FLOODPLAINS

FLOODPLAINS

*Processes and Management
for Ecosystem Services*

Jeffrey J. Opperman, Peter B. Moyle,
Eric W. Larsen, Joan L. Florsheim,
and Amber D. Manfree

UNIVERSITY OF CALIFORNIA PRESS

University of California Press, one of the most distinguished university presses in the United States, enriches lives around the world by advancing scholarship in the humanities, social sciences, and natural sciences. Its activities are supported by the UC Press Foundation and by philanthropic contributions from individuals and institutions. For more information, visit www.ucpress.edu.

University of California Press
Oakland, California

© 2017 by The Regents of the University of California

> The use of trade, product, or firm names in the publication is for descriptive purposes only and does not imply endorsement by the US government.

Library of Congress Cataloging-in-Publication Data

Names: Opperman, Jeffrey J., 1971– author. | Moyle, Peter B., author. | Larsen, Eric W., author. | Florsheim, Joan L., author. | Manfree, Amber D., author.
Title: Floodplains: processes and management for ecosystem services / Jeffrey J. Opperman, Peter B. Moyle, Eric W. Larsen, Joan L. Florsheim, and Amber D. Manfree.
Description: Oakland, California: University of California Press, [2017] | Includes bibliographical references and index. | Description based on print version record and CIP data provided by publisher; resource not viewed.
Identifiers: LCCN 2017008982 (print) | LCCN 2017012223 (ebook) | ISBN 9780520966321 (epub and ePDF) | ISBN 9780520293069 (cloth: alk. paper) | ISBN 9780520294103 (pbk.: alk. paper)
Subjects: LCSH: Floodplains. | Floodplains—California—Central Valley. | Floodplain ecology. | Floodplain management. | Ecosystem management. | Earth (Planet)—Surface—Processing.
Classification: LCC GB561 (ebook) | LCC GB561 .O66 2017 (print) | DDC 333.91/7—dc23
LC record available at https://lccn.loc.gov/2017008982

Manufactured in China

25 24 23 22 21 20 19 18 17
10 9 8 7 6 5 4 3 2 1

CONTENTS

Authors / vii

Acknowledgments / ix

1 · **INTRODUCTION TO TEMPERATE FLOODPLAINS** / 1

2 · **HYDROLOGY** / 11

3 · **GEOMORPHOLOGY** / 21

4 · **BIOGEOCHEMISTRY** / 37

5 · **ECOLOGY: INTRODUCTION** / 45

6 · **FLOODPLAIN FORESTS** / 59

7 · **PRIMARY AND SECONDARY PRODUCTION** / 69

8 · **FISHES AND OTHER VERTEBRATES** / 79

9 · **ECOSYSTEM SERVICES AND FLOODPLAIN RECONCILIATION** / 99

10 · **FLOODPLAINS AS GREEN INFRASTRUCTURE** / 115

11 · **CASE STUDIES OF FLOODPLAIN MANAGEMENT AND RECONCILIATION** / 137

12 · **CENTRAL VALLEY FLOODPLAINS: INTRODUCTION AND HISTORY** / 157

13 · **CENTRAL VALLEY FLOODPLAINS TODAY** / 177

14 · **RECONCILING CENTRAL VALLEY FLOODPLAINS** / 203

15 · **CONCLUSIONS: MANAGING TEMPERATE FLOODPLAINS FOR MULTIPLE BENEFITS** / 213

References / 219

Geospatial Data Sources / 249

Index / 251

AUTHORS

JEFFREY J. OPPERMAN is the global lead freshwater scientist for WWF and is a research associate at the Center for Watershed Sciences at the University of California, Davis.
Contact: jeff.opperman@wwfus.org.

PETER B. MOYLE is a Distinguished Professor Emeritus in the Department of Wildlife, Fish, and Conservation Biology and Associate Director of the Center for Watershed Sciences at the University of California, Davis.
Contact: pbmoyle@ucdavis.edu.

ERIC W. LARSEN is a research scientist and fluvial geomorphologist in the Department of Human Ecology at the University of California, Davis.
Contact: ewlarsen@ucdavis.edu.

JOAN L. FLORSHEIM is a researcher in fluvial geomorphology, hydrology, and earth surface processes at the Earth Research Institute, University of California, Santa Barbara.
Contact: joan.florsheim@ucsb.edu.

AMBER D. MANFREE is a postdoctoral researcher in Geography at the Center for Watershed Sciences at the University of California, Davis.
Contact: admanfree@ucdavis.edu.

ACKNOWLEDGMENTS

The authors gratefully acknowledge the financial support of the Center for Watershed Sciences, University of California, Davis, made possible by a gift from the S. D. Bechtel, Jr. Foundation. We also gratefully acknowledge the shifting array of scientists, naturalists, floodplain advocates, and other colleagues who inspired this work over the past decade or more and who enhanced our understanding of floodplains. The Center for Watershed Sciences provided (and still provides) a friendly home for floodplain research by creating an atmosphere of collegiality that makes cross-disciplinary books like this possible. We give special thanks to the following.

Jack A. Stanford and Klement Tockner, who generously reviewed the manuscript and provided much useful and encouraging feedback. We also thank the many people who reviewed sections of the book, including Dylan Ahearn, Gretchen Benjamin, Alexander Bryk, Alvaro Cabezas, Severin Hohensinner, Frans Klijn, Kris Johnson, Ron Melcer, Bryan Piazza, Jamie Pittock, Michael Reuter, Mark Reynolds, Brian Richter, Jeremy Sarrow, and John Stella. Any errors and omissions in this work are entirely those of the authors.

Jeffrey Mount, who provided insightful leadership of Cosumnes River studies that included Peter B. Moyle, Jeff J. Opperman, and Joan L. Florsheim, and led to the founding of the Center for Watershed Sciences at UC Davis. Jay Lund and Cathryn Lawrence have continued this leadership tradition at the center, where floodplain studies still flourish.

Jacob Katz of California Trout, whose research, advocacy, and contagious passion for floodplains are changing floodplain management in California.

Ted Sommer who observed the importance of floodplains to salmon and other biota in California's Central Valley and inspired others to take floodplains seriously as ecosystems.

Jeffrey J. Opperman is grateful for the support and patience of Paola, Luca, and Wren for a decade of evenings and weekends spent working on this book. Jeff also thanks the numerous scientists and practitioners at the Nature Conservancy whose work illustrates the potential for floodplain reconciliation and management for multiple benefits.

Peter B. Moyle thanks Patrick Crain for his insights into the Cosumnes River floodplain system, based on years of getting wet and muddy sampling fish, and Carson Jeffres for providing insights based on his tireless, hands-on work on both the Cosumnes and Yolo Bypass floodplains.

Joan L. Florsheim thanks the researchers and staff of the Earth Research Institute, UC Santa Barbara, for support and encouragement, as well as colleagues and students too numerous to name from UC Davis, UC Berkeley, USGS, Scripps, CalTrans, DWR, State Water Resources Control Board, UC Denver, Sonoma County, and more who shared their knowledge and opportunities to work collaboratively on California's amazing floodplain systems.

Eric W. Larsen thanks Steve Greco at UC Davis whose encouragement and infectious love of science related to the Sacramento River has profoundly influenced his work on the Sacramento River system. He also thanks Stacy Cepello, California Department of Water Resources, whose vision and singular leadership has helped protect and restore the Sacramento River for future generations, and William Dietrich for tireless enthusiasm, inspiration, and insights on the science of river geomorphology.

ONE

Introduction to Temperate Floodplains

FLOODPLAINS ARE AMONG the most dynamic, productive, diverse, and threatened ecosystems in the world (Tockner and Stanford 2002). Intact and restored floodplains generate major environmental benefits that provide significant support for local and regional economies, most notably through flood-risk management, fisheries, recreation, and seasonal agriculture (Postel and Carpenter 1997). Yet the flooding that defines floodplains—and drives their ecological productivity and diversity—is often viewed as a problem. Many floodplains sustain dense human populations and agriculture that is not compatible with inundation. Floods in industrialized countries are usually equated with disaster, prompting extensive construction projects, such as dams, levees, and channel straightening and dredging, to minimize flood impacts on the built landscape. Consequently, floodplains, particularly temperate floodplains in more developed countries, are among the most altered landscapes worldwide, most with ecosystems that are highly degraded (Tockner and Stanford 2002). Yet floodplains also present some of the best opportunities throughout the world for innovative management that reconciles human uses and environmental conservation.

In this book, we focus on floodplains in temperate regions of the world, interweaving floodplain science and management. We review fundamental processes that shape floodplains as biophysical systems and then consider new perspectives on how floodplains are managed. Thus, this book should be of interest to both scientists and managers. We strongly believe that the most promising future for temperate floodplains will arise through management solutions that allow them to function as dynamic ecosystems that are also productive and safe components of the human landscape. Achieving this future will depend on informed dialogue and collaboration among managers, scientists, and stakeholders. Our experiences with temperate floodplains lead us to believe that the long-term social and environmental sustainability of floodplains can best be guided by the interrelated concepts of "novel ecosystems" and "reconciliation ecology." *Novel ecosystems* are those that are highly altered by humans and often contain alien (nonnative) species,

FIGURE 1.1 In most of the temperate world, large rivers have become hydrologically disconnected from their floodplains by levees, bank armoring, and flow regulation as shown in this view of the Sacramento River, California, United States (photo by Christina Sloop/Bird's Eye View 2016).

such that their current biota and physical structure may differ markedly from those of the ecosystems they replaced. Nevertheless, they may be functionally quite similar to the original ecosystems (Hobbs et al. 2009, 2013, 2014; Moyle 2013). *Reconciliation ecology* is the "science of inventing, establishing, and maintaining new habitats to conserve species diversity in places where people live, work, and play" (Rosenzweig 2003, p.7). Reconciliation ecology focuses on conservation of native biodiversity while accepting the reality that virtually all habitats have a strong human presence. Thus, built landscapes, such as cities or flood-management systems, can be treated as novel ecosystems into which biodiversity conservation and ecosystem services are actively integrated in order to provide multiple benefits to human society.

The concept of novel ecosystems is generally applicable to floodplains of large temperate rivers because virtually all are modified and/or regulated. While channels and floodplain surfaces continue to be reshaped by flood frequency, extent, and duration, these forces are modified in large part by the operation of dams, levees, and drainage systems. Even most "restored" floodplains have topography that has been highly altered by agriculture and are subject to inundation regimes managed by people. Species assemblages are often in continuous flux, responding to environmental change, land and water management, and invasions of non-native species.

Because of this ecological novelty and because large areas of temperate floodplains have been converted to other land uses, reconciliation ecology provides a constructive framework for moving forward with sustainable management of floodplains for multiple purposes, including flood-risk management and conservation. People are now an integral part of nearly all temperate floodplain ecosystems and, in the chapters ahead, we will illustrate how reconciled floodplain ecosystems can provide ecological and economic benefits.

The floodplains of California's great Central Valley (figure 1.1), which inspired this book, provide a particularly cogent example of the complex and evolving relationship between people and floodplains. These floodplains have undergone ecological loss and sweeping landscape change, but they also provide examples of paths toward a more sustainable future. We draw heavily on this well-studied landscape as its lessons can be applied globally and reinforced by examples from other floodplain systems.

In short, this book is about reconciling temperate floodplain ecosystems with society's demands on the environment; it is also about how people can work proactively with novel floodplain ecosystems to generate multiple benefits. Our goal is to demonstrate that maintaining native biodiversity and natural floodplain functions can be highly compatible with a broad range of societal objectives and expectations. Innovative management approaches can produce floodplains that provide habitat for some of the world's most spectacular fish and wildlife, while also protecting communities from floods and providing clean water and open space for people. Our ambitious objective is to have the concepts, examples, and syntheses of emerging ideas in this book serve as a foundation for effective and sustainable management of temperate floodplain systems. We hope that this foundation will also provide insight for global floodplain management beyond temperate regions.

GEOGRAPHIC SCOPE AND THE DEFINITION OF FLOODPLAINS

Strictly speaking, the floodplains covered in this book occur within temperate latitudes, defined as those between the Arctic Circle and the Tropic of Cancer and between the Antarctic Circle and the Tropic of Capricorn (spanning the latitudes of 23°26′ to 66°34′ both north and south). Mountain ranges and other continental features greatly affect river flows and temperature patterns, so floodplains vary in size, character, and importance throughout the temperate regions. We primarily focus on floodplains from North America, Eurasia, and Australia, with limited mention of those in southern South America and South Africa (figure 1.2).

Floodplains support complex physical, biological, and social systems (Naiman et al. 2005). They are created by interactions between flowing water and sediment, which influence, and are influenced by, physical structures and biological processes. Floodplains are optimal sites for farms and cities because of their proximity to water, low-gradient terrain, and rich organic soils. Urban and agricultural development on floodplains is often facilitated by infrastructure such as dams and levees.

Because of these numerous and interacting features, processes, and uses of floodplains, professionals of diverse disciplines study and manage floodplain systems. As a result, hydrologists, geomorphologists, ecologists, engineers, economists, planners, and policy makers have developed a variety of definitions of floodplains (Nanson and Croke 1992; Alexander and Marriot 1999). Below we review definitions from geomorphic, hydrologic, regulatory, and ecological perspectives. We then present an ecogeomorphic definition that integrates biological and physical concepts that will be used throughout this book.

Geomorphic

Although rivers and floodplains are often viewed as distinct, separate features, from a geomorphic perspective rivers and associated floodplains are single functional systems that move water, sediment, and organic material across the landscape. The geomorphic floodplain is a set of channels and surfaces that were created by, and continue to be shaped by, fluvial processes and are inundated with some frequency. Processes that build and rework floodplains include overbank deposition and erosion, lateral migration, and channel avulsion (described in chapter 3). The resulting

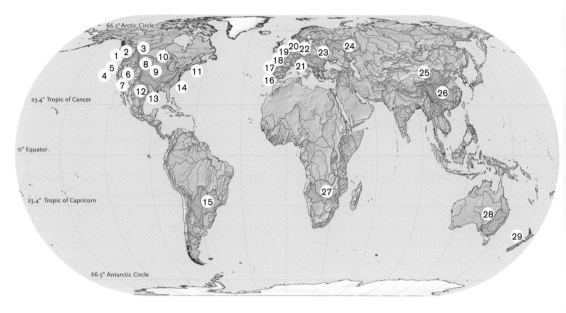

FIGURE 1.2 Global watersheds with river systems discussed in this book (Natural Earth 2016; Nature Conservancy 2016).

1 Willamette	7 San Joaquin	13 Atchafalaya	18 Ebro	24 Volga
2 Walla Walla	8 Missouri	14 Southeastern US	19 Loire	25 Yellow (Huang He)
3 Flathead	9 Illinois	Floodplain Forests	20 Meuse	26 Yangtze
4 Napa	10 Mississippi	15 Parana	21 Rhone	27 Okavango
5 Sacramento	11 Otter Creek	16 Guadalquiver	22 Rhine	28 Murray-Darling
6 Truckee	12 Brazos	17 Piedra	23 Danube	29 Waipaoa

floodplains have substrate derived from interactions among flow, sediment, and vegetation (Nanson and Croke 1992; Knighton 1998, Wohl 2010).

Hydrologic/Hydraulic

A hydraulic floodplain is a surface inundated by a defined flood recurrence interval (i.e., has a certain probability of flooding in any given year) from yearly (Wolman and Leopold 1957) to much longer periods, often 100 years or greater (Alexander and Marriot 1999). This definition is often used by hydrologists and engineers (Nanson and Croke 1992). The term "active floodplain" has been used to describe the area inundated by floods with a broad range of recurrence intervals, whether or not that surface is alluvial. An alluvial surface is created by the actions of the river, such as sediment deposition (Alexander and Marriot 1999).

Regulatory

The regulatory floodplain in the United States is essentially the hydraulic floodplain corresponding to what is termed a "100-year flood," or the surface that has the statistical probability of 1% of being inundated in any given year, based on past records of flooding. The term "100-year flood" has led to widespread confusion among citizens, who often assume that, following such an event, they are safe for the next century (Mount 1995). Because of the potential for misunderstanding, the US Army Corps of Engineers has begun to use the terminology of "percent exceedance" (Hickey et al. 2002) with the "1% exceedance floodplain" corresponding to the "100-year floodplain." However, the term "100-year flood" is well established among planning professionals and in the popular media, so it remains in common usage.

The designation of the 100-year floodplain still strongly influences patterns of development in the United States; within designated 100-year floodplains, owners of structures with federally backed mortgages are required to hold a flood insurance policy (Sheaffer et al. 2002; Pinter 2005). Such policies are administered by the National Flood Insurance Program (NFIP), which is part of the Federal Emergency Management Agency (FEMA). The process of delineating the 100-year floodplain involves considerable uncertainty due to extrapolations from short hydrological records (often less than 100 years) and because runoff patterns have high variability, which is affected by climate change and land-use changes (Milly et al. 2008). Because each new flood provides a new data point, the mapped extent of the 100-year floodplain often expands after a major flood event, as it did in regions of California following floods in 1986 (Mount 1995) and again in 1997. The extent of the 100-year floodplain will likely expand in many regions as climate change increases flood magnitudes (Kundzewicz et al. 2008).

Ecological

From an ecological perspective, floodplains are low-lying areas along rivers that support ecosystems with biological and physical characteristics that are strongly influenced by the dynamic hydrograph of the river. Floodplains serve as an interface between aquatic and terrestrial environments during inundation events (Seavy et al. 2009). Floodplain ecosystems have often been considered equivalent to riparian wetlands or "bottomland" forests, or as ecosystems that incorporate riparian forests, wetlands, and floodplain waters such as oxbow lakes (Mitsch and Gosselink 2000; Naiman et al. 2005). Temperate floodplains often support a high percentage of the biodiversity in a region, especially of vertebrates (Hauer et al., 2016).

Much of the ecological literature on floodplains emphasizes areas that are inundated frequently and for long durations, as is the case for many large river systems in the tropics (Bayley 1989; Junk et al. 1989). The timing and predictability of flooding are also important ecological variables that shape floodplain ecosystems and the extent to which species have specific adaptations for using floodplains. Flooding can be fairly predictable, as in California's seasonal climate, or highly erratic, as in southern Australia's aseasonal climate. This has led Winemiller (2004) to place floodplain ecosystems into three basic categories: *tropical seasonal, temperate seasonal,* and *temperate aseasonal.*

Tropical seasonal floodplain ecosystems have massive rainy-season floods that inundate huge tracts of forest and other habitat on a predictable basis for months at a time. Classic examples are the extensive floodplains of the Mekong and Amazon rivers and their tributaries. Due to our emphasis on temperate systems, we do not focus on this floodplain type, even though it encompasses some of the largest and most studied floodplain ecosystems (see box 1.1).

Temperate seasonal floodplain ecosystems make up the majority of floodplains in temperate regions. In these systems, flooding tends to predictably occur within a specific season although the exact timing and extent of flooding may vary tremendously among years and between river systems. In more northern areas, seasonality is driven by snowmelt, while in Mediterranean regions seasonality is driven more by rainfall, sometimes in combination with snowmelt from mountains.

Temperate aseasonal floodplain ecosystems are found in areas where large rain storms can occur during any month of the year, but often have long periods between events. Winemiller (2004) notes that the Brazos River, Texas, United States, and the Murray-Darling system in Australia are examples of aseasonal floodplain rivers.

Ecogeomorphic

In this book, we use a definition of floodplain that combines geomorphic and ecological

> **BOX 1.1 · Excluded Floodplain Systems**
>
> In some respects, it is easier to define the scope of this book by the floodplain systems that are largely (but not entirely) excluded from it: those of tropical rivers, ephemeral desert streams, boreal rivers, and alpine streams. These systems are fascinating and some are well studied. Each is briefly discussed here for clarification.
>
> *Tropical rivers* have by far the largest floodplains in the world. Because infrastructure development (dams and levees) is relatively recent in tropical zones, most tropical rivers and floodplains are still largely connected, although that is changing (Dudgeon 2000; Opperman et al. 2015). These river-floodplain systems, such as the Amazon, Mekong, Ganges, Niger, and Congo, support some of the most species-rich and productive ecosystems in the word, including a high percentage of the world's freshwater fishes and fisheries. Annual flooding typically lasts for months and floodplain forests and fisheries thrive as a result (Lowe-McConnell 1975; Goulding 1980; Hogan et al. 2004). In Africa, some subtropical rivers have a slow cycle of annual flooding and strong interactions between large mammals and the aquatic ecosystem. A prime example is the Okavango Delta, in Botswana (Mosepele et al. 2009). Many tropical and subtropical rivers are now being altered by water infrastructure development, such as large hydropower dams, which disrupt fish migrations, reduce forest flooding, and alter other ecological processes (Dudgeon 2000; Ziv et al. 2012; Winemiller et al. 2016).
>
> *Ephemeral desert streams* floodplains have been omitted largely because flooding occurs only for short periods, so floodplain dynamics are often quite different from those in other river systems. These systems have been well studied by desert-oriented stream ecologists (Grimm and Fisher 1989; Heffernan 2008).
>
> *Boreal rivers*, mostly near or above the Arctic Circle, are important floodplain rivers, but their dynamics are different enough from temperate rivers that we found it hard to include them (e.g., Yarie et al. 1998; Durand et al. 2011). Key factors in the floodplain geomorphology and ecosystem characteristics of boreal rivers include permafrost in floodplain soils, timing of ice breakup, and the effects of sudden releases of water and ice from ice dams (Peters et al. 2016).
>
> *Alpine streams* are mostly in steep canyons with confined channels providing little room for floodplain development. Mountain meadows, however, are miniature floodplains (chapter 5) that occur in lower gradient reaches. These meadows are wetland systems that have disproportionate importance compared to their surface area. For example, in the Sierra Nevada of California and Nevada, meadows are "hot spots" of biodiversity, attenuate floods, collect sediment, store water, improve water quality in streams, and sequester carbon (Purdy et al. 2012).

definitions. We consider floodplains to be features that are formed and influenced by rivers and upon which biophysical processes and ecosystems operate. The geomorphic processes influence the characteristics of the ecosystem present, while ecosystem characteristics alter the way the geomorphic processes work. We highlight seasonal floodplains along lowland rivers, generally those with broad alluvial valleys. Such floodplains often feature highly productive and diverse ecosystems and they tend to be the focus of management, both for values that depend on flooding, such as fish and waterfowl, and for keeping people and crops safe from flooding.

In the spirit of novel ecosystems and reconciliation ecology, we consider ecogeomorphic floodplains to encompass all areas that are periodically inundated by river flows, regardless of return intervals, including areas that would have been flooded prior to human changes. These areas include seasonal floodplain surfaces and features such as wetlands, oxbow lakes, ponds, and side channels as well as habitats that are often considered to be terrestrial, such as agricultural fields and riparian forests, which are periodically inundated at low frequencies. This definition approximates the definition of an active floodplain (Alexander and Marriot 1999) and is a broader definition than has usually been applied to floodplains (Bay Institute 1998; Mitsch and Gosselink 2000).

ON THE NEED TO STUDY FLOODPLAIN ECOSYSTEMS

Floodplain ecosystems received very limited study prior to the 1970s, with scientific interest confined to the relatively small community of geomorphologists—researchers who had long known that any inclusive definition of a river channel should include the floodplain (e.g., Wolman and Leopold 1957). In contrast, a large body of floodplain research related to hydrology was generated by engineers looking for ways to prevent flood waters from reaching floodplains and damaging built and agricultural landscapes. These engineering studies began in Europe in the seventeenth and eighteenth centuries and were conducted to inform the construction of flood-control infrastructure, and the Chinese have been studying and building projects to manage floods for thousands of years (Wu et al. 2016). Due to this infrastructure, most large rivers in temperate latitudes were disconnected from their floodplains prior to broader scientific study (Benke 1990; Bayley 1991; Dynesius and Nilsson 1994). Commonly, channels were straightened and armored and multichannel systems were converted to single channels, with little attention paid to effects on the biota (Ward and Stanford 1995; Brown 1998; Ward et al. 2001; Florsheim and Mount 2002, 2003; figure 1.3).

Because intact temperate river-floodplain systems are rare, it was difficult for scientists, let alone planners and decision makers, to appreciate the diversity and productivity of natural river-floodplain systems. In part because of this lack of understanding, river management and restoration efforts have seldom attempted to replicate the variability of natural processes or the full range of habitats found in complex river-floodplain systems (Poff et al. 1997; Ward et al. 2001). As a result, after decades to centuries of conversion and narrowly focused management, temperate floodplains around the world have been converted from periodically flooded ecosystems to other land uses, contributing to dramatic declines in floodplain-dependent biotas. A review by Benke (1990) concluded that the Yellowstone River is the only major river (>1000 km long) in the United States that is not significantly altered by dams, levees, and navigation infrastructure. Dynesius and Nisson (1994) reported that almost 80% of the discharges of the 139 largest rivers in North America and Europe are regulated to some extent by dams. Grantham et al. (2014) noted that in California alone there were over 1400 large dams and thousands of smaller ones that could influence the floodplains below them. In a review of trends in floodplains around the world, Tockner and Stanford (2002, p.324) describe a "dismal prognosis" because floodplains continue to be changed and degraded at a faster rate than any other wetland type.

Despite changes to nearly all rivers and floodplains in temperate systems, some rivers do retain large areas of fairly natural floodplains. For example, the lowland floodplain reaches of the Volga River (Russia) still have a fairly natural flow regime, although a cascade of upstream dams have contributed to a loss of sediment supply and to channel incision. Górski et al. (2010, 2011) studied the Volga to understand how fishes use natural floodplain habitats that have been lost elsewhere. Similarly, Moyle et al. (2007) studied floodplain use by fishes in the Cosumnes River, one of the few free-flowing rivers left in California (chapter 13), to understand how native fishes would have used the once extensive floodplain habitats of California's Central Valley. But even the Cosumnes River, with its natural flow regime, is a novel ecosystem because much of the floodplain is farmed and a high proportion of the biota is nonnative. This book draws on studies from fairly natural floodplains such as these, and also on research that illuminates the value of truly novel floodplain ecosystems, such as managed flood bypasses.

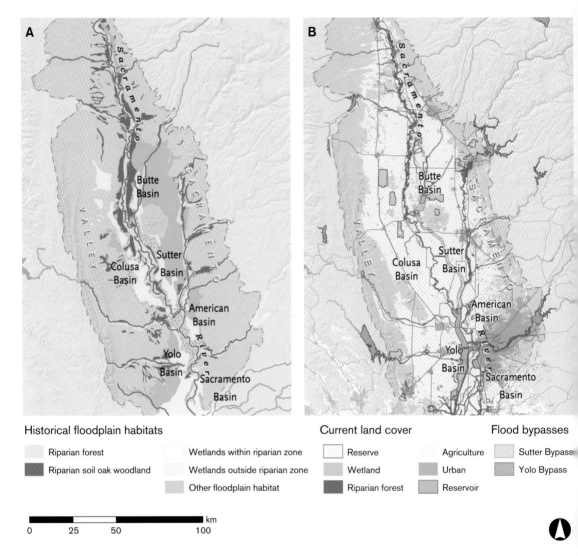

FIGURE 1.3 Distribution of floodplain habitat types in the Sacramento Valley: (A) historical and (B) current (Bay Institute 1998; Gesch et al. 2002; CalAtlas 2012).

Only in recent decades has such scientific research begun to illuminate what naturalists have known for hundreds of years, and what many rural societies living near or on floodplains have known for millennia: floodplains are complex systems of immense ecological, cultural, and economic value (Junk et al. 1989; Opperman et al. 2013). Simultaneously, hydrologists, engineers, planners, and floodplain managers have begun to question strict reliance on structural approaches to flood management (i.e., "flood control") and have begun to examine how functioning floodplains can contribute to flood-risk reduction (Haeuber and Michener 1998; Conrad 2004; Opperman et al. 2009). Because of this, floodplain restoration is emerging as a strategy not only to improve environmental health, but also as something that can reduce impacts of flooding to cities and farms. This book merges these views of floodplains, approaching them as interacting biophysical and socio-ecological systems (sensu Gunderson 2001). We posit that sustainable management of floodplains must be predicated on understanding how they function as biophysical systems. Conversely, achieving the most benefits from floodplains as biophysical systems requires understanding how biophysical processes can be reconciled with management of floodplains as socio-ecological systems.

A review of floodplain forms and functions, integrated with a discussion of management for multiple purposes, is timely because studies of temperate floodplain ecology are beginning to coalesce around common themes, and because temperate floodplains are presently the subject of large-scale restoration and new management approaches. In addition, climate change is increasing variability in the amount and timing of precipitation in most areas, increasing the likelihood of major floods. Therefore, this is an opportune time for an in-depth review that can immediately be put to use by project designers and managers.

TOPICS COVERED IN THIS BOOK

The first section of the book (chapters 2–8) focuses on the complexity of floodplains as biophysical systems, with chapters on floodplain hydrology, geomorphology, biogeochemistry, ecology, floodplain food webs, and vertebrates.

The second section of this book (chapters 9–11) examines how these biophysical processes produce a set of ecosystem services that provide value to society. In these chapters, we describe how water and floodplain management can either diminish or enhance the value of these services, with a specific focus on how floodplains can be integrated into flood-risk management, as "green infrastructure." This section concludes with a series of case studies on management challenges and solutions in temperate floodplains.

The third section of the book (chapters 12–14) provides an in-depth case study of a temperate floodplain region, California's Central Valley (figure 1.3). This section focuses on one specific floodplain region to illustrate how human activities have changed floodplains over time and how those changes affect ecosystems, physical processes, and associated risks and benefits to people. Central Valley floodplains are among the most studied temperate floodplains and they offer many examples of new approaches that managers and scientists are taking to reduce flood risk while restoring floodplain ecosystems.

CONCLUSIONS

Floodplains on large rivers in temperate regions are gaining a new appreciation by many people from natural historians and ecologists to engineers and flood managers. In many places, efforts are being made to reconnect river channels to their floodplains, although most such projects are small compared to what has been lost. There is some urgency to expand these efforts because many highly valued floodplain-dependent species, such as cranes and salmon, are in danger of extinction in the wild. Fortunately, there is also growing realization that restoring functional floodplains is often valuable from a flood-management perspective, especially in the face of climate change. These trends illustrate the potential benefits of reconciled floodplains, in which people work with natural processes rather than constantly fighting them. This book is an introduction to this new paradigm.

TWO

Hydrology

HYDROLOGY HAS BEEN described as "by far the single most important driving variable in floodplains" (Tockner and Stanford 2002). River flows control the processes of erosion and deposition that create floodplain topography—the template upon which distinctive habitats, such as riparian forests, develop (Poff et al. 1997; Ward et al. 2002). Groundwater beneath the floodplain surface strongly influences the development of vegetative communities and other ecosystem processes (Faulkner et al. 2012). Hydrological processes also structure floodplain ecosystems by controlling patterns of connectivity, residence time, and the flows that allow exchange of organisms, carbon, and nutrients between segments of the landscape (Wiens 2002). This chapter provides an overview of floodplain hydrology. Because of its underlying role in floodplain processes, hydrology is also interwoven into all subsequent chapters.

Although many temperate floodplains across the world are characterized by modified hydrology, effective floodplain management and restoration requires understanding fundamental hydrologic processes and attributes of lowland floodplain river systems. Thus, this chapter focuses primarily on unaltered hydrological processes. The effects of anthropogenic changes to hydrology are considered in subsequent chapters.

This chapter

1. defines attributes of lowland floodplain hydrology and focuses on key attributes such as connectivity, magnitude, frequency, duration, timing, rate of change, and residence time;

2. describes the various sources of floodplain water and how heterogeneous sources of water influence floodplain processes; and

3. discusses feedbacks between floodplain hydrology and geomorphic and ecological processes.

In the next chapter, we follow these interrelated themes to more fully show how hydrology affects floodplain geomorphology, and how hydrology and geomorphology create the template for ecosystem processes.

> **BOX 2.1** · Attributes of Floodplain Hydrology: A Glossary
>
> **GENERAL ATTRIBUTES**
>
> CONNECTIVITY Linkages and pathways for flows within and between channels, floodplain surfaces, and hyporheic zones.
>
> TEMPORAL VARIABILITY Hydrologic variability across a range of timescales, including single storm events, as well as climatic patterns and changes across diel, seasonal, annual, inter-annual, decadal, century, millennial, or longer scales; can be cyclic or stochastic.
>
> SPATIAL VARIABILITY Hydrological patterns and changes that vary across floodplains based on water sources and topographic and vegetation patterns.
>
> **SURFACE ATTRIBUTES (SPATIALLY AND TEMPORARILY VARIABLE)**
>
> FLOW MAGNITUDE (DISCHARGE) Volume of water per unit time; the magnitude of flow on a floodplain can be a combination of flow contributed from variable sources.
>
> FLOW STAGE Water surface elevation that varies spatially with location in channels and irregular floodplain topography and varies temporally during the passage of a flood wave.
>
> AREAL EXTENT Measure of the geographic area of flooding at a given time period.
>
> DEPTH Vertical extent of inundation.
>
> FREQUENCY Number of times over a given interval of time that a given flow magnitude or floodplain inundation occurs; recurrence interval.
>
> RESIDENCE TIME Length of time that a given unit of water remains in a given place and thus reflects the exchange rate of water at that place. Residence time can be calculated in many ways. One simple method is dividing the volume of the area of interest (e.g., floodplain extent) by the flow rate.
>
> DURATION Length of time that a given floodplain surface or water body is inundated. Duration differs from residence time in that duration refers more simply to the amount of time that a given area is inundated.
>
> TIMING Dates of occurrence of floodplain flow or inundation, such as timing of cyclic seasonal floods.
>
> **SUBSURFACE ATTRIBUTES (SPATIALLY AND TEMPORARILY VARIABLE)**
>
> HYPORHEIC FLOW Shallow groundwater flow beneath the floodplain.
>
> WATER TABLE The top of the zone of saturation; the elevation of the surface of hyporheic floodplain water.
>
> SOIL MOISTURE Water content of floodplain sediment and soil above the water table.
>
> BANKFULL FLOW Surface flow discharge in a channel when filled roughly to the top of the bank (see chapter 3).
>
> BASEFLOW Low flow discharge in channels between storms and other runoff events.

ATTRIBUTES OF FLOODPLAIN HYDROLOGY

Floodplain hydrology is characterized by a wide range of interrelated attributes that affect floodplain processes and ecology (box 2.1) and these attributes reflect the influence of diverse sources of water to the floodplain. For example, floodplain flows derived from the main river channel are largely influenced by riverine hydrological attributes, including magnitude, frequency, duration, timing, and stage (Poff et al. 1997). The attribute of connectivity facilitates this relationship between river flows and floodplain flows (Amoros and Bornette 2002; Brierley et al. 2006). Other attributes of floodplain hydrology reflect the interaction of flow with floodplain topography, such as the areal extent and spatial variability of surface water inundation, flow depth (stage), velocity, duration, and residence time.

Floodplain hydrology varies temporally at scales ranging from diel fluctuations within individual flood events to cyclic patterns (seasonal, annual, interannual, decadal, or even longer) to long-term climate change (Florsheim and Dettinger 2015). In summary, attributes of floodplain hydrology arise from interactions between floodplain topography and river flows and other sources of water; these interactions vary in complex ways within short-term events and over long-term climate patterns and changes. The resulting attributes of floodplain hydrology (defined in box 2.1) can be quantified for different purposes ranging from ecosystem restoration to flood control and management. However, these attributes are often difficult to quantify. Direct measurement over time on large, remote, and/or topographically diverse floodplains is logistically difficult. Because flow gages are sparse in much of the world, within-channel river flow attributes can also be difficult to quantify. Recent technological advances have improved the ability to visualize and quantify floodplain hydrological attributes such as inundation extent, stage, and connectivity at multiple spatial scales (Mertes 2002). For example, Landsat and other satellite-based technologies have allowed development of high-resolution models for floodplain water balance (Beighley et al. 2009) and flux (Alsdorf et al. 2000, 2007) in Amazon floodplains as well as other applications worldwide (Ashworth and Lewin 2012; Baki and Gan 2012; Gupta et al. 2012; Trigg et al. 2012; Constantine et al. 2014; Rozo et al. 2014).

Hydrologic Connectivity

Hydrologic connectivity is requisite for a floodplain that can support dynamic geomorphic processes and the ecological processes characteristic of healthy, productive floodplains. From a purely hydrologic perspective, connectivity refers to transfer of water along and between pathways linking surface and subsurface floodplain components through various fluvial and groundwater processes. This transfer of water facilitates the transfer of energy and materials—including sediment, wood, and living organisms—across the same flow pathways. Ward (1997), Ward and Stanford (1995), and Brierley et al. (2006) define riverine connectivity as energy transfer within riverine landscapes across three spatial dimensions, longitudinal (up and downstream), lateral (from channel to floodplain), and vertical (between surface and subsurface).

Longitudinal Connectivity

Longitudinal connectivity occurs when river flow is transferred along longitudinal pathways, upstream to downstream, within watersheds and is a primary driver of floodplain hydrological attributes and ecosystem processes. Longitudinal connectivity links a downstream floodplain with hydrological processes throughout the upstream watershed, such as precipitation, runoff, and discharge in tributary channels. By modifying flow regimes, dams affect this longitudinal connectivity and alter the movement of water, sediment, and organisms through river systems, changing their availability to floodplains (Ligon et al. 1995; Kingsford 2000; Fryirs et al 2006).

Lateral Connectivity

Lateral connectivity between rivers and floodplains is a dominant theme in the literature on floodplain ecosystems (Junk et al. 1989). Restoration of floodplain ecosystems often requires reestablishing lateral connectivity between main channels and floodplains to reinitiate flows of water, sediment, and organisms. The importance of lateral connectivity to geomorphic and ecological processes is a primary theme of this book and we summarize several projects to restore this connectivity.

Vertical Connectivity

Vertical connectivity is the connection between surface waters and the hyporheic zone, the heterogeneous shallow aquifer underlying floodplain-river system. Vertical connectivity features recharge or "downwelling" (movement of water from surface to subsurface) and discharge or "upwelling" (movement of water from subsurface to surface). Poole et al. (2008) expanded beyond vertical connectivity and introduced "hydrologic spiraling" to describe the three-dimensional integration of longitudinal, lateral, and vertical hyporheic flow paths. They note that the length and location of these flow paths influence the residence time of water within the hyporheic zone along with the biochemical and thermal conditions of the water and the biota within hyporheic habitats. Poole et al. (2006) suggest that common anthropogenic alterations to river systems tend to reduce complexity of hyporheic flow dynamics, and therefore simplify habitat and ecosystem processes. These simplifications occur both within the hyporheic zone and in its interactions with surface waters.

Flow Magnitude, Frequency, Duration, Timing, and Rate of Change

Geomorphic and ecological processes on floodplains are strongly influenced by attributes of the river flow regime, including the magnitude, frequency, duration, timing, and rate of change of the surface water flow (Poff et al. 1997).

Flow Magnitude

Most of the literature on flow magnitude has focused on floods that inundate the floodplain—by definition flows greater than bankfull—but river flows at or below bankfull also strongly influence floodplain processes. For example, baseflows throughout the year, including annual minimum flows, influence the water table on the floodplain with important consequences for floodplain water bodies and vegetation (Poff et al. 1997; Shafroth et al. 1998, 2000). As flow magnitude rises, flow pulses (those flows above baseflow but still below bankfull) can reinitiate flow in side channels or abandoned braids and, by raising the water table, increase the depth and surface area of floodplain water bodies (Tockner et al. 2000; Ward et al. 2001, 2002). Flow pulses also increase a river's ability to remove and transport wood or sediment. Flows large enough to inundate the floodplain are associated with many geomorphic and ecological processes on floodplains. Floods of varying magnitudes and durations are key for small- and large-scale floodplain geomorphic and ecological processes (chapters 3, 5, and 14).

Flow Frequency

Flow frequency, or how often a flow of a particular magnitude occurs, is inversely related to flow magnitude, such that small flows occur more frequently than large floods. In lowland rivers, frequent small floods, recurring every year or so, are often very important for maintaining floodplain ecosystems (Amoros and Bornette 2002). Topographically higher portions of the floodplain are inundated less frequently compared to lower areas, a hydrological gradient that strongly influences floodplain vegetation patterns. For example, on Passage Creek, Virginia, United States, the distribution of plant species depends on the frequency of exposure to destructive flooding (Hupp and Osterkamp 1996).

Flow Duration

The duration or length of time water inundates a floodplain is significant for a range of ecological and geomorphic processes. For example, predictable (e.g., annual) and long-duration overbank flows are necessary for certain floodplain ecological processes to occur—such as the accumulation of productivity on floodplains to support fish feeding—and these long-duration flows form the basis for the flood pulse concept (Junk et al. 1989), a dominant paradigm for floodplain ecology (discussed in chapter 5). Geomorphic processes are also influenced by duration; the duration of flows that exceed the magnitude required for sediment transport from river to floodplain have a strong influence on floodplain morphologic changes and development of new floodplain landforms (Florsheim et al 2006).

Flow Timing

Because many ecological processes are seasonal, such as species' reproductive cycles, the timing of floodplain flows during the year also strongly influences processes such as vegetation establishment on new floodplain surfaces and the ability of juvenile fish to access floodplain biomass. For example, long-duration inundation events in the Central Valley of California (United States) are most effective at promoting a range of ecological benefits when they occur during the spring (chapter 14; Williams et al. 2009).

Flow Rate of Change

The rate of change—or how quickly flow magnitude increases or decreases—on a floodplain affects both plants and animals. For example, terrestrial animals are more likely to find refuge during floods that rise slowly compared to a rapidly rising flood. Similarly, fish are less likely to get stranded on the floodplain with a slow rate of draining. Successful establishment of many riparian tree species, such as cottonwood (*Populus* spp.), requires that the rate of decline of the water table (controlled largely by the rate of change of flow depth in the river channel) does not exceed the ability of seedling root growth to keep pace with the declining water table (chapter 6; Mahoney and Rood 1998).

Residence Time of Water

The residence time of water—the rate at which water moves through a floodplain—exerts a strong influence on physical and ecological processes. As discussed in the chapters on ecological processes (chapters 5–8), the residence time of water on the floodplain strongly influences water quality (e.g., turbidity), biogeochemical processes, and ecological processes, such as primary and secondary productivity.

Residence time differs from duration in that duration refers more simply to the amount of time that a given area is inundated; an area can have long duration of inundation by water with either very long residence time of water (e.g., a pond) or very short residence time of water (e.g., a drainage channel with flowing water).

The residence time of water on a floodplain is controlled by characteristics of the flood wave interacting with the topography of the floodplain. The speed of the flood wave traveling down a floodplain is controlled by the valley or channel slope and by the elevation or stage of the flood water, which is a function of flow magnitude and valley dimensions. Where flood flows are topographically confined to channels or narrow floodplains, the restricted cross section increases water elevation. This increase in elevation leads to greater velocity of the flood wave and reduced residence time of the flood pulse. In contrast, where a channel is connected to a large floodplain, the expansive cross section leads to lower elevation and velocity, allowing floodplain storage, attenuation of the flood wave, and longer residence time of water on the floodplain. Floodplain vegetation can also slow water velocities and increase residence time (Welcomme 1979). Residence time tends to be longest in the topographically lowest portions of floodplains, such as depressions, wetlands, and former channels (Naiman et al. 2005). In addition to surface connections between a river's channels and floodplains, subsurface connections influence residence time (Tockner et al. 1999).

SOURCES OF FLOODPLAIN WATER

The integration of individual hydrological attributes (box 2.1) can be referred to as the floodplain's "flow regime." A floodplain's flow regime reflects the aggregation of water coming from diverse sources, including the main channel, secondary channels, tributaries, shallow groundwater or hyporheic zones, and direct precipitation. Each of these sources varies in terms of *magnitude, frequency, duration, timing, and rate of change* and so the floodplain flow regime varies based on the proportional contributions of these sources over time and interactions with floodplain landforms and vegetation. Here we draw on the *variable source area concept* of Dunne and Leopold (1978)—a concept that is well established in hydrology to describe processes and sources that contribute flow to stream channels—to illustrate the complexity of sources contributing to lowland river floodplain hydrology (figure 2.1).

In this section, we describe the various sources in the general sequence in which they contribute water to the floodplain, beginning with the input of water from direct precipitation. We then review interactions of floodplain surface and subsurface flows within the floodplain hyporheic zone and then the input of water from tributaries. Finally, we describe *the flood pulse*, the rise of water along main and secondary channel banks and onto the floodplain. These variable sources of water, and their interactions, contribute to the complexity of biophysical conditions and processes on floodplains.

Direct Precipitation

Rainfall during storms provides a source of moisture to floodplain soils and may raise the water level in floodplain wetlands. When rainfall intensity exceeds the infiltration rate in soils, or when a clay layer impedes infiltration of water into the floodplain sediment, direct precipitation can produce surface water on the floodplain prior to the entry of hyporheic water or river flows.

Hyporheic Flow Contributions

The hyporheic zone is an important source of water to floodplains and one that supports distinct ecological processes. Working on the Danube River in Austria, Tockner et al. (1999) reported that the hyporheic zone was the first source of water to contribute flow on the floodplain: water seeping from the hyporheic could begin to fill low areas of the floodplain during periods when the river lacked surface connections to the floodplain. Jones et al. (2008) found that, similar to surface flows, the direction of floodplain subsurface flows varied seasonally across the floodplain. Many factors influence the subsurface flow direction including main channel flow stage, tributary flow contributions, and the location and direction of hyporheic exchange.

Baseflow in the main river channel has a major influence on floodplain hyporheic flows, which in turn has important consequences for floodplain water bodies and vegetation (Poff et al. 1997; Shafroth et al. 1998, 2000). Flows greater than baseflow, but still below bankfull, raise the floodplain water table, increase the depth and surface area of floodplain water bodies, and can initiate flow in abandoned or side channels (Tockner et al. 2000; Ward et al. 2001; Ward et al. 2002). Conversely, when the groundwater (hyporheic) level in the floodplain is higher than the water surface in the channel, hyporheic water held in alluvial sediments flows into the river to support baseflows.

Flood Pulse

During a flood pulse, river stage rises and fills the channel until it reaches the top of the channel bank, sometimes referred to as the "floodplain lip." After filling the channel from bank to bank, a continued rise in stage results in water flowing over the bank and initiating "overbank" flow on the floodplain (Wolman and Leopold 1957; Leopold et al. 1964; Slingerland and Smith 1998). In addition to contributing water to the floodplain, the flood pulse also allows for exchange of sediment, nutrients, and aquatic biota between channels and floodplains (Junk et al. 1989). In many natural floodplain systems, overbank flows occur frequently, on the order of every year or two. However, overbank flows may occur less frequently in rivers that are highly modified anthropogenically, such as rivers where flow is regulated by dams or where channels are incised.

Once overbank flows begin, the pattern of

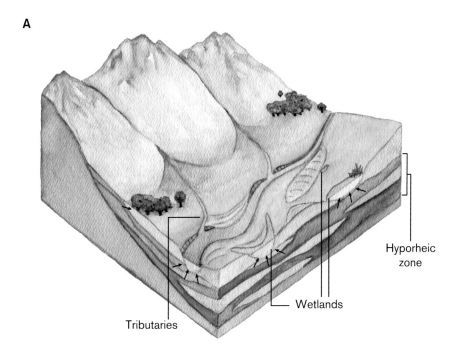
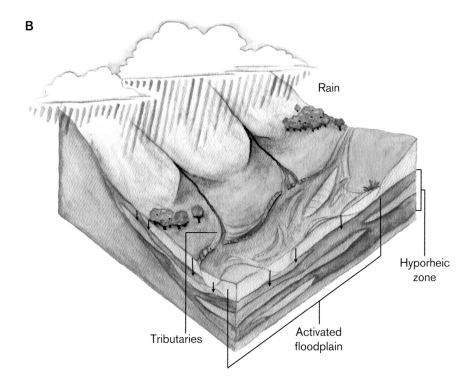

FIGURE 2.1 River floodplain and surrounding uplands, tributaries, and wetlands. Arrows show direction of exchange between floodplain water and hyporheic zone. (A) During the dry season, low flow in the main channel and tributaries and floodplain wetlands are fed by hyporheic water. (B) During the wet season, local rainfall contributes directly to floodplain wetlands and tributaries, and basin-scale runoff contributes to rising flow stage in channels until water overflows banks. During floodplain inundation, floodplain water contributes to the hyporheic zone.

flooding can be complex. In natural river-floodplain systems with irregular topography, overbank flow begins at multiple low elevation points along the floodplain-channel boundary, resulting in floodplain inundation that is initially heterogeneous spatially. This incipient flood water from the main river then mixes with local water already present on the floodplain, including water derived from direct precipitation and tributary inflows (Mertes 1997). Large natural floodplains often have multiple channels through them, increasing the complexity of inundation processes. A network of channels can cross the floodplain at various elevations and these channels convey flow from upstream and into the floodplain at a range of river flow magnitudes. Because of these complex channel forms, a given patch of floodplain may receive surface flow initially from a secondary channel that flows through the floodplain prior to receiving lateral flow from the immediately adjacent river channel. Bidirectional flow can occur even in a single floodplain channel depending on relative flow stages. For example, in floodplain rivers with complex morphology—such as the Middle Fly River (Papua New Guinea), the Lower Mississippi River and Birch Creek, Alaska (United States)—Rowland et al. (2009) describe "tie" channels that carry flow from the main channel into floodplain lakes during rising river stages; during falling river stages, the tie channels carry flow from the lake back to the main channel.

Tributary Flow

Tributaries entering alluvial valleys may interact with floodplains in several ways. Some tributaries terminate in alluvial fans at the junction with the main river such that stream flow infiltrates into the floodplain and fan sediment assemblage, becoming subsurface flow that can contribute to the floodplain hyporheic water. An example of this can be found in the Navarro River of northwestern California (United States), where small tributaries join the river's discontinuous floodplain (Florsheim, 2004). Some tributaries flow across floodplains to join main rivers and can either subsidize or take flow from the floodplain hyporheic water depending on their relative water levels. Tributaries don't always connect directly to the main river channel because a barrier, such as a natural levee, blocks the flow. In such cases, tributary flows often contribute directly to the floodplain surface flow. For example, prior to human modifications, the floodplain of the Yolo Flood Basin in Central California, United States, was separated from the Sacramento River by a natural alluvial levee (Gilbert 1917). Flows from tributaries spread across the floodplain where they merged with flood flows entering the flood basin from the main river through upstream connections (Florsheim et al. 2011).

Variable Hydrologic Sources and Floodplain Processes

The variable sources of floodplain water and their interactions contribute to the complexity of biophysical processes on floodplains. Moreover, diverse sources of water, changing over time and interacting with floodplain topography, can result in a broad range of floodplain water quality characteristics such as temperature, turbidity, and nutrient levels. Within the Alluvial Zone National Park of the Austrian Danube, for example, maximum and minimum summer water temperatures within various floodplain water bodies vary by as much as 16°C, reflecting variable sources of water (e.g., river, tributary, groundwater; Ward et al. 2002). During floods, the balance and timing of water from different sources influence floodplain water quality. In the Yolo Bypass in California, an engineered floodplain (see chapter 13), inflow from tributaries continues to provide additional nutrients to floodplain waters even as the floodplain flow begins to drain back into the primary source of water, the Sacramento River (Schemel et al. 2004).

Due to a diversity of water sources, floodplains can become inundated even during periods when river stages are not sufficient to provide lateral overflow from the main river. As discussed above, the Yolo Bypass can be partially inundated by smaller tributaries during years when it doesn't receive water from the Sacramento River (Sommer, Harrell et al. 2001). Tributary inflow, hyporheic flows, and direct precipitation strongly influence the subsequent hydrological interactions between the primary river and floodplain. Prior inundation by various sources can affect the extent to which main river flows extend laterally on to the floodplain, with important implications for geomorphic and ecological processes (Naiman et al. 2005). Inputs from groundwater and rain can maintain intermittent floodplain water bodies for much longer than can be achieved solely through surface connections with the main river (Benke 2001). Thus, these alternate sources of water increase the complexity of spatial and temporal hydrological patterns on a floodplain beyond that provided by the primary river.

FEEDBACK BETWEEN HYDROLOGY AND GEOMORPHIC AND ECOLOGIC PROCESSES

While this chapter emphasizes how riverine hydrology influences conditions and processes on the floodplain, presence of a floodplain also influences conditions and processes within a river channel. The floodplain accommodates river flows in excess of channel capacity. Overflow onto the floodplain limits the increase of stage in the channel, and thus limits the shear stress that is exerted on the channel bed and banks. In essence, the floodplain relieves pressure on the channel, slowing the increase in shear stress with increasing discharge. In some cases, this may permit a wider range of sediment grain sizes to remain on the bed than if the floodplain were not connected. When rivers are disconnected from their floodplains by levees, this "relief" function no longer occurs (unless the levee fails or is overtopped) and flow depths continue to increase with discharge, increasing erosion of the channel bed. Fine sediment fractions are more likely to be transported downstream, leading to local bed material coarsening or incision. Disconnection from the floodplain and resultant increased in-channel scour can be detrimental to riverine biota. For example, high flows can scour out salmon eggs buried in gravel (Hall and Wissmar 2004) and displace fish, invertebrates, and other aquatic organisms. Leveed rivers often incise, creating a deep channel through which water can move rapidly, with very little habitat for invertebrates, fish, and birds. This process of channel incision reduces the hydrological connection between the channel and floodplain because a larger volume of water is needed to initiate surface flow on the floodplain. Due to this process of incision, levee removal as part of a floodplain restoration project may not always lead to more frequent surface connectivity, thus increasing the challenge of reinitiating floodplain processes (Williams et al. 2009).

CONCLUSIONS

Floodplain hydrology is variable and dynamic, supporting the most diverse and productive ecosystems on Earth. There are numerous feedbacks among hydrological, geomorphic, and ecological processes on the floodplain. A floodplain's geomorphic structure is the template on which flooding occurs and shapes how each flood affects the local environment. In turn, floods can modify the geomorphic structure of the floodplain. Over time, these interactions between dynamic hydrology and floodplain morphology create the diverse ecosystems that characterize floodplains (Stanford et al. 2005). Hydrological processes also influence floodplain ecosystems through attributes of flooding (e.g., timing and duration) and through the flows of water that mediate exchange of sediment, organisms, carbon, and nutrients among segments of the landscape. Subsequent chapters focus on these connections between hydrologic, geomorphic, and ecological processes.

THREE

Geomorphology

A RIVER AND ITS FLOODPLAIN are one integrated system where the river's geomorphic processes erode, transport, and deposit sediment and organic material to create the morphology that underpins floodplain ecosystems. Floodplain geomorphology is controlled by a variety of factors including geologic setting, sediment supply rate, characteristics of sediment, transport capacity of the river, and flow regime (Nanson and Croke 1992; Poff et al. 1997; Knighton 1998; Bridge 2003). The geology and hydrology of the upstream watershed also strongly influence the downstream floodplain geomorphology.

Research on floodplain geomorphology generally addresses floodplains from the perspective of the dynamic processes that form them and has greatly expanded in the past several decades (Carling and Petts 1992; Anderson et al. 1996; Brown 1997; Petts 1998; Marriott and Alexander 1999; Bridge 2003; Freitag et al. 2009; Ritter et al. 2011; Hudson and Middlekoop 2015). This chapter provides a basic background for understanding floodplain geomorphology, primarily for large lowland floodplain rivers. As described in the section of this chapter devoted to river classification, large lowland rivers tend to have more developed floodplains than smaller upland rivers and have been more extensively studied. This chapter describes (a) floodplain surface development, (b) channel and floodplain erosion, (c) floodplain topography, and (d) floodplain classification. Box 3.1 provides a glossary of terms used in this chapter.

Here we focus primarily on unaltered geomorphic processes; subsequent chapters describe anthropogenic changes to the processes that shape floodplains and the implications for floodplain management and reconciliation. Although this book focuses on temperate floodplains, there are many more examples of unaltered large river floodplain systems in the subtropics and tropics. Because temperate and tropical floodplains are shaped by similar geomorphic processes, we draw on examples from tropical rivers when needed.

FLOODPLAIN SURFACE DEVELOPMENT

Several fluvial processes contribute to developing floodplain surfaces (Nanson and Croke

BOX 3.1 · Glossary of Terms Used in This Chapter

GENERAL ATTRIBUTES

LATERAL ACCRETION Addition of sediment to a floodplain surface laterally, usually through the sediment deposition on point bars.

VERTICAL ACCRETION Addition of sediment to a floodplain surface vertically, through settling out of sediment carried in overbank flows.

BANKFULL (STAGE) The water surface elevation that comes to the top of the river banks.

STREAM POWER A measure of the energy of flowing water often used to quantify the ability of flow to move sediment; it is defined as channel discharge (Q), multiplied by channel slope, multiplied by the specific weight of water.

CHANNEL PATTERNS

SINUOSITY A measure of the "curviness" of a meandering river; the ratio of the length of the channel centerline divided by the straight-line distance between the same upstream and downstream endpoints.

STRAIGHT Channel with little sinuosity (sinuosity ≈1.0).

MEANDERING Channel with bends that alternate from side to side; sometimes called a sinuous channel. Closely related to meandering, point bars are accumulations of bed material forming the convex bank on the inside of a meandering river bend.

BRAIDED Channel with multiple flow paths separated by unstable mid-channel bars, called braid bars, composed of the same material as channel bed material load; bars are lower than the floodplain elevation.

ANASTOMOSING Multiple channels integrated within a floodplain, with stable vegetated islands that are at the same elevation and have the same sediment composition as the surrounding floodplain.

PROCESSES

AVULSION A process through which the primary flow of a river diverts from an established channel and may form a new stable channel in a new location on the floodplain.

CREVASSE SPLAY Sediment deposited on a floodplain when sediment is transported from the main channel to the floodplain.

MEANDER MIGRATION The pattern of evolution of a meandering river as it erodes and deposits and the channel progressively moves across a floodplain.

OTHER

LEVEE A topographic high or berm adjacent to and often parallel to the channel; formed when overbank flow deposits sediment next to the channel. Natural levees slope downward away from the river channel over long distances and are often hard to detect visually, levees that are human-made usually have distinct berms.

OXBOW LAKE Body of water formed when a meander bend is cut off and abandoned.

BACKSWAMP Low-lying wetland areas, rich in organic nutrients, separated from active channels; common in the low-gradient floodplains of the southeastern United States.

FLOOD BASIN Low areas on floodplains beyond the natural alluvial levees of large river systems.

1992; Knighton 1998; Bridge 2003). Early reviews (e.g., Wolman and Leopold 1957; Leopold et al. 1964) emphasized two primary processes: (1) deposition on point bars (lateral accretion) and (2) deposition on floodplains during overbank flows (vertical accretion). On the rivers that they studied, generally relatively small meandering rivers, Wolman and Leopold (1957) found that much of the floodplain development resulted from lateral accretion linked with within-channel deposits. On larger, lowland rivers, floodplain development results from more diverse processes such as avulsion or deposition within channels that later become incorporated into the floodplain.

Floodplain geomorphic processes are closely associated with the processes that shape a river's channel pattern. Therefore, before examining floodplain development processes, we provide a brief review of the major types of river channel patterns. An early approach classified channels as straight, meandering, and braided (Leopold and Wolman 1957). More recent classifications consider additional forms and transitional forms between these basic patterns (Knighton 1998). For the classification we describe here, we add "anastomosing" rivers to the earlier list. Although this type of lowland multichannel river is often considered a subset of the river type called anabranching, here we refer to that type as "anastomosing," following Knighton (1998).

Lowland rivers may be broadly divided into single-channel and multiple-channel rivers (figure 3.1). Within those classified as single-channel rivers, meandering rivers have alternating bends and are sinuous, and so-called straight rivers have very little or no sinuosity; however, naturally straight rivers are rare. Braided rivers have multiple flow paths separated by *unstable* mid-channel *bars*, while anastomosing rivers have multiple channels separated by vegetated *stable islands*. The bars in a braided system are lower in elevation than the floodplain, composed of the same sediment as the channel bed material load, and are inundated by bankfull flow stage. In contrast, the islands in an anastomosing river are at the same elevation or higher than the floodplain, composed of the same sediment as the surrounding floodplain sediment assemblages, and are inundated during overbank flows. Anastomosed-channel floodplain islands tend to be much wider and longer than braided channel bars. Individual channels in an anastomosed system can themselves be braided, meandering, or straight. Figure 3.2 shows a section of a river that includes examples of all the patterns.

Lateral Accretion and Channel Migration

Rivers can migrate across their floodplains (figure 3.3). Under equilibrium conditions, this process of channel migration deposits sediment on one side of the channel in roughly equal proportion to the sediment eroded on the opposite side such that channel width remains relatively constant. As a channel meanders, a point bar is built on the inside of the bend that continues to extend into the stream toward the outside cut bank; this bank recedes through erosion. As the bar grows toward the cutbank, older locations of the point bar become progressively farther from the main channel, and also become higher in elevation as sediment is deposited. Consequently, more distal areas of the point bar are inundated less frequently and deposition is mainly of finer sediment, layered on top of the older and coarser deposits. This process of deposition during meander migration is a specific type of lateral accretion, and is one of the primary processes for building floodplain surfaces in meandering rivers. The establishment of vegetation on a point bar augments accretion because vegetation has high hydraulic roughness (Knighton 1998). Coarse channel deposits left behind as a channel migrates become incorporated into the floodplain and can eventually serve as preferential flow pathways for groundwater (see chapter 2; Naiman et al. 2005). The sediment and geomorphic forms created by river migration contribute to floodplain topographic features, described in a section later in this chapter.

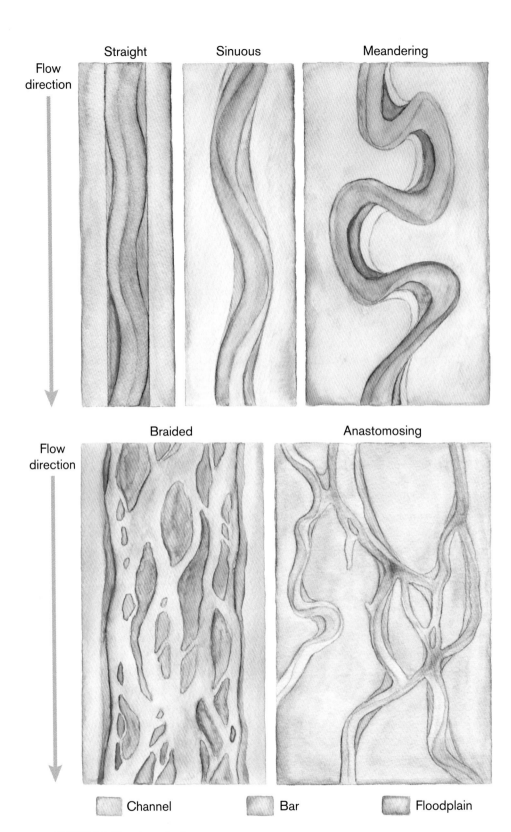

FIGURE 3.1 River channel patterns.

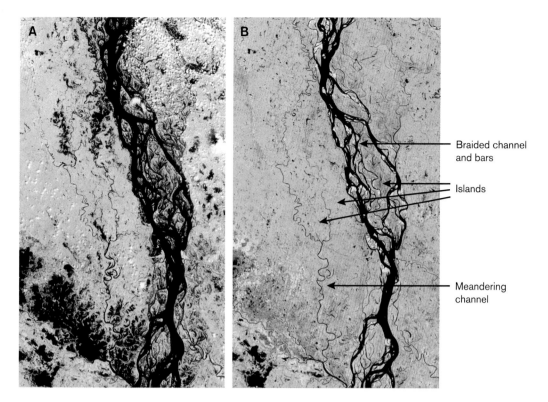

FIGURE 3.2 Anastamosing Brahmaputra River system in central Bangladesh showing channel patterns in flood conditions in September 2014 (A) and lower flow conditions in October 2014 (B). Individual channels in an anastomosing system can themselves be braided, meandering, or straight, as indicated in the figure (after Bridge 2003; USGS 2014a).

Lateral accretion (i.e., deposition) and meander migration are two different perspectives from which to describe the same phenomenon. The rate at which lateral accretion (migration) occurs is a function of stream power and bank composition (Nanson and Hickin 1986; Larsen, Fremier, and Greco 2006). This process is described further later in this chapter when we discuss the effect of flooding on lateral accretion. The lateral extent of the floodplain, across which meander migration can occur, can be limited by restraints such as natural topographic features or constructed levees, bridges, and other infrastructure. A river's migration rate varies greatly among years based on individual flow events; therefore, migration rates are usually expressed as long-term averages. To compare between rivers of different sizes, it is useful to express migration rates in terms of channel widths. The Mississippi River migrated at a long-term rate of 45–60 m/year (roughly 0.02–0.05 channel widths per year) prior to man-made levees and bank armoring (Hudson and Kessel 2000). Micheli et al. (2004) reported that long-term migration rates on the Sacramento River ranged from 2.8 m/year (0.008 channel widths per year) in the years 1896–1946, to 4.2 m/year (0.011 channel widths per year) in the years 1946–1997. The processes of meander migration are described further when we discuss channel and floodplain erosion during flooding.

Floodplain development by lateral accretion and bar building also occurs on braided rivers. Braid-channel accretion is caused by several different processes that collectively develop floodplains on braided rivers. For example, braid bars can become stabilized if the primary braid channels shift laterally away from the bar (figure 3.4). Large flood events often deposit braid bars at high elevations; the bars are then not susceptible to being reworked by

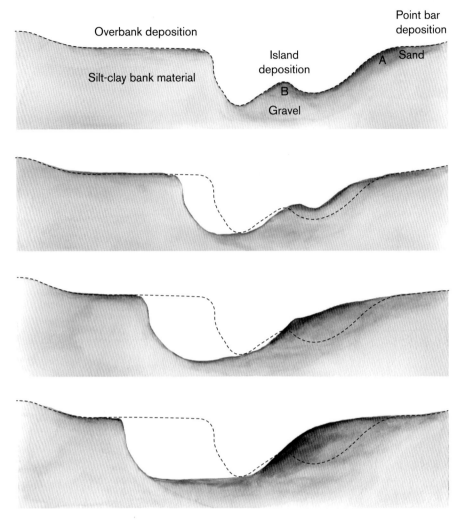

FIGURE 3.3 Floodplain development through lateral accretion. The channel is migrating to the left. Dashed lines indicate initial channel cross section. Note the upward fining in the bar progressively building on the right (A) and the mid-channel bar (B) that becomes incorporated into the floodplain (after Knighton 1998).

subsequent smaller floods (Nanson and Croke 1992). Large wood can induce deposition downstream of the wood, leading to development of a mid-channel bar (Abbe and Montgomery 1996) or island (Edwards et al. 1999; Gurnell et al., 2001). Growth of vegetation in the sediment deposited behind the large wood can lead to further deposition of wood and sediment and the formation of a stable island.

Through processes of bank erosion and deposition, rivers can rework their entire floodplains over a period of centuries (e.g., 250 years for Waimakariri River, a braided river in New Zealand [Reinfelds and Nanson 1993]) to millennia (e.g., 1300 years for the River Exe in Great Britain [Hooke 1980]). These processes are tightly coupled to ecosystem development. For example, meander migration creates an age gradient of floodplain surfaces upon which habitats such as forests can develop (e.g., figure 3.5). The floodplain surfaces of different ages support forests in different stages of ecological succession (chapter 6; Salo et al. 1986; Stanford et al. 2005; Greco et al. 2007).

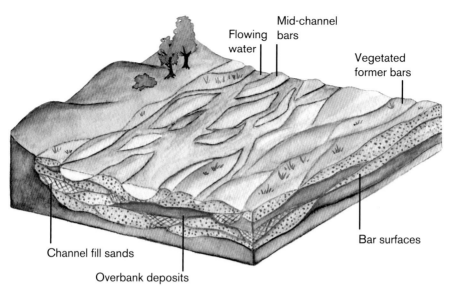

FIGURE 3.4 Floodplain development in braided channels. As the channel shifts, bars are abandoned and become part of the floodplain.

Vertical Accretion

Overbank flood waters carrying suspended sediment will typically leave deposits of silt and sand immediately adjacent to the channel where the water velocity first slows, while finer-grained sediment is deposited further away from the channel. Vegetation and topography influence this basic pattern of preferential sediment size deposition (Knighton 1998). Vertical accretion can be a dominant mode of deposition and floodplain development in streams with frequent overbank flooding, a high load of fine sediment, and laterally stable banks that resist channel migration (Ritter et al. 1973). The contributions of vertical accretion may have been overlooked by early researchers because channel migration can rework vertically accreted deposits, thus removing evidence of the overbank deposition (Knighton 1998).

Rates of overbank deposition can vary between years and spatially across a floodplain based on characteristics of river flows, such as the frequency of overbank flows, and sediment characteristics. Rates also depend on floodplain characteristics including topography, vegetation, or the interactions of flood waters from diverse sources that vary in their suspended sediment load (Knighton 1998; Mertes 2000). By slowing water velocity, vegetation growing on floodplain surfaces increases the vertical accretion of fine sediment.

Bridge and Leeder (1979) reviewed existing literature for the long-term rates of vertical accretion in rivers, mostly in the United States, that showed a large range of variability—from 0.0 to >3000 mm/year. For example, the Mississippi River, over a period of 30,000 years, had an average rate of vertical deposition of about 1 mm/year (Fisk 1944). Mertes (1994) estimated rates of vertical accretion in the Amazon to be 0.8–1.6 cm/day corresponding to annual rates of centimeters to meters of deposition. On the floodplain of the Cosumnes River (California), predisturbance Holocene deposition rates of up to about 3.0 mm/year kept pace with sea level rise and tectonic basin subsidence; anthropogenic disturbances caused a rapid increase in floodplain sedimentation rates up to 25 mm/year between 1849 and about 1920 (Florsheim and Mount 2003).

Deposition of sediment on floodplains can be important to the sediment dynamics of the entire river system. Floodplain sedimentation

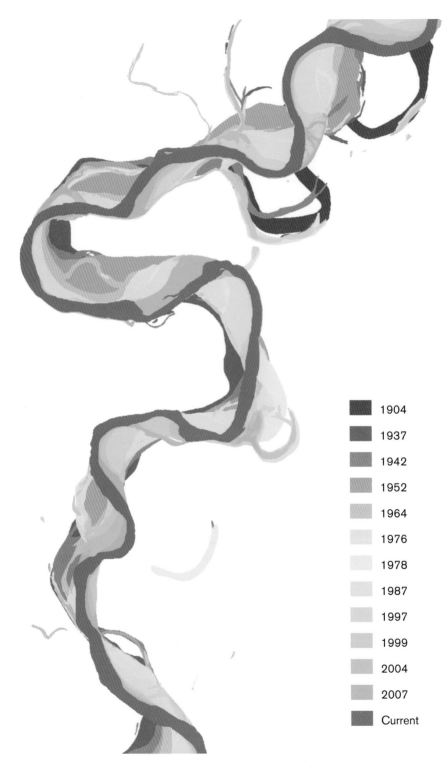

FIGURE 3.5 Meander migration and floodplain surfaces along a reach of the Sacramento River, California. Floodplain surfaces represent historic channel locations. Colors represent floodplain age and show extensive reworking of the floodplain by progressive migration (CDWR 2012).

can decrease a river's suspended sediment load. For example, floodplain deposition has been shown to be responsible for a decrease of 20–25% in suspended sediment loads in rivers (Walling et al. 1996). In another example, Valett et al. (2005) reported that over 90% of the suspended sediment in flood waters that entered a floodplain of the Middle Rio Grande (New Mexico, United States) was deposited. Removing suspended sediment from the water through floodplain deposition may improve water quality in the main channel and may improve soil development on the floodplain itself. When sediment accumulation rates increase due to upstream erosion from land use, the sediment trapping function of hydrologically connected floodplains may help protect downstream estuaries from excess sedimentation (Noe and Hupp 2005). The suspended sediment deposited on the floodplain can also provide nutrients to promote growth of riparian forests (Yarie et al. 1998). However, once a floodplain's accommodation space for sediment is filled, less of a river's load is deposited on the floodplain (Swanson et al. 2008; Constantine et al. 2014).

Floodplains are composed of sediment assemblages and layers that reflect the processes that form them and can therefore provide insight into paleoenvironmental factors that existed during floodplain formation (Brown 1996). Floodplain stratigraphy provides a record of climatic variation (Knox 1987; Aalto et al. 2003; Malamud-Roam et al. 2006, 2007) and land-use changes. For example, a layer of "post-settlement alluvium" is found in many temperate floodplain rivers. These layers were deposited during periods of rapid land-use changes, such as the onset of agriculture, deforestation, or mining, which resulted in elevated rates of upland erosion and deposition in lowlands (Jacobson and Coleman 1986; Lewin 1996; Lecce 1997; Hupp et al. 2015). The increase in overbank deposition on the Cosumnes River floodplain after upstream land-use changes, described above, provides an example. Trimble (1999) found that much of the sediment eroded from farm fields in the 1920s in a Wisconsin (United States) watershed was stored downstream in floodplains within the channel network. Thus, floodplains are effective in sequestering sediment and thereby buffering effects of land-use disturbances (Gomez et al. 1999; Hupp et al. 2015). Residence time of sediment in floodplain storage is variable and depends on how frequently rivers can rework their entire floodplain through channel migration or other erosional processes. Floodplain storage of alluvium tends to increase with valley width and decreasing slope (Ritter et al. 2011).

How much sediment is or can be stored on a floodplain is an important question for land-use managers (Dietrich and Dunne 1978; Brown 1987) because floodplains can accumulate large amounts of sediment (Bendix 1992; Walling et al. 1992). Various tracers related to anthropogenic activities are utilized to document floodplain sedimentation in lowland rivers. For example, sediment from hydraulic mining in the Sierra Nevada in California can be identified by its high quartz content (Mount 1995). Singer and Aalto (2009) used ^{210}Pb (an isotope of lead related to human activity) to assess overbank sedimentation rates in the Sacramento River's highly altered floodplain system. Walling et al. (1992, 1996) used ^{137}Cs, a radionuclide found as fallout from testing of nuclear weapons during the middle twentieth century, as a marker to aid in dating sediment strata to quantify sedimentation rates.

Avulsion

Avulsions are dynamic processes through which the primary flow of a river diverts from an established channel and creates a new stable channel across the floodplain. Avulsions are triggered by a combination of erosion and deposition and can result in a new channel or reoccupation of an older channel along with the creation of new floodplain surfaces. Avulsions occur in all types of rivers, meandering, braided, anastomosing, and even "straight" ones. The recurrence interval for avulsion events can vary widely, from once every few decades (Kosi River, India) to more than a thousand years (Mississippi River, United States; Slingerland and Smith 2004). Avulsions often occur when sediment deposition in the channel or an obstruction, such as a jam of large wood, impedes or blocks flow and forces it onto the floodplain (Montgomery et al. 2003).

Avulsion is the dominant process for floodplain evolution in some types of rivers. For example, anastomosing rivers commonly have low gradients, low width-to-depth ratios, and cohesive, stable banks. Therefore, channel migration rates are low and are a minor contributor to floodplain formation; the dominant processes include vertical accretion during overbank flow and avulsion. In anastomosing rivers, avulsion is characterized by the switching of channel location during a natural levee break as water and sediment from the main channel enter the floodplain (Slingerland and Smith 2004).

Avulsions both erode and deposit sediment, influencing floodplain geomorphology. During avulsion, sediment is transported from the channel onto the adjoining floodplain, often creating a crevasse splay, a fan-shaped sediment deposit formed when sand and silt are transported through a levee break and then deposited by the spreading water. Below we illustrate how the erosional and depositional processes associated with avulsions, including the formation of crevasse splays, influence floodplain development and topography.

The Cosumnes River (California, United States) is the site of a floodplain restoration project where management interventions initiated processes similar to those that would occur during a natural avulsion. Managers excavated a breach in the levee, allowing the river to access the floodplain through the breach during high flows. When flood waters flowed through the breach, erosion occurred on the floodplain near the breach where flow velocity was highest, while deposition (vertical accretion) of a crevasse splay occurred further from the breach. While an avulsion (i.e., channel switching) has not occurred at the site, floodplain development that occurred after the breach illustrated the types of geomorphic processes associated with avulsions and how these processes combine to create complex floodplain topography.

Florsheim and Mount (2002) documented the morphology of splay complexes and concluded that (a) sediment on splay complexes is organized into diverse landforms that include lateral levees, lobes, and new floodplain channels that form in the newly deposited sediment; (b) the juxtaposition of breach scour and sediment deposition creates relatively high floodplain relief that decreases with distance from a breach; (c) relief becomes more pronounced over time as higher-magnitude floods scour the old floodplain sediment and add new sand and silt onto the surface of the splay deposit; and (d) sediment is transported in main and secondary channels that extend topographic variability further down the floodplain. Florsheim et al. (2006) identified a discharge threshold for connectivity and sediment transfer from the channel to the floodplain and predicted the volume of sand deposited in a sand-splay complex using a model that assumes floodplain sedimentation is proportional to the concentration of sediment and volume of water that flows across the floodplain.

In anastomosing rivers, avulsion is the dominant process that creates the multiple channel form. Channels separate and rejoin, creating islands with surfaces at the same elevation as the floodplain (Richards et al. 1993). Main, secondary, or tertiary channels may be present at different elevations; lower elevation channels convey flow before higher-elevation channels do, such that different channels carry different portions of the total flow as stage rises and falls. Over time, natural levees that form adjacent to individual channel segments isolate lower elevation islands that may become wetlands (Smith and Perez-Arlucea 1994).

In meandering rivers, channel cutoffs are a common process of geomorphic adjustment that results in changes of channel planform (Fuller et al. 2003; Micheli and Larsen 2011). While some researchers consider cutoffs as a type of avulsion (e.g., Rowntree and Dollar 1999; Shields et al. 2000), others consider them separate processes (e.g., Brizga and Finlayson 1990) and distinguish between cutoffs as events that connect the ends of a meander bend and avulsions as larger-scale events that result in the switching of the primary channel into a new location on the floodplain (Brooks and Brierley 2002). Below we discuss cutoffs as well as avulsions that result in switching of the primary channel as mechanisms for floodplain development in meandering rivers.

Channel cutoffs can be classified as either a neck or a chute cutoff. In either case, the cutoff results in a new channel forming across a floodplain surface and abandonment of the previous channel. Cutoffs increase local channel gradient and thus the ability to transport sediment (Knighton 1998). A neck cutoff involves the erosion of the narrow strip of land between two limbs of a meander bend, while a chute cutoff is characterized by the incision of an incipient channel across a point bar or floodplain interior to the bend. Following either form of cutoff, the abandoned channel can become a water body on the floodplain, known as an oxbow lake (Greco and Plant 2003; Larsen, Girvetz et al. 2006; Larsen et al. 2007; Michalková et al. 2011).

Smith et al. (1989) describe an avulsion event on the meandering Saskatchewan River (Canada) that resulted in the switching of the main channel into a lower area of the floodplain. Brooks and Brierly (2002) studied floodplain development on the Thurra River in southeastern Australia and similarly found that avulsions result in the channel relocating to a lower part of the floodplain, within which a new meander belt forms. In this system, cutoffs occur within these meander belts with greater frequency (once every 1000 years) compared to avulsions (once every 5000 years).

CHANNEL AND FLOODPLAIN EROSION DURING FLOODING

In the previous section, we discussed development of floodplain topography, in which we described how erosion and deposition form floodplain surfaces. In this section, we focus on erosion of floodplain surfaces, which can be eroded both by channel migration across floodplains and by erosive overbank flows. The rate and type of erosion vary with channel type, flow hydraulics, sediment load in transport, characteristics of the sediment composing the banks and floodplain, and vegetation.

Channels respond to floods through changes in bed level and planform. Channel changes during high flows are a function of (1) initial channel form, (2) flow hydrology and hydraulics, (3) channel boundary resistance, (4) sediment supply, and (5) the historical sequence of previous floods (Wohl 2000). The ability of flows to alter channel boundaries is a function of flow *magnitude* and *duration* (Costa and O'Connor 1995; Larsen, Fremier, and Greco 2006) as well as the channel-boundary resistance from sediment and vegetation (Baker and Costa 1987; Larsen, Fremier, and Greco 2006). The most effective flows in performing geomorphic work are those that have the greatest cumulative energy over a given time period, which is a product of stream power and duration. Total energy expended by flood flows has been directly correlated with total sediment transport over time in a river channel (e.g., Wolman and Miller 1959). Stream power is the product of three values: the channel discharge (Q), the channel slope (downstream gradient), and the specific weight of water. *Cumulative* stream power is defined as the sum of stream power over a specified time interval. Meander migration rates can be directly correlated with cumulative stream power (Larsen, Fremier, and Greco 2006). Although in this section we focus on the contribution of floods to meander migration, a range of flow magnitudes contribute to meander migration. Indeed, over time, most meander migration work is done by flows of moderate magnitude. This is because, while large floods have high magnitude, they often have short durations; flows with smaller magnitudes but far longer durations can have a greater contribution to cumulative stream power.

During overbank flows, broad floodplains can dissipate much of the energy that flood flows could otherwise exert against channel banks. As a result, large floods that inundate the floodplain do not exert substantially more shear stress on river channel banks than does a bankfull flow (Baker and Costa 1987; Larsen, Fremier, and Greco 2006). This is because such flows do not significantly increase the water depth in the main channel. One corollary to this effect is that increased force *is* exerted on a stream bank when the boundary is heightened by levees or other means because, absent the "relief" provided by overbank flows, the flood waters are confined in the channel and shear stress continues to increase with stage and discharge.

As flows erode the banks, the river channel migrates into the floodplain and thus erodes floodplain surfaces. The rate and type of channel erosion are influenced by the type of channel, and, within types, the specific channel form. Braided-channel rivers are characterized by easily eroded banks with multiple flow paths separated by bars, and braided rivers tend to be laterally unstable (Knighton 1998). In meandering channels, channel migration occurs through erosion of a cutbank with simultaneous deposition on a point bar. The spatial pattern of bank erosion is determined by the flow path, which in turn is influenced by the channel curvature (Johannesson and Parker 1989). The rate of bank erosion and channel migration varies as a function of stream power and channel resistance (Micheli et al. 2004; Larsen, Fremier, and Greco 2006). Meandering channels of a midrange curvature tend to migrate most rapidly (Hickin and Nanson 1984). As described above, braided and meandering streams can rework their entire floodplain through progressive channel migration and periodic cutoffs (avulsions).

In addition to channel migration and avulsion, erosion of surfaces across the floodplain occurs during very high magnitude flood events, particularly flood events with a low sediment load (Knighton 1998). Wohl (2000) provides several examples of severe localized floodplain erosion. For example, coastal rivers in New South Wales (Australia) that have high energy but are laterally stable undergo gradual vertical accretion due to overbank deposition, over a period of hundreds to thousands of years, with episodic extreme floods eroding the accumulated alluvium. Inbar (1987) reported the erosion of 1 m of sediment from a 100–400 m wide floodplain for a distance of 2000 m in the Jordan River basin from a single event. When significant erosion of floodplain surfaces happens on a large scale, it can introduce massive amounts of sediment into the river. Episodic large-scale floodplain erosion greatly complicates the predictions of sediment transport rates in such basins (Nanson 1986).

TOPOGRAPHIC FEATURES OF FLOODPLAINS

Processes such as channel migration, avulsion, and sediment deposition and erosion combine to create diverse and irregular topographic features on floodplains (figure 3.6). These topographic features are significant because they constitute the physical structure of floodplain ecosystems (Florsheim and Mount 2002; Ward et al. 2002). Although often erroneously considered to be flat, floodplains often display considerable topographic heterogeneity, with floodplains along large rivers having relief on the order of ~10 m. Even relatively small differences in topography (<1 m) can have significant influences on soil characteristics (Thoms 2003), plant communities (Hupp and Osterkamp 1985; Salo et al. 1986), and biodiversity (Poff et al. 1997). For example, Thoms (2003) found that, on the floodplain of the lower Balonne River (Australia), channel islands, banks, and natural levees had between 70% and 123% more total organic carbon in the

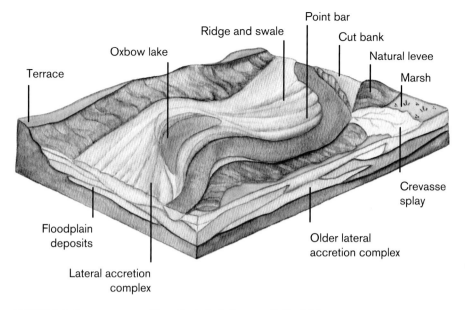

FIGURE 3.6 Common topographic and hydrologic features of a floodplain.

topsoil than did abandoned channels and flat floodplain surfaces. Due to differences in inundation frequency and duration, plant communities can differ dramatically on surfaces with relatively small differences in elevation (Hupp and Osterkamp 1996).

Sediment deposits adjacent to the channel often form low ridges that become natural alluvial levees on riverbanks (Klasz et al. 2014; figure 3.7). Natural alluvial levees next to the channel can be the topographic high point within a floodplain (Lewin 1996) because rivers deposit their coarsest load immediately adjacent to the channel during overbank flow (Knighton 1998). Natural levees then slope gradually to lower-elevation areas within the floodplain (Hudson and Heitmuller 2003). Natural levees flanking river channels generally display considerable variability in height and width, which influences inundation dynamics as overbank flows can preferentially enter the floodplain through low spots in the natural levee (see chapter 2).

Crevasse splays also create characteristic floodplain topography. After deposition, subsequent flows can cut channels through a splay, increasing topographic heterogeneity of the floodplain and influencing patterns of vegetation establishment (Florsheim and Mount 2002). Over time, crevasse splays evolve toward a set of related topographic features, including channels, wetlands, and levees (Smith et al. 1989; Smith and Perez-Arlucea 1994).

Floodplains of meandering rivers are frequently characterized by "ridge and swale" topography (figure 3.6). Subtle ridges form due to (a) the sequential development of chute channels across a point bar, (b) the migration of transverse sand bars onto point bars, or (c) the deposition of suspended sediment over a point bar, often initiated by wood on the bar. The formation of vegetation on the ridges can promote further accretion and maintenance of the ridge-swale topography (Nanson and Croke 1992). Wetlands or ponds may form in the swales between ridges (Ward et al. 2002). A variety of fluvial geomorphic processes create diverse water bodies on the floodplain (e.g., box 3.2). In large floodplains of tropical rivers, the proportion of permanent water on a floodplain can vary from 5% to 60%, with most systems having 10–20% covered in permanent water

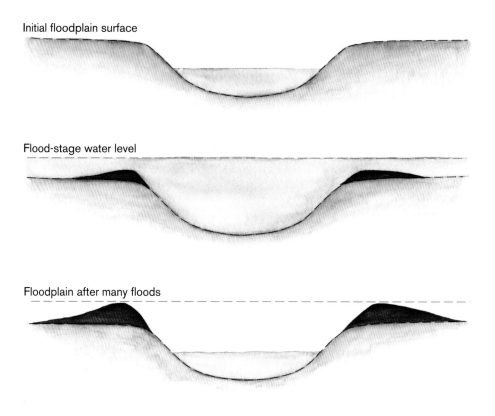

FIGURE 3.7 Natural levees form along the banks of a river connected to its floodplain. Energy dissipates as flows spread across the floodplain and slow down. Coarser, heavier materials are deposited nearest to the channel, forming natural levees (indicated in red).

> **BOX 3.2 · Build Your Floodplain Vocabulary**
>
> Scientists in Europe developed a system of naming floodplain water bodies (Amoros 1991). *Eupotamon* describes channel features with permanent running water; *parapotamon* are off-channel water bodies, or side arms, connected to the main channel at their downstream end; *plesiopotamon* are backwaters that lack permanent connections to the main channel but are strongly influenced by river discharge; *paleopotamon* describes isolated, often senescing, floodplain water bodies, such as backswamps. For those who enjoy relatively obscure linguistics, we recommend linking this classification system with a similar one for fish (chapter 8), allowing the floodplain scientist to discuss eurytopic fishes that inhabit the parapotamon!

(Welcomme 1979). In addition to permanent water bodies, floodplains can contain seasonal wetlands and channels that contain surface water in response to rising river stages, groundwater, or direct precipitation.

A meandering river can experience both "neck" and "chute" cutoffs, as described above, which result in the formation of oxbow lakes—the remnant channel portions that become isolated from the river after the cutoff event (Knighton 1998). Oxbow lakes tend to have water year-round and so function like lakes, whereas side channels, abandoned channels, and abandoned braids retain the characteristics of a flowing-water feature because they contain water primarily when flooding (Ward et al. 2002). Side channels can be connected to the river at one, both, or neither end during low flow, and begin to carry more flow as river stage rises (Amoros 1991). Floodplains can also contain a variety of other still-water, flowing-water, and semi-flowing-water bodies, including ponds formed when tributaries are dammed by large wood (Ward et al. 2002) or beavers.

Flood basins are low areas beyond the natural alluvial levees of large river systems and may be the lowest elevation areas of a floodplain. They commonly contain internal features such as channels, natural levees, and lakes or marshes that may be seasonally dry (Gilbert 1917; Bryan 1923; Allen 1965; Bridge 2003). Flood basins are often segmented into secondary basins by crevasse splays or natural alluvial levees of tributaries (Bridge, 2003). The flood basins of the Sacramento Valley (California) are described at length in chapter 13.

The term "backswamp" refers to portions of flood basins, or entire flood basins, that are nutrient rich with organic deposits. Backswamps tend to be located in the low areas of floodplains not subject to channel migration where vertical accretion from overbank flows occurs infrequently and at low rates over time (Nanson and Croke 1992; Lewin 1996) and deposition is dominated by organic matter (Aalto et al. 2008). Backswamps, better known as just "swamps," are common in the low-gradient floodplains of the southeastern United States, and tend to be covered with dense forests.

CLASSIFICATION OF FLOODPLAINS

Classification systems of floodplains can help inform restoration and management efforts to ensure that actions are appropriate for a given type of floodplain. For example, different floodplain types will respond differently to flow and sediment regime changes. In order to inform management actions, classifications are often based on the key processes that form the floodplain, such as fluvial processes (e.g., meandering, avulsions) and patterns of deposition and erosion. Additional parameters often used in floodplain classification schemes include the following: (a) large-scale geometry (such as width); (b) sediment type (e.g., silts, sand, etc.); (c) micro forms (scroll bars, splays); and (d) some measure of hydraulic energy (e.g., stream power).

Common *river* channel classification systems provide some insight into the characteristics of the *floodplain* classification schemes. Schumm (1977) broadly classified river channel networks into erosion, transport, and deposition reaches, with deposition reaches having the potential to develop well-defined and broad floodplains. The basic channel pattern classification described earlier in this chapter (straight, meandering, braided, and anastomosing) provides insights into the primary processes, such as meander migration and channel avulsion, which build and erode floodplain surfaces.

Early studies of floodplains formed the basis for much of the early theory of floodplain development, such as the observed predominance of lateral accretion and presumed relative unimportance of vertical accretion. Subsequent studies of a broader set of floodplains suggested a more diverse range of formational processes. To impose some order on this complexity, Nanson and Croke (1992) proposed a classification system for floodplains based on these development processes. Their classification is based primarily on energy (stream power) and sediment type, an indicator of the ability of a river's banks to resist erosion (cohesive vs. noncohesive banks). Their classes have been characterized as disequilibrium, equilibrium, and low-gradient floodplains (Ward et al. 2002). Note that this is almost identical to Schumm's river classification into erosion, transport, and deposition reaches. Phillips and Desloges (2015) classified alluvial floodplains in areas influenced by glacial processes in southern Canada near the Great Lakes based on stream-power resistance and floodplain sedimentology (essentially percentage of sand), which are quite similar to the two main parameters used by Nanson and Croke (1992).

Welcomme (1979), focusing primarily on how floodplains function for fisheries, defined three primary types of floodplain. First, *fringing floodplains* are strips of land flanking rivers, with floodplain width generally varying inversely with channel slope. Second, *internal deltas* are formed where one or more rivers flow over a geological/topographic feature that causes lateral spreading over an alluvial plain. River channels generally divide into braided channels that flow through such a plain. Because these areas are often very expansive and extremely flat, they can flood due to direct precipitation in addition to flooding from high river flows. Examples include the Gran Pantanal of the Paraguay River, the Apure-Arauca tributaries to the Orinoco, and the Okavango River Delta, Botswana. Third, *coastal deltaic floodplains* occur near where a river enters the ocean where the main river channel divides into multiple distributaries that flow through a fan-shaped delta produced from the deposition of river-borne sediments. These are generally not considered to be river floodplains because they are also part of estuaries, with strong tidal influences.

Geographic Information Systems (GIS), remote sensing, and computer modeling technologies are in the process of revolutionizing how scientists analyze geomorphic forms and processes. GIS analyses can provide a much greater amount of information to inform classification systems, and allow the classification of extremely large areas. New technologies make it possible to understand patterns at much greater spatial and temporal scales. Examples from around the world include the following:

- Analyzing the entire channel network of the Rhone valley in France using GIS methods to describe and test the environmental factors that influence floodplain width (Notebaert and Piégay 2013)
- Using remote sensing to analyze geomorphology in order to distinguish floodplain ecosystems in tributaries to the Amazon River in Peru (Hamilton et al. 2007)
- Using Landsat imagery to study oxbow lake formation and the detailed hydrodynamics of the floodplain in the Bolivian and Colombian Amazon (Trigg et al. 2012; Constantine et al. 2014; Rozo et al. 2014)
- Using Landsat imagery to analyze the channel and floodplain types on the largest rivers in the world (Ashworth and Lewin 2012)
- Analyzing channel dynamics and floodplain processes of the Ganges and Brahmaputra rivers (Baki and Gan 2012; Gupta et al. 2012)

Lóczy et al. (2012) used digital elevation models (DEMs) to classify floodplain types in foothill regions of Hungary. Their method is promising, as is the work in the Rhone River valley mentioned earlier (Notebaert and Piégay 2013), because it can be used to analyze floodplain types on a large scale, based on digital information. Their classification system was able to successfully describe changes in floodplain character from variables derived from DEMs, including floodplain width, local valley confinement, channel pattern, and valley floor gradient.

CONCLUSIONS

Floodplains are as diverse as the landscapes that surround them and, despite classification attempts, they exist as a continuum of types and forms. The diversity and dynamism of geomorphic processes and forms on floodplains create a shifting stage of habitats on which diverse organisms can act (Stanford et al. 2005). The relationships among hydrology, geomorphic processes and patterns, and ecological responses are the focus of chapters 4–8. By altering the basic hydrological and geomorphic processes that maintain this shifting mosaic, traditional management of rivers and floodplains has generally simplified ecological processes and communities. As a result, there has been a considerable decline in the extent, diversity, and productivity of floodplain ecosystems. Fortunately, new models of river and floodplain management are emerging that seek to accommodate natural geomorphic processes and complex floodplain ecosystems. These management approaches are the focus of the second and third sections of the book.

FOUR

Biogeochemistry

BIOGEOCHEMICAL PROCESSES CONSIST of the interactions among chemical, geological, and biological processes that determine availability of essential nutrients and other materials that sustain life on floodplains. Three floodplain processes in particular influence the biogeochemistry and nutrient dynamics of water and soil: sedimentation, biological assimilation, and biogeochemical transformations. *Sedimentation* is important because water moving on a floodplain tends to decrease in velocity and depth, allowing nutrients adsorbed to sediment to settle out onto the floodplain. *Biological assimilation* is performed by plants ranging from phytoplankton to riparian trees that directly assimilate dissolved nutrients. *Biogeochemical transformations* occur because conditions on inundated floodplains often promote biogeochemical reactions, such as the transformation of dissolved organic carbon (DOC) and nutrients into particulate organic carbon (POC). Through these processes, floodplains can remove nutrients from river water and transform them into various organic forms, including forms that are readily biologically available.

Floodplains are often "hot spots" of biogeochemical activity, defined as locations that have disproportionately high rates of biogeochemical reactions relative to the surrounding landscape due to the presence of appropriate environmental conditions and key reactants (McClain et al. 2003; Burt and Pinay 2005). The biogeochemical activity of floodplains provides some key ecosystem services, such as nutrient retention, and is an important motivation for restoring and conserving floodplains (Mitsch et al. 2001; Brettar et al. 2002; Noe and Hupp 2005; Ahearn et al. 2006).

Biogeochemical activity is not evenly distributed across floodplains because of their complex and diverse structure, including variable substrate sizes and characteristics, variable flow paths for groundwater, and multiple interfaces between terrestrial and aquatic systems (Mitch and Gosselink 2000; Sheibley et al. 2006). In addition to being "hot spots" of biogeochemical activity, a floodplain's variable flow regime results in "hot moments," defined as "short periods of time that exhibit disproportionately high reaction rates relative to longer intervening time periods" (McClain et al. 2003). Hot moments occur when hydrological processes reactivate or bring into contact biogeochemical reactants and/or create the necessary environmental conditions, such as anoxia or long hydraulic residence time, to promote biogeochemical processes (McClain et al. 2003).

Due to the spatial complexity of floodplains and inter- and intra-annual variability of flows, biogeochemical processes can vary spatially and over time; floodplains can act as sinks, sources, or transformers of nutrients—as well as heavy metals and pollutants—and these roles can vary annually, seasonally, or even within a single flood event (Mitch and Gosselink 2000; Valett et al. 2005). The same floodplain can act as a sink for an inorganic form of a nutrient and a source of the organic form of that nutrient; a wetland or riparian forest can be a sink for a nutrient during the growing season and a source of the nutrient through detritus during periods of hydrological connection following the growing season (Mitch and Gosselink 2000). The biogeochemical activity of a floodplain, and whether it will be a source, sink, or transformer of nutrients, is based on river water chemistry, characteristics of the floodplain hydrology (timing, duration, and spatial pattern of inundation), and characteristics of the floodplain, including soil types and the available pool of organic matter (Valett et al. 2005). Very few studies have tried to directly quantify fluxes to and from floodplains during the timescale of a flood event to document when, and under what conditions, floodplains are sinks, sources, or transformers of biogeochemical reactants (Tockner et al. 1999; Ahearn et al. 2006). Here we briefly review some of the better understood phenomena including dissolved oxygen (DO) flux, nutrient transformation in soils, mercury flux, nutrient retention, and, finally, temporal patterns of sequences and fluxes.

DISSOLVED OXYGEN FLUX

DO in floodplain waters is generally initially high upon first inundation, due to high levels in river water and turbulent mixing as flood waters enter the floodplain. Levels of DO then tend to drop as floodplain vegetation begins to decay, with important implications for aquatic fauna. For example, on an experimental floodplain along the Rio Grande River, DO in floodplain water dropped from an initial level of 3 mg/L DO (40% saturation) to approximately 0.5 mg/L DO (Valett et al. 2005), which is below the tolerance levels of most fishes and other aquatic organisms. Levels of DO can vary across a floodplain based on the amount of available organic matter and rates of decomposition (Valett et al. 2005), source of water, temperature, and hydraulic residence time. On the Cosumnes River, a storm event resulted in the transport of a large amount of algal biomass to a portion of the floodplain that, due to topography, had a long residence time. The influx in algal biomass raised respiration rates and dropped DO levels from 6.2 mg/L (60% saturation) to 3.0 mg/L (30% saturation) over the course of a few days (Ahearn et al. 2006). This drop in DO resulted in 100% mortality of juvenile Chinook salmon (*Oncorhynchus tshawytscha*) within an experimental enclosure; juvenile salmon in an enclosure less than 100 m away remained in good health (Jeffres et al. 2008), illustrating the potential for considerable spatial heterogeneity in biogeochemical processes and water quality on floodplains.

The photosynthetic activity of phytoplankton can influence DO levels, with DO rising during the day with photosynthetic activity and then dropping at night due to respiration and the biological oxygen demand of detritus and suspended organic solids. During periods of disconnection or long hydraulic residence time, floodplain water bodies can stratify with higher DO near the surface and low DO at the bottom. The Rio Grande floodplain, described above, had levels near 0 mg/L DO at the bottom, approximately 2 mg/L DO at the surface, and 4–5 mg/L DO at depths between 50 and 100 cm (the depth where DO was highest reflected the depth at which river water entered the floodplain, due to thermal stratification; Valett et al. 2005).

Rapid destratification, due to abrupt changes in surface temperature or high winds, can cause rapid mixing of the water, drops in DO, and fish kills. Floating mats of vegetation generally contribute to low DO levels because they suppress phytoplankton, reduce aeration by reducing wind and wave action on the water surface, and use up oxygen when they decay (Welcomme 1979).

SOILS AND NUTRIENT TRANSFORMATION

Much of the biogeochemical activity of floodplains, including denitrification and methanogenesis (see below), depends on anoxic conditions and the interface between well-oxygenated and anoxic zones in floodplain soils and at the soil-water boundary (Mitch and Gosselink 2000; Ford et al. 2002). During periods of disconnection between river and floodplain, these biogeochemical hot spots occur primarily within wetlands, at interfaces between groundwater flow paths and riparian vegetation, and at the interface between the channel and the floodplain, including the hyporheic zone. During periods of connection and inundation (the "hot moments") much larger areas of the floodplain can contribute to high rates of biogeochemical activity as the interface between oxygenated and anoxic conditions expands and contracts.

As described above, DO in floodplain waters is often initially high upon first inundation and low DO or anoxic conditions subsequently occur as organic matter begins to decay and floodplain soils become saturated (Welcomme 1979). Soil texture strongly influences soil oxygen conditions and biogeochemical reaction rates. Soil texture can be highly heterogeneous across a floodplain due to variable patterns of deposition and erosion and varying amounts of organic material (Pinay et al. 2000). Floodplain soils can range from being primarily mineral (e.g., recent channel deposits) to containing high proportions (>20–35%) of organic material (Naiman et al. 2005, 2010). Fine-textured soils are more likely to support anoxic conditions because they are poorly drained and diffusion of oxygen through saturated soil can be orders of magnitude slower than diffusion through a porous and well-drained soil. Floodplain soils subject to extended inundation can be characterized as *hydric*, meaning that they formed during periods of inundation sufficient to cause anaerobic conditions. Coarse floodplain soils, such as abandoned channel deposits with gravel, may lack hydric characteristics because they do not become sufficiently waterlogged to become anaerobic (Mitch and Gosselink 2000).

Within floodplain soils, a sequence of oxidation-reduction, or redox, reactions proceeds based on the availability of oxygen and other electron acceptors. Oxidation is broadly defined as a molecule's loss of hydrogen or an electron (and an increase in oxidation number), whereas reduction is the gaining of hydrogen or an electron (and a decrease in oxidation number). The decomposition of organic matter proceeds fastest when oxygen is available to act as a terminal electron acceptor for oxidation (Mitch and Gosselink 2000).

Anaerobic conditions can form in fine-textured floodplain soils within hours to days after inundation, although a thin oxidized layer can remain at the surface of the soil at the soil-water interface. Oxidized ions (e.g., Fe^{3+}, Mn^{4+}, NO^-_3, and SO^-_4) may remain in this layer, while reduced forms are found in lower layers (e.g., ferrous and manganous salts, ammonia, and sulfides). As oxygen is depleted, the soil's redox potential (a measure of electron availability in a solution) drops and a sequence of other chemicals act as the electron acceptor. For example, one of the initial reactions that occurs following oxygen depletion is the reduction of nitrate (NO^-_3) to nitrite (NO^-_2), creating N_2O or N_2 gas (figure 4.1). With increasingly lower redox potential, a series of other ions become electron acceptors, such as manganese (manganic compounds reduced to manganous compounds), then iron (ferric ions are reduced to ferrous ions), then sulfur (sulfates are reduced to sulfites). Under the lowest redox potentials (the most reduced conditions), carbon dioxide or organic matter itself becomes the electron acceptor and is reduced to methane gas (CH_4) and other low-molecular-weight organic compounds. This is referred to as methanogenesis and produces what is commonly called "swamp gas" (Mitch and Gosselink 2000), a major contributor to global warming.

OTHER FACTORS INFLUENCING BIOGEOCHEMICAL PROCESSES

In addition to hydroperiod, floodplain sediment (soil) characteristics, and oxygen dynamics, floodplain biogeochemical activity is also influenced by the water chemistry of surface and groundwater inflows. The nutrient load and ionic composition of river and floodplain water is a function of characteristics of the precipitation, geology, and land use of the contributing watershed (Welcomme 1979). Spink et al. (1998) sampled floodplain soils along seven temperate rivers in Europe and North America and observed that soil mineralization rates and nutrient availability were positively correlated with levels of nutrients in river water.

Other properties of the floodplain itself also influence biogeochemical activity, including the available pool of organic matter, which is a function of vegetative productivity, time since last flood (Valett et al. 2005), and land-use practices on the floodplain. For example, widespread grazing in tropical and subtropical floodplains results in major contributions of manure, which can locally influence chemical and nutrient conditions (Welcomme 1979). Flooding history can influence biogeochemical processes as prolonged absence of flooding can lead to greater accumulations of organic matter. Valett et al. (2005) found that a floodplain that had been disconnected from the river for over 50 years (long "inter-flood interval" or IFI) had three times the standing stock of large wood and an order of magnitude greater standing stock of leaf litter than a floodplain that was still connected and thus frequently flooded (short IFI). When the long IFI floodplain was experimentally inundated, respiration rates were 2–4 times greater than those of the short IFI floodplain, even as inundation increased respiration by 50 times on the short IFI floodplain. The high respiration rates led to low DO on the long IFI floodplain but not on the short IFI floodplain (although the hydraulic residence time also likely influenced respiration rates).

MERCURY

Recently, researchers have examined how mercury cycles in the Central Valley river system in California, United States, illustrating how floodplains influence the cycling of minerals and nutrients. Mercury was used during historic gold mining, which, along with mercury mines, introduced large quantities of mercury into rivers flowing into the Central Valley. Much of this mercury is still present in the Central Valley hydrologic system. Wetlands, including floodplains, provide the necessary conditions for converting inorganic mercury into methylmercury, a form that is toxic to animals including humans. Thus, mercury cycling is an important consideration for the management and restoration of California floodplains (Alpers et al. 2008).

Under low oxygen or anaerobic conditions, inorganic mercury can undergo methylation to form methylmercury, a toxin that can bioaccumulate through a food web. Methylation first requires the conversion of various forms of inorganic mercury into a reactive form ("reactive Hg(II)"), a process that is facilitated by high concentrations of dissolved organic matter (DOM). Second, reactive Hg(II) is converted to methylmercury by anaerobic sulfate-reducing bacteria. Because wetlands generally contain both high levels of DOM and low-oxygen or anoxic conditions, they tend to have much higher rates of methylation compared to open-water or other environments. Episodic wetting and drying—particularly complete drying between inundation events—can promote the formation of reactive Hg(II), and thus methylation. Because floodplains contain soils with high DOM, episodic anoxic conditions, and complete drying between inundation events, they are of particular concern for mercury cycling (Alpers et al. 2008).

Phytoplankton and other algae can assimilate aqueous methylmercury, thus allowing the toxin to enter the food web. Bioaccumulation can then occur within invertebrates, fish, mammals, and other organisms, with increasing concentrations at higher levels in the food web. Henery et al. (2010) found that juvenile Chinook salmon rearing in the Yolo Bypass, California, a seasonally inundated floodplain environment (see chapter 13), accumulated methylmercury faster in their tissues than did juvenile salmon rearing in the Sacramento River. They noted that this differential uptake of methylmercury as juveniles would become insignificant, in terms of adult tissue concentration, because juveniles rear in these different environments for only relatively short time and adults are roughly three orders of magnitude larger than juveniles.

RETENTION OF NUTRIENTS

Riparian forests and riverine wetlands have been recognized as filters that reduce the loading of nutrients, including nitrate and phosphorous, into rivers and streams from adjacent agricultural fields and uplands. The ability of riparian vegetation and wetlands to reduce nutrient loading into surface waters has been observed at scales ranging from the field (Peterjohn and Correll 1984; Woltemade 2000; Forshay and Stanley 2005;) to the landscape (Omernik et al. 1981; Lowrance et al. 1984; Osborne and Kovacic 1993). In general, the presence of wetlands, including riparian forests and floodplain wetlands, will result in a watershed retaining more nutrients, and exporting more organic material, than a watershed lacking these wetlands (Mitch and Gosselink 2000).

Floodplains are frequently sinks for nitrate (Tockner et al. 1999; Forshay and Stanley 2005; Valett et al. 2005) either through plant uptake, which may lead to temporary retention, or through loss of nitrogen gas following denitrification (figure 4.1). For example, approximately two-thirds of the nitrate load entering an experimental floodplain along the Rio Grande, Texas, United States, was retained by the floodplain (Valett et al. 2005), while a Danube River floodplain in Europe retained 45% of the nitrate it received (Tockner et al. 1999). Woltemade (2000) reported that wetlands could remove up to 68% of nitrate from water draining agricultural land. As described below, residence time strongly influences the ability of a floodplain to process and assimilate nitrogen. For example, due to low residence time of water in floodplains during a flood event in 2011, BryantMason et al. (2013) found that the Atchafalaya River Basin, which flows into the Gulf of Mexico, retained only 7% of the nitrate that flowed into it. For this system to remove more nitrogen, Piazza (2014) recommends increasing connectivity between rivers and floodplains and other biogeochemical "hot spots" where longer residence can promote sedimentation and denitrification. For the Mississippi River basin as a whole, Mitsch et al. (2001) recommend large-scale restoration of wetlands and floodplain forests (10–25 million ha within the 300 million ha basin) to significantly reduce the export of nitrogen to the Gulf of Mexico, which is impaired by excess nutrients contributing to a hypoxic zone (see chapter 11).

Denitrification is a critical step in the nitrogen cycle as it is the primary way that nitrogen is lost from terrestrial and aquatic systems and returned to the atmosphere. Riparian and floodplain soils can serve as locations of denitrification, which requires a source of nitrogen oxides (such as nitrates), anaerobic conditions, heterotrophic denitrifying bacteria, and organic material that can serve as an electron donor (figure 4.1). Denitrification can occur when water with nitrates intercepts an anoxic zone in which microbial denitrification can take place, such as in wetlands and in the hyporheic zone (Burt and Pinay 2005). Denitrification can occur across broad areas during extensive floodplain inundation as the spatial extent of saturated, anoxic

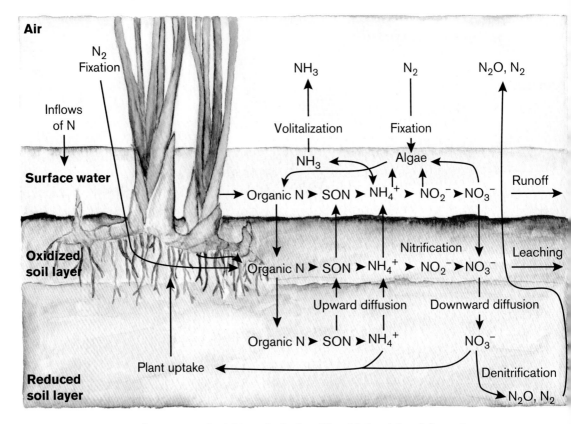

FIGURE 4.1 Nitrogen transformations in a floodplain wetland (adapted from Mitch and Gosselink 2000).

soils expands. High residence time is important for this to occur (Woltemade 2000), and thus greater rates of denitrification occur on broad floodplains than on narrow floodplains along high gradient rivers (Burt and Pinay 2005).

Sheibley et al. (2006) used laboratory techniques and mesocosms to investigate the potential for the Cosumnes River floodplain (see figure 13.2) to remove nitrate. They found that the denitrification potential of the floodplain soil varied widely and was influenced by soil properties such as the percentages of organic matter, total nitrogen, and clay in the soil. Extrapolating from the rates of denitrification within mesocosms on the floodplain, Sheibley et al. (2006) estimated that the Cosumnes floodplain could remove a small proportion of the total annual flux of nitrate in a wet year (1–4%) but a relatively high proportion in a dry year (24%) because of differences in residence times and biogeochemical processing between the large floods of a wet year and the small floods in a dry year. These estimates may be high because the mesocosm experiments provided optimal conditions for removing nitrate.

By monitoring actual flood events, Ahearn et al. (2006) reported that the Cosumnes River floodplain retained 17% of total inorganic nitrogen during a flood event. The varying retention rates are strongly influenced by the size of the floodplain relative to the contributing watershed and other factors that influence hydraulic residence time (Woltemade 2000).

Several other nitrogen transformations occur in floodplains including nitrification, ammonification, and nitrogen fixation (Mitch and Gosselink 2000; figure 4.1). Nitrogen fixation entails the conversion of gaseous N_2 to organic nitrogen by bacteria or cyanobacteria (blue-green algae). Vascular plants, such as the riparian tree white alder (*Alnus rhombifolia*), can fix nitrogen through a symbiotic relationship with nitrogen-fixing bacteria of the genus *Rhizobium* (Mitch and Gosselink 2000; Naiman et al. 2010). Within riparian forests of red alder (*A. rubra*) in the Pacific Northwest, rates of nitrogen fixation can exceed 300 kg/ha/year (Naiman et al. 2010). Mitsch and Gosselink (2000) provide detailed descriptions of the biogeochemical transformations of nitrogen and several other nutrients.

Because phosphorous is often adsorbed to sediment particles, it is usually retained by floodplains through deposition of sediment, and phosphorous retention rates are often correlated with sediment deposition rates (Noe and Hupp 2005). Thus, it is the hydraulic characteristics of riparian vegetation (high roughness) and wetlands (low velocity) that promote phosphorous retention through sediment deposition (Burt and Pinay 2005). A riverine swamp along the Illinois River, United States, received 10 times as much phosphorous as was released from the swamp, with the majority of phosphorous retention occurring through sediment deposition during a flood (Mitsch et al. 1979). Retention of phosphorous has also been observed in experimental riverine wetlands in Illinois (Mitsch et al. 1995) and in a restored and managed floodplain along the Rio Grande (Valett et al. 2005).

Hydraulic residence time (i.e., the age of water on a floodplain) exerts a strong control over biogeochemical fluxes. Deposition rates of particulate matter, sediment, and the nutrients adsorbed to sediment increase with residence time. Similarly, longer residence time allows for more opportunities for biological assimilation (e.g., phytoplankton taking up nutrients from the water column) and biogeochemical transformation. The residence time during a specific flood event can determine whether a floodplain is a source or sink for a given constituent.

Infrastructure that disrupts upstream and local hydrological connectivity can greatly influence biogeochemical activity, both by reducing direct inundation of floodplains and by reducing residence time (Gergel et al. 2005). River-floodplain systems in which the river and channel are rarely connected have much lower rates of sediment and nutrient retention than hydrologically connected river-floodplain systems. Thus, reconnection of floodplains has been recommended as a strategy for reducing sediment and nutrient loading to downstream systems such as estuaries (Noe and Hupp 2005).

TEMPORAL SEQUENCES AND FLUXES

Tockner et al. (1999) describe three phases of river-floodplain connectivity that govern floodplain biogeochemistry and productivity, based on their study of a floodplain along the Danube River (Austria). During Phase I, the floodplain is hydrologically decoupled from the river; nutrients in floodplain water bodies, such as oxbow lakes, are low as phytoplankton compete for resources and undergo high grazing pressure from zooplankton. This can be described hydrologically as the "disconnection" phase and ecologically as the "biotic interaction phase." During Phase II, floodplain water bodies receive nutrient-rich groundwater inflows due to rising river stage and, hydrologically, this phase is referred to as "seepage inflow." The increased nutrient availability and long residence time spurs development of phytoplankton and, thus, biologically this is the "primary production phase." This phase can also be characterized as a "flow pulse" (Tockner et al. 2000) with below-bankfull connectivity between river and floodplain.

During flood pulses (Phase III), large volumes of river water enter the floodplain, resulting in low residence time of water on the floodplain and greatly increased discharge from the floodplain. Much of the transport of material to and from the floodplain occurs during these connections; thus, Tockner et al. (1999) refer to this period as the "transport phase." For example, 84% of the discharge from the floodplain and greater than 90% of the flux of suspended solids (SS), POC, fine particulate organic matter (FPOM), and coarse particulate organic matter (CPOM) occurred during the transport phase. All of this flux occurred during only 3% of the sample period (Tockner et al. 1999).

Based on their research on the Cosumnes River floodplain (California, United States), Ahearn et al. (2006) and subsequent unpublished research expanded on Tockner's conceptual model of floodplain biogeochemistry dynamics and further partitioned the flood pulse/transport phase into three subphases: the flushing, transport, and draining phases. A flood pulse can begin with a *flushing phase*, as river water displaces antecedent floodplain water, characterized by the initial export, or "flushing," from the floodplain of water with a chemical signature distinct from that of the river water. During flushing, solutes, sediment, and organic matter are exported from the floodplain as they are pushed out by incoming river water. The flushing phase ends when floodplain outflow and inflow have similar chemical signatures. Thus, the flushing phase is defined by chemical as well as hydrological parameters and will not occur during all flood events; a flushing phase requires antecedent water on the floodplain that arises from a nonriverine source (e.g., direct precipitation) or, if riverine in origin, has developed distinct characteristics in terms of chemistry, turbidity, or temperature. During the flushing phase the Cosumnes floodplain exported DOC and chlorophyll *a* (Chl *a*).

The *transport phase* begins when floodplain inflows and outflows have similar water characteristics. Residence time is low during the transport phase, with mean residence time estimated at only 1.2 hours on the Cosumnes floodplain. Due to low residence time, little processing occurs on the floodplain and so the floodplain is not a source of products produced by floodplain processes, such as Chl *a*. However, because the floodplain has lower velocities and greater roughness than the river, deposition and retention of sediment and nutrients does occur during the transport phase.

The transport phase ends when floodplain outflow exceeds inflow, and the *draining phase* commences. During the draining phase, hydrological residence time increases and, during the event described by Ahearn et al. (2006) the Cosumnes floodplain again became a net exporter of DOC and Chl *a*. In large floods, fluxes of biogeochemical constituents are dominated by the transport phase, but the flushing and draining phases can be relatively important during small floods. Because small floods have proportionately more flow occurring during the long-residence time draining phase, during which processing can occur, a sequence of small floods are more effective at absorbing Nitrogen and producing Chl *a* (phytoplankton) than a single large flood event (for the same volume of water).

The hydraulic residence time of the transport phase has a particularly strong influence on biogeochemical processes on a floodplain. The relative size of a floodplain will strongly influence the residence time during transport phases, which will tend to dominate overall fluxes, and thus this relative size plays a large role in the biogeochemical behavior of the floodplain. For a floodplain that is small relative to its upstream watershed, such as the Cosumnes, the transport phase will be characterized by a very short residence time and minimal floodplain processing. In contrast, a floodplain that is large relative to its upstream watershed can have greater range of residence times during flood events, including the transport phase. These phases, and their influence on floodplain primary productivity and the export of biologically available carbon, are discussed further in chapter 7.

CONCLUSIONS

Just as flood waters change the floodplain, floodplains change flood waters through biogeochemical processes. The influence of these processes depends on connectivity and residence time of water. Floods that include long periods of high residence time are most likely to result in high biogeochemical activity, especially biological assimilation and biogeochemical transformations. The potential for floodplains to "filter" water of undesirable nutrients and chemicals is receiving greater attention and is an additional justification for restoring and managing floodplains (chapter 9).

FIVE

Ecology

INTRODUCTION

FLOODPLAINS SUPPORT COMPLEX and diverse ecosystems that are among the most productive on the planet. They can support levels of primary and secondary productivity that exceed that of many purely terrestrial or aquatic ecosystems (Tockner and Stanford 2002). Both the productivity and diversity of floodplains are strongly controlled by dynamic and variable connectivity with river and hyporheic flows (chapter 2). The intermittent pulse of flood waters is largely responsible for high floodplain productivity (chapter 7), while high flows induce erosion and deposition, which increases habitat heterogeneity (chapter 3). This heterogeneity in turn contributes to high levels of biodiversity (Salo et al. 1986). Riverine faunas, ranging from invertebrates to freshwater dolphins (Martin and da Silva 2004), use and often depend upon flooded habitats. Numerous fish species use floodplains for spawning, rearing, and feeding, and as refuge from harsh conditions found in flooding rivers, such as high turbidity and turbulent water. Waterfowl flock to flooded lands to feed, rest, and escape predators, especially during migration. Flooding renews riparian forests through enrichment of soils and creation of optimal conditions for recruitment of many trees and shrubs. Flooding of land often creates a flush of nutrients and detritus that supports blooms of zooplankton and other invertebrates, which are fed upon by fish, waterfowl, and other animals. Not surprisingly, floodplains support some of the largest freshwater fisheries in the world, especially in Southeast Asia (Welcomme 1979).

This chapter has two overarching themes. First, temperate floodplain ecosystems are intricately linked to the ecosystems of rivers that flow through them; both riverine and floodplain ecosystems are strongly influenced by the river's flow regime. Second, hydrological connectivity between river and floodplain across a broad range of flows is required to maintain floodplain ecosystem processes and river-floodplain interactions. We begin by discussing these two key concepts—flow regime and connectivity—before discussing how they together underpin processes that create and maintain dynamic habitats on floodplains. We conclude with a review of several major conceptual frameworks for river-floodplain ecosystems that prominently feature connectivity and dynamic processes.

FLOW REGIME AND CONNECTIVITY

Floodplain ecosystems are intimately linked to river flow characteristics, including magnitude, frequency, duration, seasonality, and rates of change (Poff et al. 1997; Yarie et al. 1998; Trush et al. 2000). As a result, floodplain ecosystems respond to a broad range of flows. For example, the water table within the floodplain hyporheic zone is influenced by the stage in the adjacent river and so periods of low river flow correspond to a lowering of the floodplain water table; floodplain vegetation and water bodies, such as ponds and wetlands, are influenced by the fluctuating level of the water table over time. Even during periods of low flow, subsurface connectivity between rivers and floodplains can help maintain water-loving shrubs and trees such as willows and cottonwoods. River flows rising to just below bankfull reinitiate flow into secondary channels and, through the rising water table, provide subsurface inflow to floodplain water bodies (chapter 2). Frequent, long-duration overbank flooding supports and maintains numerous ecological processes (Junk et al. 1989), while rare, high-magnitude events induce erosion and deposition on the floodplain, generating a shifting mosaic of topographic features and habitat types (Stanford et al. 2005, chapter 3).

Due in large part to hydrologic variability, the physical characteristics of active floodplains are constantly changing in time and space. Floodplain patches often cycle through stages, sequentially displaying characteristics of terrestrial, wetland, lentic, and lotic habitats within a single year. In some systems, a flood pulse occurs in most years with predictable timing and often inundates large portions of the floodplain. In other rivers, such as the Brazos River, Texas, United States, flood timing is unpredictable in terms of season, and the extent of inundation varies dramatically (Zeug et al. 2005).

The typical timing of peak flows in rivers varies from region to region. Arthington (2012) describes 15 distinct flow regime groups from around the world based on Haines et al. (1988), which are distinguished in part by timing of peak flows. Eleven of the groups occur in temperate regions. Flood seasonality is relatively predictable in most rivers, except those that are categorized as temperate aseasonal rivers (chapter 1). For example, in regions with a Mediterranean climate, flooding occurs in winter and early spring. However, even in rivers with predictable seasonal flooding, event timing, extent, and duration can vary within a region and between years due to shifting precipitation patterns, topography, and other factors. The biota of floodplains reflects these differences in variability. Native fishes in regions with fairly predictable flooding, such as the Central Valley of California, are often adapted to use floodplain habitats, while such adaptations are largely lacking in fishes of highly unpredictable systems such as the Brazos River, Texas, and Murray-Darling River, Victoria, Australia.

Floodplains are subject to processes of erosion and deposition from both in-channel flows (e.g., erosion of a cutbank) and overbank flows and other processes such as channel avulsion (chapter 3). Thus, patches of substrate and vegetation are continually being destroyed, reworked, and created, resulting in a shifting mosaic of habitat types (Salo et al. 1986; Richards et al. 2002; Ward et al. 2002, Stanford et al. 2005). These geomorphic processes also lead to subtle differences in floodplain topography; relatively small differences in elevation can result in biologically significant differences in distance to the water table during the nonflood season and in differences in depth and duration of flooding. These differences exert strong influences on plant community composition, also contributing to the mosaic of vegetation types on the floodplain (Poff et al. 1997; Stella et al. 2011).

Floodplain ecosystem processes depend on connectivity and a dynamic flow regime and this dependence explains why floodplains—and particularly those in temperate regions—

have been extensively degraded or converted to other land uses: most water-management activities alter connectivity and river flow regimes. Dams affect longitudinal connectivity and disrupt downstream movement of sediment and nutrients, reducing or eliminating their deposition on floodplains. Dams also disrupt or eliminate upstream and downstream movement of organisms, such as fish species that spawn in headwaters and rear in floodplains. Dams with large storage capacities alter river flow regimes, especially those dams managed to capture flood waters and reduce flood crests. Similarly, levees and floodwalls disconnect floodplains from river flows to protect land uses that are not compatible with inundation.

The dependence of floodplain ecosystems on flow and connectivity means that floodplain restoration often focuses either on reconnecting floodplains or on restoring components of the flow regime, or both (Opperman et al. 2013). On regulated rivers, Yarnell et al. (2015) recommend combining these two restoration needs by integrating functional flows with management of floodplain surfaces to promote connectivity. The functional flows approach recognizes that reinstating the full natural flow regime is unrealistic in managed systems, and instead proposes to manage reservoir releases in order to recreate targeted components of the natural hydrograph. The targeted components support the most essential geomorphic, ecological, or biogeochemical processes, such as peak flows that can drive geomorphic change on floodplains. Integrating functional floodplains within water-management systems is a major focus of chapters 10–15 of this book.

DIVERSITY, DISTURBANCE, AND SUCCESSION

Diversity

Floodplains are often described as centers or hotspots of biodiversity (Hauer et al. 2016), although data for global patterns of floodplain-dependent or floodplain-associated species are lacking (Tockner and Stanford 2002). Floodplains are hypothesized to support high levels of biodiversity for three main reasons. First, floodplains connected to natural rivers experience frequent ecological disturbance events (floods, erosion, deposition) that alter existing habitats and thereby release resources or create new space for colonization. Over time, this dynamism results in an ever-shifting array of habitat types that support high levels of biodiversity (Salo et al. 1986). Second, as floodplains undergo dramatic seasonal changes, they can be temporarily occupied by species from adjacent ecosystems, including both terrestrial (forests on higher terraces) and aquatic (river channel) ecosystems. For example, insects such as grasshoppers will colonize floodplains during extended dry periods. They are forced to move to higher elevations at the onset of flooding, attracting large, diverse flocks of birds to feast on this easily available prey. Third, temperate floodplains are subject to invasions by alien (nonnative) species (chapter 9), which may temporarily increase species richness but can contribute to extirpations of some native species in the long run. Increasingly, temperate floodplain ecosystems support food webs containing both native and alien species, functioning as what are best described as novel ecosystems (chapter 9).

Disturbance

The general pattern of frequent but variable flooding appears to provide an intermediate level of disturbance that can increase levels of plant species richness in a region (i.e., the intermediate disturbance hypothesis [IDH]; Connell 1978). Within active floodplains, disturbance is frequent enough to prevent competitive exclusion by a few dominant species of plants (Salo et al. 1986; Pollock et al. 1998), while enough refugia from frequent disturbance exist to ensure that the system contains more than just "weedy" plant species (Grime 1979, 1988). Pollock et al. (1998) found that patterns of species richness of plants on floodplains corresponded to predictions of Huston's dynamic-equilibrium model

(Huston 1979), which considers productivity as well as disturbance to predict species richness. This model predicts that the greatest plant species richness will occur on sites with intermediate levels of disturbance (e.g., flooding) and intermediate levels of productivity. Whether IDH and the Huston model apply to animals using floodplains remains to be tested. Dodds (2009) notes that there is little direct evidence for IDH in floodplain and stream ecosystems overall because so many other factors, including magnitude, timing, and length of floods, also affect biodiversity. The number of potential influencing factors makes it hard to define precisely what "disturbance" or "intermediate disturbance" means to floodplain organisms other than plants. Regardless, research on riverine food webs has found that frequent flooding can increase food-web length and increase the flow of energy to higher trophic levels (Wooton et al. 1996), which potentially increases the diversity of organisms within a floodplain. During a disturbance event, such as very large flood that can scour the floodplain, the heterogeneous topography and vegetation on the floodplain can contain low-velocity refugia for aquatic species, such as within jams of large wood. Following a disturbance, recolonization follows the same three axes of connectivity described in chapter 2: longitudinal, lateral, and vertical (Peters et al. 2016).

If flooding can increase biodiversity, the absence of flooding should, over time, reduce floodplain biodiversity (Sparks et al. 1990). This reduction occurs through loss of physical processes that create diverse habitat mosaics and because plant species that do not depend on flooding are favored, including those that can form monospecific stands (e.g., tamarisk). Thus, from an ecological perspective, floods are not destructive events, but rather they are the driver needed to create and maintain diverse habitats in which floodplain biotas thrive (Stanford et al. 2005). Prevention of floods is consequently the greatest stressor to floodplain ecosystems in the modern world (see the discussion of succession below).

Various studies have described the importance of floodplains or floodplain-associated habitat types (e.g., wetlands and riparian forests) to biodiversity (Hauer et al. 2016). For example, Stanford et al. (2005) reported that nearly 70% of regional vascular plant species were found on one floodplain site, the Nyack floodplain on the Middle Fork of the Flathead River, Montana. Although riparian/floodplain habitats represent less than 1% of the land area in the Inyo Mountains of southeastern California, approximately three quarters of vertebrate species in this region use riparian corridors at some point during their life history (Kondolf et al. 1987). Similarly, wetlands, many of which are on floodplains, represent approximately 5% of US land surface but are critical habitat for 35% of endangered species (Kusler 1996). Aquatic species are imperiled at far higher rates than terrestrial species (Ricciardi and Rasmussen 1999) and this is due, in part, to losses of floodplains and, more generally, to widespread hydrological alteration, including loss of heterogeneity in aquatic habitats (Poff et al. 1997; Richter et al. 1997; Tockner and Stanford 2002). Remnant floodplains often contain habitat types that have elsewhere been converted to other land uses, such as forests within a landscape dominated by agriculture. They thus harbor species that have been extirpated elsewhere in the region. In general, the importance of riparian vegetation and other floodplain habitats is magnified in dry areas or in areas with highly variable precipitation patterns (Jolly 1996).

Due to the processes of channel migration and floodplain erosion and deposition, a mosaic of floodplain landscape elements (i.e., habitat or patch types) is continuously being created and destroyed (Stanford et al. 2005). When a patch of habitat is destroyed at one location, that same habitat type has likely been created by other processes at other locations along the river corridor; the overall proportion of a given element may thus remain relatively stable in the landscape over time (Galat et al. 1998; Naiman et al. 2010). For example, floodplain lakes and other water bodies are continually being created through channel migration and avulsion. Through successional processes over time, these lakes become marsh and eventually revert to terrestrial habitat. While individual lakes come and go, the overall proportion of lake to marsh to terrestrial plain may remain relatively constant if the floodplain is active (Welcomme 1979), similar to the concept of a "shifting-mosaic steady state" described by Bormann and Likens (1979). Floodplain ecosystems must be quite large and connected to rivers with natural flow regimes to sustain these steady-state dynamics without continuous human intervention (Ward et al. 2002).

Succession

Floodplain habitats can undergo succession, the processes through which an ecosystem undergoes changes in species composition and structure over time (Salo et al. 1986; Greco and Plant 2003). The relatively frequent disturbance regime of floodplains results in the creation of floodplain surfaces (e.g., bare ground) or features (e.g., an oxbow lake) where ecosystems can develop through succession. In the absence of disturbance, successional processes can lead to forests that are more uniformly mature and less diverse in terms of plant species, age classes, and structure (Decamps et al. 1988). Through the hydrarch succession process, floodplain water bodies gradually fill with silt and organic matter and transition from open water to marsh to terrestrial habitat (Amoros 1991).

Dynamic hydrologic and geomorphic processes are required to counteract the tendency for successional processes to homogenize channel forms, water bodies, and floodplain plant communities (Welcomme 1979; Ward et al. 2002). High-magnitude flows reset the successional stage of plant communities by removing mature forest through erosion while creating new areas for regeneration through sediment deposition. Similarly, high-magnitude flows can reset the successional stage of an oxbow lake by removing a sediment or vegetation plug or scouring out the lake bed (Amoros 1991). Further, very large flows can lead to widespread erosion of the floodplain, creating areas of bare ground and new floodplain water bodies, or to channel avulsion and associated floodplain sedimentation and new channel formation (Ward et al. 2002).

River-management actions, such as adding levees and regulating flow, generally reduce dynamic processes and connectivity, also reducing the frequency of disturbance events. Thus, over time, river management generally allows successional processes to progress toward a more homogenous and less complex mosaic of habitats (Bravard et al. 1986; Decamps et al. 1988; Amoros 1991; Barnes 1997; Ward et al. 2001). For example, because flow regulation of the Danube River has reduced the frequency of large floods, existing water bodies are being converted to terrestrial habitats within the floodplain below Vienna (Tockner and Schiemer 1997). Hale (2004) found that floodplain reserves on the Wisconsin River that were not connected to rivers with dynamic hydrographs lost early successional communities. Successional processes in floodplain forests are discussed further in chapter 6.

The processes of island formation and destruction in braided rivers effectively illustrate the concepts of disturbance, succession, and habitat mosaics (figure 5.1). High-flow events erode and reshape or destroy islands while incipient islands are simultaneously being created, often behind wood jams in the channel (Abbe and Montgomery 1996; Edwards et al. 1999; Gurnell et al. 2005). Shifting channels can lead to islands becoming connected to the riparian forest and, conversely, channel avulsion can lead to a portion of the riparian forest becoming an island. These dynamic processes result in a diversity of forest ages and structure (Gurnell and Petts 2002). Rivers with flow regulation generally have fewer islands and tend to transition to single-thread channels flanked by mature riparian forests.

The processes of island formation, evolution, and destruction integrate attributes of hydrology, geomorphology, and vegetation, and require a dynamic riverine environment and

FIGURE 5.1 Island forming near the floodplain in the braided Chilcotin River, BC, Canada. Large wood being transported by the river becomes stranded on gravel deposits in the shifting channel. The wood can trap other pieces of wood, causing additional gravel deposition on the bar. Trees and shrubs colonize the protected gravel, and under the right conditions, the island can grow to a fairly large size and support a diverse floodplain forest, as the island in the background does. In this photo, the stranded wood is on the upstream end of the bar, while new vegetation is established on the downstream end (photo by Carson Jeffres).

connectivity between river and floodplain. Ward et al. (2002) suggest using islands (e.g., number per river kilometer) as landscape-level indicators of river corridor condition and of the extent to which natural processes are operating, including biological processes. Similarly, Dykaar and Wigington (2000) suggest using the area of islands and bars as a geomorphic indicator of river-floodplain integrity on the Willamette River in Oregon. They reported an 80% decline in island and bar area between 1910 and 1988 due to flow regulation and bank armoring. In Central California, flow regulation and bank armoring of major rivers have contributed to loss of islands and the simplification of river channels toward single-thread channels (The Bay Institute 1998). Such simplification results in loss of ecosystem complexity and biodiversity.

MAJOR CONCEPTUAL FRAMEWORKS FOR RIVER-FLOODPLAIN ECOLOGY

While floodplains are intimately connected to rivers, conceptual models developed for river ecology have not always reflected that floodplains and rivers function as an integrated biophysical system. One reason for this is that nearly all large rivers in the temperate world were heavily altered prior to scientific study (Benke 1990; Bayley 1991; Dynesius and Nilsson 1994; Nilsson et al. 2005). Initially, aquatic ecologists did not fully recognize the extent of this alteration and so did not appreciate the importance of floodplains to riverine ecosystems (Bayley 1991). Below we review major conceptual frameworks for river-floodplain ecology; the evolution of these frameworks over time reflects the growing appreciation for the importance of floodplains.

RIVER CONTINUUM CONCEPT

The central concept guiding river ecology in the temperate world in recent decades has been the River Continuum Concept (RCC; Vannote et al. 1980). As a model, the RCC describes how riverine biota, sources of energy, turbidity, and temperature regime change in a downstream direction with increasing stream order (figure 5.2).

A. River Continuum Concept (RCC)

Longitudinal gradient with local effects

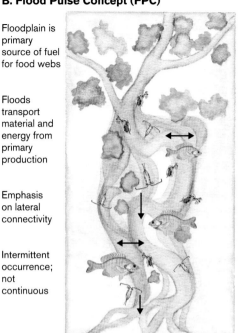

Shredders
Predators
Collectors
Grazers

Flow direction

B. Flood Pulse Concept (FPC)

Floodplain is primary source of fuel for food webs

Floods transport material and energy from primary production

Emphasis on lateral connectivity

Intermittent occurrence; not continuous

C. Riverine Productivity Model (RPM)

Emphasis on contribution of edges in food-web structure

Local primary production

Leaves and wood are sources of carbon

Addresses gaps in river continuum and flood pulse concepts

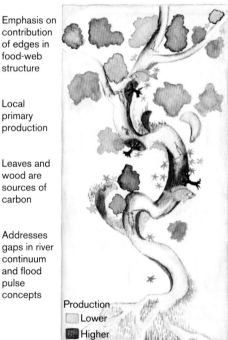

Production
Lower
Higher

D. River Wave Concept

High floodplain autochthonous production and allochthonous inputs (FPC explains best)

High upstream allochthonous inputs (RCC explains best)

High local instream autochthonous production and allochthonous inputs (RPM explains best)

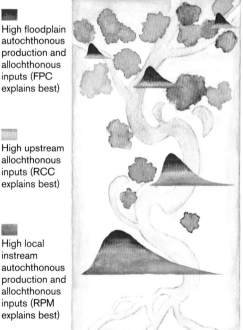

FIGURE 5.2 River ecosystem models describe energy sources, biotic production, and change through space and time.
(A) The river continuum concept describes a longitudinally linked gradient punctuated by local effects.
(B) The flood pulse concept emphasizes lateral connectivity with the floodplain.
(C) The riverine productivity model focuses on the contribution of edges.
(D) The river wave concept integrates components from preceding models, describing energy flows as waves in time and/or space. (See Humphries et al. 2014.)

The RCC predicts that invertebrate diversity is maximized in middle-order reaches and that low-order (headwaters) and high-order reaches (large rivers) will have lower diversities of invertebrates. For large rivers, the RCC predicts that most productivity will be derived from upstream sources because depth and turbidity will limit autochthonous sources (internally produced, such as algae and aquatic plants) and the small ratio of channel perimeter to channel width will limit allochthonous (externally produced) sources, such as organic material from riparian forests. Thus, the RCC predicts that nearly all carbon available to drive large-river food webs comes from upstream, suggesting that downstream reaches depend on the "leakiness" of upstream reaches.

The RCC has proved to be a useful model for discussion and several studies promptly tested the concept, providing various examples of confirmation, disagreement, and refinement (e.g., Cuffney 1988). Examining several large rivers for consistency with the RCC, Sedell et al. (1989) reported that confined rivers with minimal floodplains conformed relatively well with the RCC, while rivers with extensive floodplains conformed less closely, noting that the most productive areas in rivers are those connected to floodplains.

FLOOD PULSE CONCEPT

Junk et al. (1989) recognized the absence of floodplains in the RCC and, in response, developed the Flood Pulse Concept (FPC). Based on research conducted on unregulated rivers, including both tropical (the Amazon) and temperate (the Illinois) rivers, the FPC introduced a lateral component to the longitudinal, headwaters-to-mouth, paradigm of the RCC. The FPC posited that, within reaches with extensive floodplains, ecosystem productivity was largely controlled by lateral interactions of river and floodplain and not by longitudinal linkages of headwaters to main-stem river (figure 5.2). Junk et al. (1989, p.112) state:

The pulsing of the river discharge, the flood pulse, is the major force controlling biota in river-floodplains. Lateral exchange between floodplain and river channel, and nutrient recycling within the floodplain have more direct impact on biota than the nutrient spiraling discussed in the RCC (Vannote et al. 1980). We postulate that in unaltered large river systems with floodplains in the temperate, subtropical, or tropical belt, the overwhelming bulk of the riverine animal biomass derives directly or indirectly from production within the floodplains and not from downstream transport of organic matter produced elsewhere in the basin.

Thus, the FPC posits that the position of a patch of habitat along a lateral transect from the river channel through a floodplain has greater influence on ecosystem processes within that patch than does its position along a longitudinal transect, from headwater to mouth, as predicted by the RCC.

Many large lowland rivers feature a predictable annual flood pulse event that connects the river with its floodplain and allows for exchange of water, energy, nutrients, and organisms. This connectivity provides an energetic subsidy to the riverine ecosystem in three ways. First, river organisms can follow the expanding edge of water as it moves across the floodplain and directly access food resources on newly flooded land, such as terrestrial insects and plants. Junk et al. (1989) refer to the floodplain inundated during the flood pulse as the "aquatic-terrestrial transition zone." For example, fish utilizing the floodplain can eat leaves, fruits, and nuts of floodplain trees, as well as terrestrial invertebrates and other organisms caught in flood waters (Welcomme 1979; Chick et al. 2003). Second, during sustained flood pulses, floodplains become areas of high productivity in the water column (chapter 7) with food webs driven by autochthonous production (phytoplankton, periphyton, and aquatic macrophytes) and decomposition of terrestrial vegetation. Rates of decomposition and nutrient cycling may be increased by sequential wetting and drying (Junk et al. 1989). Although the

FPC emphasizes the importance of terrestrial resources to the floodplain subsidy, recent research suggests that this may have been overemphasized and that carbon derived from autochthonous sources within the inundated floodplain, particularly various forms of algae, is more labile (biologically available) than carbon derived from terrestrial plant sources (Winemiller 2004; Junk and Bayley 2008). These two mechanisms for energy subsidy hinge on riverine organisms accessing expanded sources of productivity on the floodplain. A third process that subsidizes riverine ecosystems is transport of carbon, nutrients, materials, and organisms back into the river as the floodplain drains during flood recession. In this way, floodplain productivity can benefit organisms that do not actually move onto the floodplain.

Sequential terrestrial and aquatic productivity, and its exchange through dynamic flow regimes and connectivity, results in rapid recycling of organic matter and the characteristic high productivity of river-floodplain systems (Bayley 1991). For example, production of animal biomass on floodplains is estimated to be 3.5 times that of terrestrial habitats (Tockner and Stanford 2002). Bayley (1995) describes floodplain ecosystems as having a "flood-pulse advantage." This advantage refers to the significantly greater per-unit-area production of fish within rivers with a dynamic flow regime and connectivity with a floodplain compared to rivers or reservoirs lacking such connection via a flood pulse. Supporting the flood-pulse advantage, Risotto and Turner (1985) found that the amount of bottomland hardwood forest, a proxy for extent of floodplain, was a significant predictor of fish biomass within a given reach of the Mississippi River.

Several studies have found that floodplain productivity can be exported to riverine ecosystems, consistent with the FPC. In Georgia's Ogeechee River, the flux of organic matter from floodplains to the river exceeded primary productivity in the river by a factor of seven (Benke 2001). Furst et al. (2014) found that, during the 2010–2011 flood in the Murray-Darling system, the Chowchilla floodplain (67 km^2) exported a large quantity of zooplankton into the Murray River (approximately 6 tonnes/day dry weight), increasing the abundance of zooplankton downstream of the floodplain compared to upstream. Junk et al. (1989) state that large, piscivorous fishes that are adapted to main-channel habitats, and do not themselves move onto floodplains, often rely on prey species that do feed and grow on floodplains. Floodplains can continue to export nutrients to rivers during baseflow periods through hyporheic flow (Stanford and Ward 1988). Floodplain productivity can also be exported to adjacent terrestrial ecosystems, as when spiders feed upon emergent insects produced in floodplain habitats (Ward 1989). However, other studies have found that floodplains are relatively closed systems that retain most of the productivity generated or processed within them (Hamilton and Lewis 1987; Lewis 1988). Thus, the extent to which floodplain productivity supports downstream riverine ecosystems or adjacent terrestrial ecosystems requires further study (Ballinger and Lake 2006).

The FPC was derived from systems with floods that occur relatively predictably and have considerable duration, during a time of year that allows high productivity because of high light levels and warm temperatures (Johnson et al. 1995). The predictability of flooding likely influences the extent to which biota adapts to floodplain habitats. In general, species in regions with unpredictable flood pulses do not exhibit behavioral or life-history adaptations that take advantage of floodplains (Johnson et al. 1995; King et al. 2003). In fact, unpredictable floods may be harmful to species (Junk et al. 1989).

Johnson et al. (1995) posit that floods of intermediate duration maximize productivity. If floods are of short duration, autochthonous production will be minimal and many species will not be able to complete the floodplain-dependent portions of their life histories. If flood pulses are very long, floodplains can take on characteristics of a lotic system and experience stagnation and low levels of dissolved oxygen. Further, a very long flood pulse will limit the terrestrial growth phase of plants (Ahn et al. 2004), which is a key part of the productivity advantage of river-floodplain systems. Intermediate levels of flooding have also been posited to maximize diversity of habitats and species on floodplains (Richards et al. 2002) although the relationship between flood flows and biodiversity is complex and thus rarely fits simple models (Dodds 2009).

In the decades after publication of the FPC, researchers have attempted to adapt it to a broad range of temperate systems, including those in which flooding may be less predictable, of shorter duration, and/or occurs during times of the year less conducive to productivity (e.g., winter, with short daylight and low temperatures; Thorp et al. 1998; Gutreuter et al. 1999; Tockner et al. 2000; Sommer, Harrell et al. 2001). Although the transferability of specific aspects of the FPC to some temperate systems has been questioned (Thorp et al. 1998), the general concept of a flood-pulse advantage for species, especially fishes, has been confirmed in numerous temperate systems (e.g., Gutreuter et al. 1999; Sommer, Nobriga et al. 2001).

Tockner et al. (2000) extended the FPC to a broad range of river systems by expanding beyond the concept's original focus on predictable, long-duration events. They emphasized the importance of a diversity of flows, ranging from below-bankfull "flow pulses" to rare major flood events. They also drew a distinction between unpredictable and high-magnitude "erosive flooding" events and the long-duration flood pulses of the FPC. Erosive flooding shapes the geomorphology of rivers and floodplains and thus provides the physical template for habitat heterogeneity and biodiversity, while flood pulses promote productivity on the floodplain and exchange of material between rivers and floodplains. In addition to the overbank flood pulses emphasized by the FPC, below-bankfull events, termed "flow pulses," also have ecological importance for both the river and the floodplain by influencing habitat conditions within floodplain channels, the water table, and hydrology of floodplain water bodies. This emphasis on the ecological importance of a broad range of flows largely parallels the concept of the natural flow regime (Poff et al. 1997).

Tockner et al. (2000) also expanded on the FPC to emphasize the importance of diverse sources of flood water, including direct rainfall, tributary water, overland runoff, soil water, and rising groundwater, in contrast to the FPC's original emphasis on lateral overflow from a large river. The source of water can have an important effect on ecological responses to flooding due to differences in temperature, nutrients, and hydraulic properties. For example, overbank flow can have high velocities, while hyporheic inundation has very low or no velocity (see chapter 2).

SHIFTING HABITAT MOSAIC

Tockner, Lorang et al. (2010, p.77) state that the Shifting Habitat Mosaic (SHM) concept "recognizes that the interaction of physical and biological processes in a floodplain produces a continually changing spatial pattern of habitats fostering high biodiversity." Stanford et al. (2005) suggest that the SHM can unify many of the diverse conceptual models and theories of how river floodplains function as biophysical systems. The SHM focuses on how hydrologic processes create, maintain, and change diverse patches of habitat across three dimensions on a floodplain (longitudinal, lateral, and vertical).

Flooding drives geomorphic processes such as meander migration and avulsion that contribute to various channel forms, water bodies (e.g., from cutoff events and localized scour on the floodplain), and diverse patches of varying age and vegetation composition. In addition to contributing to heterogeneity of surface habitat patches, these geomorphic processes also contribute to heterogeneity of subsurface patches. For example, the process of channel abandonment results in coarse subsurface deposits that provide preferential pathways for groundwater through a floodplain. During periods of low flow, the interaction of water from various sources—surface, upwelling from the subsurface, and water entering from other sources such as tributaries—contributes to a range of water quality, temperature, and productivity across the floodplain. For example, primary productivity is much higher in zones of upwelling because the water contains greater nutrients.

The habitat mosaic of a floodplain is composed of patches that vary in their successional stage, productivity, water quality, species composition, and other characteristics; emergent properties of floodplains occur through the interactions between these patches (Tockner, Lorang et al. 2010). The proportion and location of these patches change through time in response to rapid events, such as a flood clearing established vegetation, and processes that occur over longer timescales, such as vegetation establishing and maturing on fresh alluvial deposits (Stanford et al. 2005).

Tockner et al. (2010) suggest that the SHM and FPC can collectively serve as the "two fundamental concepts" that explain the development and evolution of floodplain ecosystems and processes. They indicate that the FPC applies primarily to processes that occur during a period of inundation with little attention to the physical turnover of biophysical structure or habitats of the floodplain. In contrast, the SHM emphasizes the physical turnover of structures and habitats in response to flood events and ecosystem responses (e.g., forest regeneration) over time.

A conceptual model for Central Valley floodplains (Opperman 2012) provides a similar division, with processes split among submodels. The first submodel focuses on processes that develop and maintain the complex physical template of the floodplain, corresponding to the SHM. Other submodels focus on processes that occur during inundation events, such as increased primary productivity and exchange of organisms between patches, corresponding to the FPC.

RIVERINE PRODUCTIVITY MODEL

Thorp and Delong (1994) introduced the Riverine Productivity Model (RPM) as a response to both the RCC and FPC. While the RCC predicts that large-river food webs are driven by the export of organic matter from the upstream watershed and the FPC posits that large-river food webs are dominated by floodplain productivity, the RPM holds that large-river food webs are driven by local autochthonous production, including phytoplankton, periphyton, and macrophytes, and direct inputs from the riparian zone, not limited to periods of floodplain inundation. Thorp and Delong (1994) argue that although organic carbon derived from upstream sources may represent the largest pool of carbon within a river reach, this imported organic matter tends to break down slowly and be less biologically available than the smaller pool of more labile carbon from autochthonous and local allochthonous sources, such as that contained in fresh leaves rather than leaf fragments that have already been highly processed by bacteria and insects upstream. Thus, the carbon assimilated into food webs can largely originate from internal or local sources even though total ecosystem carbon is dominated by detritus from upstream sources.

A later iteration of the RPM (Thorp and Delong 2002) maintained the emphasis on the importance of autochthonous production by algae and de-emphasized the importance of local riparian inputs. Thorp and Delong (2002) cite numerous studies that demonstrate the importance of algal pathways in food webs, even where algae represented a relatively small proportion of the overall available carbon. The relative importance of various carbon sources, including from upstream, local riparian, and within-floodplain sources, is elaborated in chapter 7.

RIVER WAVE CONCEPT

Humphries et al. (2014) offered the River Wave Concept as a single conceptual model that

> **BOX 5.1 · Miniature Floodplains: Unappreciated Phenomena?**
>
> The focus of most temperate floodplain studies is on the expansive seasonal floodplains of large rivers and on the extensive flooding that often occurs there. However, floodplain processes also occur at much smaller scales and, as on larger floodplains, these processes depend on flow regime and connectivity. The various river conceptual models discussed previously suggest that small-scale floodplains can cumulatively export productivity downstream to rivers. Here, we discuss different manifestations of "miniature floodplains," including those that flank small streams and also the edges of larger streams or rivers that are activated during minor increases in flow. This discussion also applies to narrow floodplains along the edges of channels that are constrained and therefore lack access to more expansive floodplains.
>
> Halyk and Balon (1983) reported that floodplain pools along a small stream in Ontario, Canada, provided an important contribution to fish production in the system. Similarly, a mountain meadow is essentially a miniature floodplain flanking the stream flowing through it and providing important habitat for birds, amphibians, and fishes (Purdy et al. 2012). Mountain meadow streams typically spill over their low banks onto the meadows during snowmelt events. Meadows can retain a large volume of water that is slowly released during the summer months, maintaining mesic soil conditions for plant growth in the meadow and maintaining streamflow during dry periods. In some mountain regions, meadows can be quite extensive along streams in mid-elevation valleys and these not-so-miniature floodplains can provide important habitat for salmonids, including migratory forms (Bellmore et al. 2012).
>
> Channel edges, including the edges of large rivers with steep banks, can also serve as miniature floodplains. Flow pulses that inundate edge habitat are often associated with successful reproduction of certain fish species, because this edge habitat provides many of the same benefits for fish as do large, inundated floodplains. These benefits include cover, lower water velocities, warmer temperatures, and higher flood densities. For example, a six-week flow increase in Putah Creek, California, in spring greatly improves spawning success of native fishes (Kiernan et al. 2012). While most of Putah Creek is incised and thus not connected to extensive floodplains, this flow pulse "connects" the river to edge habitats where larval fish can rear in a 1–2 m wide band of flooded vegetation along the stream edge (P. Moyle, unpublished observations).

draws on three of the concepts discussed above (RCC, FPC, and RPM). The River Wave Concept considers streamflow as a series of waves travelling longitudinally (downstream) and laterally (on and off floodplains), with the waves varying in shape, amplitude, wavelength, and frequency (figure 5.2). There are two basic types of river waves: those characterized by flood peaks in response to precipitation events in the watershed, which show high crests and move rapidly through the system, and those created by baseflow conditions, with long wavelengths. The shapes of these waves are further influenced by the interactions of climate, hydrogeology (e.g., groundwater storage), and geomorphology (e.g., extent of connected floodplains).

The River Wave Concept describes a lowland flood pulse characteristic of the FPC as a long, low-amplitude wave at the lower end of the river that is the summation of shorter, higher-amplitude waves from upstream tributaries, which may each have somewhat different timing of high-flow events. As predicted by other models, upper tributaries provide organic matter inputs that accumulate in the stream and riparian zone during low-flow periods. This organic matter is carried downstream by the crest of the wave of flow and can be deposited on floodplains as the wave amplitude declines.

Humphries et al. (2014) argue that, because stream biotas evolved in response to the flow characteristics of their local rivers, they are sensitive to the shape, amplitude, wavelength, and frequency of river waves. These characteristics are analogous to the ecological importance of flow magnitude, duration, and frequency described by the natural flow regime (Poff et al. 1997). Understanding these adaptive responses therefore allows biologists and managers to better predict what will happen as flow regimes change, whether from natural events, flow management, or climate change.

Conceptual frameworks presented here share a common emphasis on the underlying importance of flow regime and connectivity for maintaining floodplain processes and the benefits they provide. The importance of flow regime and connectivity hold true even for small and unusual floodplains (box 5.1).

CONCLUSIONS

Floodplain ecosystems are intricately linked to the rivers that flow through them and are structured by hydrologic (chapter 2), geomorphic (chapter 3), and biogeochemical processes (chapter 4). Both riverine and floodplain ecosystems are strongly influenced by the river's flow regime. Hydrological connectivity between river and floodplain across a broad range of flow levels is required to maintain floodplain ecosystems and river-floodplain interactions. These basic attributes of floodplain ecosystems are common themes within the broader conceptual models that explain how riverine ecosystems function. Flooding translates downstream through longitudinal connectivity, integrating basin-scale inputs of runoff, sediment, nutrients, and organic matter. As a rule, flooding can only drive geomorphic and ecological processes on floodplains where they are hydrologically connected to the river. In a given region, there is likely to be a direct, positive relationship between total floodplain area connected to rivers and levels of productivity, biodiversity, and ecosystem services supported by floodplains.

Floodplain-dependent species and communities are therefore vulnerable to the numerous management activities that reduce longitudinal connectivity and flow variability (e.g., dams and diversions) and lateral connectivity with floodplains (e.g., riprap and levees). Because this water-management infrastructure is widespread, particularly within temperate river basins, floodplain ecosystems and their associated species have been greatly reduced in extent across the temperate world (Tockner and Stanford 2002). Further, floodplains that remain connected are often highly degraded and feature biota with a high proportion of nonnative plants and animals. They can consequently be characterized as novel ecosystems (chapter 1). River and flood managers are increasingly recognizing the value of remnant floodplains, not just for their environmental values but also as part of flood-risk-reduction strategies. This recognition is also catalyzing the reconnection of portions of previously disconnected floodplains (Opperman et al. 2009). Using floodplains to store and convey flood waters allows them to function as "green infrastructure" within flood-management strategies. Additionally, these connected floodplains can be managed to provide multiple benefits, including restoration of biodiversity and ecosystem services (Tockner et al. 2010); this multipurpose management is the focus of chapter 10.

Designing systems that deliver multiple benefits requires that both scientists and managers understand the effects of ongoing and potential management actions on basic floodplain processes and on the novel ecosystems that most temperate floodplains now support. In the next three chapters, we first examine more closely the ecological importance of complex plant communities on floodplains (chapter 6) and how floodplain processes affect ecosystems through primary and secondary production (chapter 7). We then assess how vertebrates, especially fishes, use both natural and modified floodplains (chapter 8).

SIX

Floodplain Forests

FLOODPLAINS SUPPORT HIGH productivity of valued species, especially fishes and waterfowl. These species are supported by floodplain food webs that are fueled by primary productivity from sources both local and distant and by species from single-celled algae to massive trees. In chapters 6–8, we focus on various components of these food webs. This chapter begins with a discussion of the zonation of plant species on floodplains, from aquatic to terrestrial, before focusing on vascular plants that grow on land surfaces that are only intermittently flooded. For convenience, we refer to these plants as "terrestrial" vegetation. Terrestrial sources of primary productivity—including trees, shrubs, and grasses—grow on floodplain surfaces that are only temporarily or occasionally flooded and they photosynthesize primarily between periods of inundation, although some grasses may start to grow while flooded. Due to their importance on temperate floodplains, the chapter emphasizes floodplain forests (often referred to as riparian forests).

ZONATION

Floodplain vascular plant communities are patterned into zones that reflect the interactions of soil characteristics, hydroperiod, and flood disturbance (Stella et al. 2013). Floodplain soils are heterogeneous due to variable sediment sizes as well as to variable organic matter and nutrient content (Naiman et al. 2005). The typical hydroperiod—the repetitive seasonal patterns of inundation and/or water table level—and soil characteristics create regular, predictable patterns of vegetation zonation, while intermittent disturbance from high-energy flood events introduces stochasticity into these patterns by opening new space for plant colonization (Spink et al. 1998; Stanford et al. 2005).

In this chapter, we use the term "floodplain forest," but intend for this term to be equivalent to "riparian forest." Generally, riparian forests are regarded as being characteristic of river edges, while floodplain forests are characteristic of wide reaches of valley floor, with multiple channels. However, the two types of forest are in reality difficult to separate in areas with extensive floodplains.

Welcomme (1979) described five primary vascular plant zones on floodplains:

1. Permanent water with submerged vegetation only (e.g., oxbow lakes)
2. Permanent flooded areas with rooted or floating emergent vegetation
3. Seasonally flooded areas with emergent vegetation
4. Areas that are occasionally flooded (between mean flood and highest flood levels)
5. Areas that are not flooded, but whose water table is influenced by the flood regime

Zones 4 and 5 include riparian/floodplain forest, grassland, or savanna, depending on the hydroperiod, soil type, and time since disturbance. The local climate and depth to groundwater also determine the vegetation composition of the floodplain. In regions with sufficient precipitation year-round, floodplains tend to be forested, although in many parts of the world previously forested floodplains have become grasslands or savannas through clearing and grazing (Welcomme 1979).

FLOODPLAIN FORESTS AND RIVER FLOW REGIMES

Temperate floodplains are often covered in a mosaic of patches of forest with different ages, species composition, and structure (Stanford et al. 2005). Many trees characteristic of floodplains display specific adaptations to the dynamic floodplain environment (Kozlowski et al. 2002). For example, they may have seeds that can float with seed release synchronized with the natural flow regime so that high flows can distribute seeds to appropriate germination sites (Stella et al. 2006). Reflecting their position on surfaces frequently exposed to geomorphic processes, floodplain trees often have flexible stems and the ability to grow adventitious roots, such as roots that sprout from a trunk or stem that has been buried in sediment (Naiman and Descamps 1997). Through the growth of adventitious roots, new trees can establish from vegetative propagules—including stems, branches, or even whole live trees—that have been moved and deposited by a flood (Francis and Gurnell 2006; Opperman et al. 2008).

Due to factors such as nutrient-rich soils and access to water, floodplain perennial plants, especially trees, often have high growth rates. As a result, temperate floodplain forests tend to have higher basal area and stem density than adjacent upland forests (Naiman and Descamps 1997). The high growth rate and productivity of floodplain forests is due in part to periodic inundation events that deposit sediment, nutrients, and organic matter (Yarie et al. 1998). In semiarid regions, flooding contributes to this productivity by recharging groundwater and thus increasing the water available during the growing season, particularly if flooding occurs during or relatively soon before a period of drought stress (Robertson et al. 2001; Stella et al. 2013). Extended periods of inundation, due to extended flood events, management actions, or beavers, can reduce growth of floodplain forests and lead to mortality of floodplain trees (Mitsch et al. 1979; Conner et al. 1981). For example, flood-control operations on the Roanoke River, Virginia (United States) produced artificially long overbank flows, causing mortality of saplings within the floodplain forest (Pearsall et al. 2005). In fact, Jackson and Pringle (2010) caution that in highly modified river systems, *reduced* connectivity between river and floodplain may be desirable at times, such as where "novel" riparian vegetation has established following alteration of natural flood regimes and become a refuge for rare species.

Floodplain forest processes and patterns—including regeneration, succession, and heterogeneity—are linked to various components of the hydrograph and the dynamic hydrologic and geomorphic processes of the floodplain environment (Decamps et al. 1988; Trush et al. 2000; Bendix and Stella 2013; Marks et al. 2014). The relationship between floodplain forests and a variety of flow characteristics represents one of the best-studied examples of linkages between hydrological variability and floodplain ecosystems. Multiple flow levels have significance for riparian forests, including those ranging from periodic high-energy floods to minimum baseflows. Forest dynamics are also influenced by other characteristics of the hydrograph, such as recession rate and interannual variability.

Cottonwoods and poplars (*Populus* spp.) provide an illustrative example of the relationship between riparian/floodplain tree population dynamics and river flow regimes (Lytle and Merritt 2004); they have a life history that is characteristic of many temperate riparian/floodplain tree species, such as willows (*Salix* spp.). Cottonwoods release seeds in spring that are transported by either wind or water (Braatne et al. 1996). To establish, cottonwood seedlings require bare, mineral soil with abundant sunlight and soil moisture. Thus, cottonwoods regenerate most effectively on recently deposited alluvial substrate, such as sand and gravel on point bars. Strahan (1981) found that cottonwoods and willows established only on fully exposed alluvium and no cottonwood or willow seedlings were found below the canopy of mature trees. To establish, survive, and grow, seedlings' roots require contact with moist soil. As river stage declines, the roots must grow downward to keep up with declining water tables. Seedlings experience high mortality if the water table drops faster than the rate of downward extension of root systems (Kranjcec et al. 1998; Stella et al. 2010). On point bars, the water table is strongly influenced by river levels and so cottonwood establishment is strongly influenced by the rate of recession of river flows as they transition from high spring flows to low summer flows (Mahoney and Rood 1998).

Because seedling establishment depends on presence of bare, moist substrate, cottonwoods have evolved to release seeds during periods of high runoff when appropriate establishment sites are most likely to have been created. For example, on the Tuolumne River, California, Stella et al. (2006) found that the period of maximum cottonwood seed release could be predicted by a degree-day model which, combined with historical hydrological and temperature analysis, indicated that cottonwoods release their seeds to coincide with peak runoff during spring snowmelt.

Although cottonwoods are tolerant of inundation, seedlings and small saplings can suffer high mortality due to physical scouring or deposition during subsequent floods (Wilcox and Shafroth 2013). Therefore, conditions for large pulses of cottonwood recruitment are characterized by specific patterns of interannual variability: a large flood in the year of seedling establishment, which provides for extensive areas of freshly deposited alluvium for establishment, followed by several years of smaller floods that minimize scour mortality of saplings. These relationships between hydrological variability and riparian tree dynamics have been summarized in conceptual models such as the Recruitment Box Model (Mahoney and Rood 1998; figure 6.1), which Rood et al. (2005) recommend as a basis for restoring flood flows in diverse floodplain river systems.

Because of these specific flow-related requirements for regeneration, establishment of cottonwoods and other riparian vegetation is tightly coupled with flow regime. On regulated rivers, capture of floods by reservoirs can diminish availability of fresh alluvium and thus diminish rates of seedling establishment. Reservoir management can also lead to rapid recession rates that can desiccate seedlings (Mahoney and Rood 1998). For example, in the western United States, reservoirs that store water for irrigation often capture much of the snowmelt flood pulse, which can change the recession limb of the spring flood pulse below the dam from a gradual decline to a sharp decline, resulting in very rapid lowering of river levels and water table elevation. Along regulated rivers, regeneration may be restricted to a narrow riparian band where the water table is close enough to the surface to allow roots to reach it during early growth stages. However,

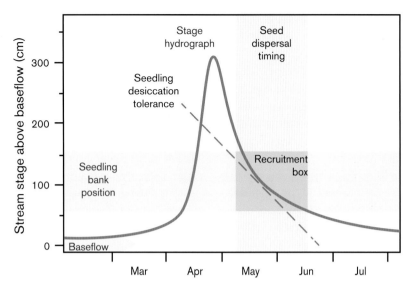

FIGURE 6.1 Recruitment box model for cottonwood trees. The box represents the window of time, defined by the hydrograph and period of seed dispersal, when recruitment is possible (after Mahoney and Rood 1998).

this band is often very close to the active channel where seedlings may be exposed to prolonged inundation or may become scoured or buried during high flows (Wilcox and Shafroth 2013; Kui et al. 2014). Regeneration of cottonwoods and similar species is thus often greatly diminished below dams, and riparian forests can become dominated by older trees that established under a previous flow regime (Bradley and Smith 1986; Rood and Mahoney 1990; Snyder and Miller 1991; Stella et al. 2003). The loss of dynamic flows and disturbance on the floodplain can result in a greater proportion of riparian vegetation reproducing through vegetative propagation or clonal growth, rather than through sexual reproduction through seed germination (Douhovnikoff et al. 2005).

In addition to establishment along point bars, abandoned channels provide important sites for regeneration of riparian vegetation. Stella et al. (2011) found that nearly half of the area of floodplain forest along a reach of the Sacramento River was associated with abandoned channels. Regeneration associated with abandoned channels, or other off-channel floodplain features, may be particularly important in rivers in which flow regulation or climate change has negatively impacted primary regeneration processes along point bars.

FOREST SUCCESSION AND SPATIAL PATTERNS

Floodplain forests are structured by both internal (autogenic) successional processes, such as vegetative competition and facilitation, and external (allogenic) successional processes, such as the riverine-controlled processes of scour, inundation, sediment deposition, and depth to groundwater (Edwards et al. 1999; see chapter 5 for a more general discussion of succession). Successional sequences are often initiated following a flood as seedlings become established on the bare mineral soil deposited by the flood; in fact, tree regeneration on newly deposited point bars is sometimes used as an example of primary succession, the ecosystem development that starts on bare soil and goes through a series of progressively more diverse stages (Clements 1936; Morris and Stanford 2011). Small floods can initiate successional processes in localized patches, whereas large, floodplain resetting floods (chapter 14) can do so across much of the floodplain (Lytle and Merritt 2004).

In addition to creating conditions that initiate successional processes, flood events interact with forest community dynamics throughout the successional sequence. For example, periodic sediment deposition from floods leads to aggradation of the forest floor, resulting in surfaces that become inundated less frequently, allowing different species to colonize. Further, some mid-successional species require mineral soil for effective regeneration, and are thus prevented from establishing beneath trees until a flood buries the accumulated organic litter layer with alluvium. The buried organic layer can then provide nutrient sources for developing trees (Yarie et al. 1998). While flooding and sediment deposition influences plant regeneration, plant growth affects weathering and nutrient availability of deposited sediment. Plant roots and decomposing organic matter dramatically increase rates of weathering of deposited sediments, increasing the availability of nutrients such as phosphorous (Naiman et al. 2010).

Within floodplains that experience natural, or near-natural, hydrologic and geomorphic processes, the distribution of trees and other plants, in terms of age, size, and species, is a function of the dynamic hydrologic processes of the river—which create and continue to influence the landforms where stands of trees grow—and autogenic successional processes that operate within the stands (Strahan 1981; Hupp and Osterkamp 1985; Fierke and Kauffman 2006). A meandering river provides the most apparent expression of the relationship between geomorphic processes and forest dynamics and distribution. Meandering rivers remove mature vegetation as they cut into banks while depositing substrate for new vegetation communities on the associated point bar, usually across the river (chapter 3). As a result of these processes, floodplains of a meandering river can display zonation of plant species' composition along a lateral gradient moving away from the river along a point bar to the beginning of upland vegetation (figure 6.2). This gradient represents both time and elevation; locations further away from the channel are older and higher elevation. Point bars and other depositional features near the active channel are dominated by early successional species such as cottonwoods and willows. Further away from the river channel is mid-successional forest containing species such as mature cottonwood and willow that had established earlier when that surface was bare alluvium. Additional trees species present are those that are capable of establishing and growing under the willow-cottonwood overstory. Even further from the river (and thus an older surface) are forests dominated by later successional species—those that established under the original overstory but now dominate, while the short-lived early successional species have diminished or disappeared from the stand.

Beyond zonation along lateral gradients, floodplain forests display other spatial patterns. While the discussion above emphasized an orderly pattern of forest along a meandering river, a variety of processes—including channel cutoffs and human disturbance—also influence the distribution of riparian tree species and age classes, thus introducing patchiness into the lateral gradients. Anastomosing rivers are dominated by geomorphic processes, such as avulsion, that are more episodic and less consistent than meander migration and therefore floodplain vegetation along anastomosing river channels is patchy rather than displaying distinct zonation of species or age classes (Ward et al. 2002).

At larger spatial scales, at the river landscape scale, floodplain forests can be described by the Shifting Habitat Mosaic concept (SHM; chapter 5), which holds that while specific floodplain plant communities may appear and disappear at the site level, largely driven by flood events, the total amount of each community will remain relatively constant at the landscape scale (Tockner et al. 2000; Lyle and Merritt 2004; Stanford et al. 2005). On the Flathead River, Montana, United States, the mosaic

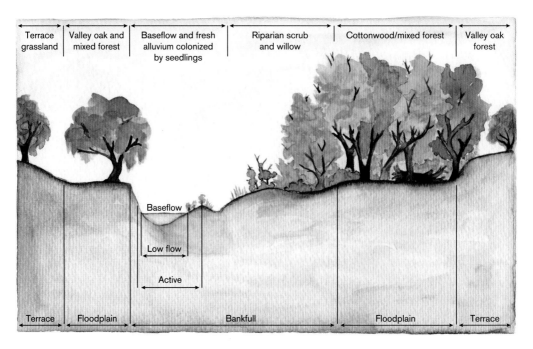

FIGURE 6.2 Generalized cross section of a historic Central Valley, California, floodplain, showing dominant woody vegetation. Along meandering rivers in California's Central Valley, point bars and early successional forests are dominated by Fremont cottonwood (*Populus fremontii*), red willow (*Salix lasiandra*), and arroyo willow (*S. lasiolepis*). Mid-successional forests include box elder (*Acer negundo*), black walnut (*Juglans hindsii*), and Oregon ash (*Fraxinus latifolia*). Higher on the floodplain are late successional forests that are rarely inundated and dominated by valley oak (*Quercus lobata*; Strahan 1981; Stella et al. 2003). Note that this figure illustrates a riparian corridor along a river with a natural flow regime; along rivers lacking regular flooding, mature forest can occupy more of the floodplain (chapter 5).

contained 10 distinct vegetation types or phases, ranging from an initial phase of bare substrate to mature forest (Egger et al. 2013). As described in chapter 5, river regulation can reduce the frequency and/or magnitude of flood flows that underpin this pattern, and so forests along regulated rivers may become more homogenous than a patchy mosaic. Geerling et al. (2013) emphasize the importance of integrating the SHM into floodplain restoration and propose *cyclic floodplain rejuvenation* as a basic concept to guide restoration projects. For this concept to operate, it requires having a large enough area of floodplain so that natural processes can work or, in the absence of adequate flows, enabling managers to manipulate the floodplain to mimic natural processes of disturbance and succession.

TREES AND GEOMORPHIC PROCESSES

The discussion above emphasizes how riparian plant dynamics are influenced by geomorphic processes on the floodplain. In turn, trees and shrubs can influence floodplain geomorphology (Gurnell and Petts 2002; Bendix and Stella 2013). For example, willows can grow on an alluvial plug, a deposit that forms between a side channel and the main channel. Willow growth can then promote further deposition on the alluvial plug, which can eventually lead to hydrologic isolation of the side channel, except during high flows. Very high flows can restore the connectivity by blowing out the alluvial plug and vegetation (Amoros 1991).

Floodplain forests are the source of large wood—both dead wood and live wood (Opperman et al. 2008)—that can exert considerable

FIGURE 6.3 A floodplain forest in the Atchafalaya basin (left) and the Lumber River, North Carolina (right) (left photo by Joseph Baustian and right photo by Amber Manfree).

influence on geomorphic processes (Naiman et al. 2010), ranging from influencing small-scale topographic variation on the floodplain (Florsheim and Mount 2002) to creating massive wood jams that lead to channel avulsion (Montgomery et al. 2003). Wood can be transported on to the floodplain from the upstream watershed, although much of the wood is derived from local inputs from channel migration and cutoff events (Naiman et al. 2010). Wood deposited on the floodplain during flood flows can promote deposition of fine sediment, which serves as substrate for riparian seedlings. Seedlings and saplings growing in this alluvium can also be protected from scour during subsequent high flows by the upstream wood. This process can lead to formation of forested islands in the floodplain (figure 5.1; Fetherston et al. 1995; Abbe and Montgomery 1996). Along the Tagliamento River, Italy, much of the initial regeneration on such nuclear islands comes from sprouts from partially buried trees, usually willows and poplars (Edwards et al. 1999; Francis and Gurnell 2006; Opperman et al. 2008).

Historically, volumes of wood in temperate rivers and floodplains were much higher than today, with much wood removed through "snagging" to improve navigation. The Atchafalaya once featured a logjam that was 170 km long, a physical structure that greatly increased the hydrological connectivity between the river channel and floodplains. Nearly a half century of effort was required to remove sufficient wood to make the Atchafalaya navigable, in 1880 (Wohl 2011). Similarly, the influence of wood from beaver activity is also reduced from previous levels. Throughout North America, beaver once numbered between 60 and 400 million and, by using wood to make dams, it is estimated they created 200 million acres of wetlands. By dramatically slowing drainage and runoff, this use of wood by beavers would have had widespread influences on hydrology and geomorphic processes in North America. On floodplains, beaver dams create ponds (Pollock et al. 2004) and affect vegetation through raising the water level. The influence of beavers may have been particularly important in arid regions where their dams raised the water table and effectively expanded the extent of wetland and riparian habitat (Andersen and Shafroth 2010).

HARDWOOD BOTTOMLAND FORESTS OF THE SOUTHEASTERN UNITED STATES

It is easy to forget how extensive temperate floodplain forests once were because today's remnant forests are mostly quite small. One of the most spectacular examples of little-appreciated floodplain forest is the hardwood bottomland forest of the southeastern United States (figure 6.3), which once covered about 40–50

> **BOX 6.1** · Can Fishes Help Forests?
>
> That fishes can benefit from riparian forests is fairly obvious: trees shade the water, making it cooler; provide complex cover when they fall into a river; provide substrate for invertebrate prey; and are important in shaping the topography of floodplains and stream channels. But there is growing realization that fish can benefit riparian forests as well. The classic studies of Goulding (1980) showed that fishes are major dispersers of seeds of tropical floodplain trees in the Amazon. Chick et al. (2003) have shown fish can also disperse seeds in temperate floodplains. They noted red mulberry (*Morus rubra*) and swamp privet (*Forestiera acuminata*) fruit were found in the stomachs of channel catfish (*Ictalurus punctatus*) in the floodplain of the Mississippi River, United States. They then demonstrated that, not only did the seeds germinate after being passed through a catfish digestive system, but also the process improved germination success. The importance of this phenomenon is not known, but this study suggests that dispersing plants by fish may be more common than realized.
>
> Another example of where fish benefit forests has been documented along rivers with healthy salmon runs. In these rivers, salmon, in the form of decaying carcasses, deliver marine-derived nutrients to riparian trees and floodplain forests (Naiman et al. 2010; Morris and Stanford 2011). Helfield and Naiman (2001) report that Sitka spruce (*Picea sitchensis*) in the riparian forests of rivers with salmon runs derived approximately a quarter of their foliar (leaf) nitrogen from marine sources, delivered through the decomposing bodies of salmon or by the processing of salmon through the digestive tracts of carnivores and scavengers. Trees supported by marine-derived nitrogen grew significantly larger than trees in riparian forests of rivers lacking salmon runs. Because larger trees benefit salmon habitat—particularly in the form of large instream wood—this represents a positive feedback loop between salmon and floodplains. Merz and Moyle (2006) found that salmon carcasses were also providing nutrients to an agricultural crop growing along the Mokelumne River (California, United States). Up to a quarter of the foliar nitrogen in *Vitis vinifera* plants was marine derived, an ecological pairing of wine and salmon.

million ha of coastal plain from Texas to Virginia. The forests were associated with the lower reaches of many rivers meandering across the low-gradient landscape (Stanturf et al. 2001). Today, most of that forest is gone, either converted to agriculture and/or isolated behind flood-control levees. Stanturf et al. (2001) estimate up to 200,000 ha could be restored to floodplain forest with a high degree of ecosystem function, although not necessarily identical to the historic forests that once existed.

These hardwood forests exist in an extremely flat alluvial landscape and so small differences in topography, measured in centimeters, can create large differences in plant communities. The most productive regions of the floodplain, featuring the largest trees, are on higher ground, such as the natural levees and other low ridges that are characterized by sandy or silty loams (Stanturf et al. 2001). Three additional forest types are recognized: cypress-tupelo, elm-ash-hackberry, and riverfront (Hodges 1997; Stanturf et al. 2001). The cypress-tupelo forest is associated with low areas that are permanently flooded (swamps). This forest type can remain unchanged for hundreds of years and some bald cypress stands are 200–300 years old. The elm-ash-hackberry forest is the most abundant forest type, occurring on sites with poorly drained soils that are typically flooded annually (Stanturf et al. 2001). The riverfront forest is characterized by trees such as cottonwood and shrubs such as willow as the early successional species; they are replaced by shade-tolerant maples (*Acer* spp.) and hackberry/sugarberry (*Celtis* spp.) if no tree-removing flood occurs.

These forests provide important habitat for wildlife and are especially renowned for supporting high diversity of both resident and migratory birds. They can also provide refuges for large mammals in the region. For example, in Louisiana, black bears (*Ursus americanus*) are now primarily found in remnant bottomland hardwood forest, along the Mississippi and Tensas Rivers and in the Atchafalaya Basin (Louisiana Department of Wildlife and Fisheries 2015). The forests are drained by bayous, essentially slow-moving streams that can be nearly static during late summer but that often support distinctive fish faunas (Chen et al. 2015). These bayous are the main delivery and exit system for flood waters during the winter wet season, so they can function more like rivers during periods of high precipitation and runoff. Their high biodiversity has made hardwood bottom forests a target of large-scale restoration programs, which have had varying degrees of success (Stanturf et al. 2000, 2001). However, there is growing realization that restoration also has benefits for flood-risk management, because farmland restored to forest does not have to be protected by levees.

CONCLUSIONS

Floodplain forests are productive and dynamic, with forest establishment and successional processes strongly influenced by hydrologic and geomorphic processes driven by river flows. In turn, floodplain trees—both alive and dead—affect geomorphic processes including avulsion and sediment deposition. Floodplain forests are important sources of habitat, food, and organic material for the ecological processes and species described in the next two chapters. As discussed in more detail in chapter 8, floodplain forests often support high levels of animal biodiversity, in part due to their diverse tree species, size classes, and structures (Naiman et al. 2005). In addition to supporting terrestrial species, when trees fall into the water they provide habitat and structure for fish and other aquatic species (box 6.1). Floodplain trees and other vascular plants also contribute large amounts of detritus that are processed within aquatic food webs, the focus of the next chapter.

The importance of such forests for biodiversity is magnified in regions where forests are otherwise rare due to climate (e.g., semiarid regions) or because of conversion to agriculture or other land uses (Jolly 1996). In areas such as the southeastern United States, floodplain forests often provide the last remaining large expanses of native forest in a region, important as refuges for a wide array of species. Restoration of these forests generally requires restoration of floodplain processes (Stanturf et al. 2001).

SEVEN

Primary and Secondary Production

FLOODPLAINS ARE TYPICALLY highly productive of organic matter and so floodplain food webs often support high concentrations of fish and wildlife. In this chapter, we review the foundations of food webs for higher trophic levels: primary production (from photosynthesis) and secondary production (from consumption of primary production). The study of primary and secondary production is important in stream ecology (Allan 1995; Cushing and Allan 2001), stimulated in part by the development of the River Continuum Concept (Vannote et al. 1980; chapter 5). However, most studies focused on streams when they are not flooding and on areas upstream of floodplains. Junk et al. (1989) proposed the Flood Pulse Concept (FPC), in part because the River Continuum Concept (RCC) did not adequately explain the sources of productivity in the lower reaches of rivers, especially on floodplains. For temperate regions, the study of primary and secondary production on floodplains was further constrained by the extreme alteration of river-floodplain systems. As a result of limited study of healthy systems and incomplete conceptual models, scientists held a general view that in situ primary production on inundated floodplains was not very important in riverine food webs. This view is changing.

In this chapter, we first look at terrestrial and aquatic sources of primary production that contribute to aquatic food webs on floodplains. We emphasize both terrestrial contributions, primarily through detritus, and the major role of in situ aquatic production. We then discuss secondary production, primarily from zooplankton and aquatic macroinvertebrates. Fish and other members of higher trophic levels are discussed in chapter 8. Microbial production is mentioned only briefly because of general lack of floodplain-specific information.

SOURCES OF PRIMARY PRODUCTION

Floodplain food webs are driven by carbon produced by plants both on the floodplain and imported from elsewhere in the river network (chapter 5). While the previous chapter discussed the ecology of "terrestrial" primary producers, such as floodplain trees, in this chapter we focus on sources of primary productivity that contribute to aquatic food webs on the floodplain (e.g., trees are discussed in this chapter only in terms of their contributions to food webs in the river or within the inundated floodplain).

Carbon from both terrestrial and aquatic sources can enter floodplain food webs in dissolved and particulate forms. Particulate and dissolved forms of organic matter are functionally distinguished by size; dissolved forms can pass through a 1.0 μm filter (Sobczak et al. 2002). Particulate organic matter (POM) is composed of phytoplankton, bacteria, protozoa, detritus, and other organic matter adsorbed to mineral sediment particles. Particulate forms are much more biologically available because organisms such as zooplankton can ingest them directly. Dissolved forms must first be converted to particulate forms by passing through the microbial loop, transforming dissolved organic carbon into particulate carbon in the form of bacterial or protozoan biomass. This additional step results in large losses in energy from respiration (Sobczak et al. 2002; Jassby et al. 2003).

During inundation, carbon sources to aquatic food webs include both external (allochthonous) and internal (autochthonous) sources. External sources include matter from the river, from floodplain forests, and from other terrestrial vegetation that grows when the floodplain is not inundated. Internal sources include phytoplankton, aquatic macrophytes, and emergent plants that grow following inundation.

DETRITUS IN FOOD WEBS

Detritus, generally defined as dead organic matter along with associated microorganisms, is a major contributor to the pool of POM and is derived both from riverine inputs and from vegetation growing in the floodplain, such as macrophytes, periphyton, and leaf litter from floodplain trees (Muller-Solger et al. 2002). Early studies of floodplain food webs emphasized the importance of detritus as the main source of carbon to metazoan food webs (Junk et al. 1989), in part because vascular plant material, both living and dead, seemed to dominate the carbon budget of floodplains (Hamilton et al. 1992). For example, Bayley (1989a) reported that the primary productivity of an Amazon floodplain was dominated by aquatic macrophytes (69% of total carbon) and forest litter (24%) with relatively small proportion represented by phytoplankton (5%) and periphyton (2%) although he acknowledged that he likely underestimated periphyton due to incomplete sampling.

Although detrital carbon from vascular plants can dominate the total pool of carbon available to floodplain food webs, such carbon is highly resistant to biological processing (refractory), compared to algal carbon, which can be processed quickly (labile; Thorp and Delong 1994). Earlier studies may have overemphasized the importance of detrital pathways because results gleaned from gut samples of detritivorous fish and other organisms assumed that ingestion of detritus was equivalent to assimilation. However, evidence from gut samples can be misleading because organisms may be ingesting detritus primarily for the value of the associated biofilm—periphyton, fungus, and bacteria that can coat the surface of detritus—and are primarily assimilating the labile components of the biofilm. Research with chemical markers (stable isotopes and fatty acids [FA]) has indicated that detritus is less important to the food webs that support higher trophic levels than previously thought and, in most floodplains, algae is the primary contributor to food webs (Hamilton et al. 1992; Thorp et al. 1998; Sobczak et al. 2005). We explore the question of sources of carbon for floodplain food webs in a later section of this chapter.

AQUATIC PRIMARY PRODUCTION

Aquatic primary production on floodplains comes from aquatic macrophytes and algae, both phytoplankton and periphyton. Macrophytes and algae can remain in permanent aquatic habitats, such as oxbow lakes or river channels, and respond to inundation by growing and reproducing within temporally flooded habitats. Phytoplankton are unicellular algae suspended in the water column, while periphyton consist of algae attached to inundated vegetation or other substrates. Major classes of algae include diatoms, green algae, and cyanobacteria (blue-green algae; Cushing and Allan 2001). Increasingly it appears that carbon derived from algal sources is the main source of carbon for floodplain food webs (Araujo-Lima et al. 1986; Hamilton et al. 1992; Thorp et al. 1998). Algae can be a major contributor to primary productivity in permanent floodplain water bodies and on broad expanses of floodplain during extended periods of inundation. The productivity of algae is regulated by four primary factors: light, nutrients, grazing by zooplankton, and hydrology.

Light

Photosynthetic activity increases with light levels so algal productivity is reduced in turbid waters but increases in flooded areas where fine sediments have dropped out due to lower water velocity. Due to higher light levels and warmer temperatures, algal growth is greater during spring or summer flooding compared to winter (Cushing and Allan 2001). Thus, the productivity described in the FPC (chapter 5) is most relevant for rivers where flooding generally coincides with periods of the year with warmer temperatures and greater light levels (Schramm and Eggleton 2006).

Nutrients

Algal growth depends on the uptake of dissolved nutrients from the water column. These nutrients are supplied through river inflow or from the processing of organic matter through biogeochemical pathways (chapter 4). Algal productivity can be limited by low nutrient levels, particularly nitrogen (Welcomme 1979).

On a California floodplain, Ahearn et al. (2006) found that phytoplankton growth was initially limited by nitrogen availability. Later in the inundation season, the phytoplankton community shifted from diatoms and green algae to nitrogen-fixing cyanobacteria, and further algal growth became limited by phosphorus availability (Grosholz and Gallo 2006). Phytoplankton blooms can deplete the water of nutrients leading to declines in productivity. Subsequent inundation, in situ mineralization of organic matter, or inflows of nutrients from other sources (e.g., tributaries) can replenish nutrients in the water column and continue to maintain phytoplankton growth (Schemel et al. 2004).

Grazing

In lakes, grazing by zooplankton has long been recognized as a major driver of phytoplankton abundance. When floodplain conditions allow zooplankton populations to establish and persist (e.g., periods with low water velocity and high residence time), zooplankton grazing pressure can lead to declines of phytoplankton (Grosholz and Gallo 2006). A similar relationship may exist for grazing on periphyton attached to aquatic macrophytes by herbivorous insects and snails.

Hydrology

Hydrology (flow) is the master variable that causes the algal community to shift in species and abundance during the course of an inundation event. Sampling in the Yolo Bypass, California, indicated that the initial phytoplankton response to inundation was from fast growing and small species, adapted to higher velocity, turbid environments, including picoplankton (*Synechococcus*), nanoflagellates, small diatoms, and filamentous cyanobacteria (Sommer, Harrell, Solger et al. 2004). Sommer, Harrell, Solger et al. (2004) characterized such phytoplankton species as "first responders" that help provide rapid initial primary productivity to support floodplain food webs. As residence time of the water increases in response to decreased flows, the phytoplankton community becomes increasingly dominated by larger species, such as diatoms and green algae. In the next sections, we explore how the flow regime and other factors affect primary productivity by phytoplankton, periphyton, and aquatic plants.

PHYTOPLANKTON

The density of planktonic algae in flowing water systems is generally inversely proportional to flow velocity because of downstream displacement of cells. High densities of phytoplankton are generally restricted to slow-moving waters and thus during initial phases of flooding are primarily found in floodplain water bodies, in backwaters or low-velocity river channel margins (Reckendorfer et al. 1999; Schiemer et al. 2001). Biomass accumulation of phytoplankton will not occur if the hydrological residence time is shorter than phytoplankton growth rates (Schemel et al. 2004). Thus, phytoplankton productivity generally increases with increasing hydrological residence time (Hein et al. 2004). Concentration of chlorophyll-a, an indicator of phytoplankton productivity, was positively correlated with residence time of floodplain water in California's Central Valley (Schemel et al. 2004; Ahearn et al. 2006). However, the relationship between residence time and productivity is not linear because with increasing residence time, phytoplankton can decline from nutrient depletion and grazing by zooplankton (Grosholz and Gallo 2006).

Based on work on a Danube floodplain, Tockner et al. (1999) partitioned floodplain aquatic primary productivity dynamics into three phases of ecological processes that correspond to the three hydrologic phases described in chapter 4. Here we focus on the third phase ("transport"), which features surface connections between the river and floodplain. Ahearn et al. (2006) expanded this concept further by partitioning the transport phase into three sub-phases: flushing, transport, and draining phases. During the flushing stage, flood waters push antecedent water off the floodplain, potentially exporting algal biomass from floodplain water bodies into the river. During the transport phase, phytoplankton productivity on the floodplain is limited by low residence time and may be restricted to shallow, littoral zones (analogous to the "inshore retention concept" of Reckendorfer et al. [1999] and Schiemer et al. [2001]). Finally, phytoplankton biomass increases and peaks during the draining phase as water velocity decreases and residence time, water temperature, and water clarity all increase. However, draining is often interrupted by additional flood pulses. During draining periods with long hydrological residence time, phytoplankton productivity can increase with water depth, primarily because deeper water lacks emergent vegetation. Emergent vegetation can reduce phytoplankton productivity due to (1) shading; (2) allelopathic chemicals from macrophytes, and (3) higher densities of zooplankton associated with vegetation that can exert grazing pressure (Ahearn et al. 2006).

Ahearn et al. (2006) noted that the peak in chlorophyll-a in water leaving the floodplain during the draining phase can be attributed to both an increase in overall residence time of floodplain water and because more peripheral parts of the floodplain begin to drain. These more distal areas included shallow littoral zones that had greater residence time and thus greater algal productivity and biomass; this productivity is exported as the distal areas drained. Therefore, the peak of algal biomass is due to both temporal and spatial heterogeneity of the water. The falling limb or draining stage of the hydrograph produces more algal biomass. By draining, the areas that have produced high algal biomass during previous parts of the flood cycle mix with the rest of the floodplain water for overall export. Both the Cosumnes River and Yolo Bypass floodplains (in the Central Valley of California, United States) produce peak concentrations of phytoplankton during the

FIGURE 7.1 Attached filamentous algae mats growing on aquatic macrophytes and emergent vegetation, Cosumnes River floodplain, California, United States (photo by Jeff Opperman).

draining phase that can greatly exceed the threshold required for growth of zooplankton populations (Muller-Solger et al. 2002). Because a long residence time generally promotes the greatest phytoplankton productivity, Ahearn et al. (2006) suggest that maximum export from the floodplain to river would occur through several flood pulses, with sufficient temporal intervals between them. Managers could achieve this pattern in areas where flooding could be managed (e.g., through gates or reservoir releases).

PERIPHYTON

In addition to phytoplankton, periphyton can be an important source of algal productivity (Douglas et al. 2005). Periphyton consists of algal films and filamentous growth on fixed surfaces, such as logs, terrestrial vegetation, and aquatic macrophytes (figure 7.1). Periphyton is generally difficult to measure accurately and so there are few quantitative studies of their role; one study that did attempt to measure productivity of epiphytic (attached) periphyton found approximately 10 times as many attached algal cells as floating cells (Welcomme 1979). Though difficult to measure, some evidence suggests that periphyton may contribute more to food webs than do phytoplankton (Welcomme 1979). For example, Bunn et al. (2003) found that filamentous algae were the primary source of energy for aquatic consumers in Australian floodplain water bodies. Lopes et al. (2015) found that periphyton provided the primary source of energy for bottom feeding fish sampled in floodplain habitats of the Paraná River (Brazil).

In addition to providing a food source, periphyton contribute to habitat complexity;

long strands or clumps of algae can provide habitat for invertebrates that is rich in food resources and sheltered from predation (Power 1995).

AQUATIC MACROPHYTES

Aquatic macrophytes are often abundant on floodplains, with their importance and species composition directly tied to the amount of permanent water on the floodplain (lakes, ponds, ditches) and the frequency with which these waters are hydraulically connected to the flooding river. On a Danube River floodplain, for example, Janauer et al. (2013) found that macrophyte abundance and diversity were associated with depth and connectivity of floodplain water bodies in complex ways. Thus, some species had their highest abundance in deeper lakes with low frequency of connection to the main channel, while others were associated with areas with more frequent connections. Puijalon and Bornette (2013) noted that the degree of scouring by flood flows determined in part what species are present, with some early successional species (e.g., *Elodea canadensis*) becoming established most quickly in flooded areas with long durations of inundation.

Despite their patchy patterns of occurrence, aquatic macrophytes often provide most of the carbon from primary productivity within inundated natural floodplains. A large biomass of macrophytes can be produced during extended periods of flooding, especially in tropical floodplain systems. Subsequent death and decomposition of this biomass can lead to hypoxic conditions within the remaining aquatic habitat (Winemiller 2004).

Although aquatic macrophytes can dominate plant biomass in some floodplain systems, recent research suggests that they contribute little carbon to floodplain food webs. Macrophytes enter food webs primarily as detritus and, because macrophyte detritus is resistant to biological processing, it is processed primarily by microbes (Winemiller 2004). Few studies have quantified the amount of processed carbon that exits the microbial loop to enter higher trophic levels; Bunn et al. (2003), using isotope analyses, suggest the proportion is likely small. Although aquatic macrophytes may not provide a dominant portion of the energy for aquatic organisms, they provide important shelter and structure for fish (Hayse and Wissing 1996) and invertebrates (Welcomme 1979) and also can provide a structural substrate for periphyton (Lewis et al. 2001), which is more available to biological processing (Hamilton et al. 1992; Lopes et al. 2015). Though most studies emphasize the importance of algal carbon to floodplain food webs, Magana (2013) reported that aquatic macrophytes in the floodplain of the Rio Grande (United States) were an important source of food for chironomids and other aquatic invertebrates.

SOURCES OF CARBON TO SUPPORT FOOD WEBS

Understanding how energy flows through trophic levels, the sources of primary productivity that underpin food webs, and how those flows and sources change through time is fundamental to understanding how floodplain ecosystems function and how to manage and conserve them. These questions, particularly the sources of organic matter at the base of metazoan food webs in rivers and floodplains, have become the subject of a scientific debate (Roach 2013). In fact, three of the conceptual frameworks reviewed in chapter 5—the RCC, the FPC, and the Riverine Productivity Model (RPM)—focus to a great extent on the sources and flows of organic matter through food webs. For large river-floodplain systems, the RCC suggested that the most important source of carbon is detritus imported from upstream, while the FPC emphasized the importance of flooded terrestrial vegetation. With the RPM, Thorp and Delong (1994, 2002) pointed out that both of those sources were fairly recalcitrant and they suggested that local aquatic (periphyton, phytoplankton, and aquatic macrophytes) and riparian (e.g., freshly deposited leaf litter from channel margins) sources provide carbon that is more biologically accessible and therefore more important to river food webs in some rivers. Rather than suggesting that the RPM replace earlier models, they noted that the RCC may be most appropriate for headwater streams and small rivers, the FPC for rivers with extensive floodplains, and the RPM applicable to large rivers with confined channels, such as the Ohio River.

A vast literature has accumulated as researchers have examined river-floodplain food webs to test, refine, and combine these conceptual frameworks. As noted above, initial work that focused on what organisms were ingesting, as opposed to what they were assimilating, may have overemphasized the importance of detritus.

Moving beyond studies of ingestion, researchers began to report that algae—though often a relatively small portion of the total carbon pool—was the most important source of carbon for higher trophic levels. For example, Sobczak et al. (2005) found that although phytoplankton represented less than 10% of the organic matter at a site in the Sacramento River (United States), it provided more than 90% of the carbon available to zooplankton (detritus dominated the carbon pool, but very little was biologically available). Studies using isotopic signatures or FA biomarkers within organisms such as fish have further improved understanding of sources of carbon because these chemical markers reveal assimilation, not just ingestion. Hamilton et al. (1992) found that most fish in an Amazonian floodplain had isotopic signatures indicating that they relied on food sources supported by algal carbon. The floodplain they studied was dominated by grass mats (C_4). While many organisms were found within the grass mats, few of them were actually obtaining carbon produced by the grasses; this included species that ingested grass detritus but primarily assimilated the associated bacteria and algae. FA within animal tissue can also be used to indicate sources of food because terrestrial and aquatic vegetation have different FA profiles. Rude et al.(2016), studying bluegills (*Lepomis macrochirus*) collected in the Illinois River (United States) and floodplain habitats, found that fish collected from the floodplain had FA profiles consistent with consumption of productivity derived from aquatic sources, while fish from the river had FA profiles indicating greater consumption of productivity derived from terrestrial sources. Young et al. (2014) found similar results between the FA profiles of channel catfish within river and floodplain habitats of the Kaskaskia River (Illinois, United States). Based on this evidence, researchers in both tropical (Hamilton et al. 1992) and temperate (Thorp et al. 1998; Sobczak et al. 2005) floodplains have concluded that, despite the dominance of living and detrital vascular plant material within the overall carbon pool, aquatic food webs in floodplains were supported primarily by algal carbon.

However, more recently, researchers have begun to question the generalization that algal carbon underpins floodplain food webs, noting that the relative importance of algal and detrital sources can shift over time and is strongly influenced by hydrology and associated parameters, such as turbidity (Pingram et al. 2014; Cook et al. 2015). For example, during periods of flooding with high turbidity, detrital sources (allochthonous inputs of terrestrial vegetation or macrophytes) can grow in importance relative to algae (Zeug and Winemiller 2008a). Reflecting that food webs can shift based on hydrology, Magana (2013) found that the Rio Grande silvery minnow, *Hybognathus amarus*, an endangered cyprinid, consumed algae during periods of low flow but switched to eating chironomids during floods and the chironomids fed primarily on aquatic macrophytes.

Additionally, detritus can become covered by biofilms composed of periphyton as well as microbial and fungal communities that assimilate detrital carbon into the biofilm. Consumption of biofilm, such as by invertebrate "scrapers," provides a route from detritus to the mesofauna food web via the microbial loop (Cook et al. 2015).

Roach (2013) undertook a review of studies that used chemical markers (isotope and FA analyses) to track sources of productivity for river and floodplain food webs. She found that C_4 grasses contributed to floodplain food webs in only 1 of 26 studies reviewed, while C_3 plants were significant contributors in 5 studies. Algae were the primary contributor to food webs in most of the studies (21 of 26). In her conclusions, Roach wrote that "rather than debate which conceptual model is most accurate," it would be most useful to understand how various factors—such as hydrology, sediment load, and nutrient availability—influence the relative importance of sources of productivity and how these factors and sources change over time" (p.295). For example, she found that carbon from C_3 plants (detritus) tended to be an important source of productivity within rivers that were relatively turbid and that detrital carbon increased its importance during periods of high flow and high turbidity. Further, sampling period may be affecting conclusions as Roach (2013) found that many of the studies that concluded that algae were predominant sources of productivity only sampled during low-flow periods. Thus, both Roach (2013) and Zeug and Winemiller (2008a, p.1741) suggest that researchers not seek "a single conceptual model of carbon dynamics (that) can apply to all rivers." Instead, they should seek to understand the range of factors that influence sources of carbon, how those factors vary over time and space, and how their sampling methods can account for that variability.

SECONDARY PRODUCTION: ZOOPLANKTON AND AQUATIC MACROINVERTEBRATES

Zooplankton and other invertebrates are the primary trophic linkages between the various sources of primary productivity and higher-level consumers such as fish (Keckeis et al. 2003). The zooplankton fauna of floodplains is often dominated by rotifers, copepods, and cladocerans. Common floodplain macroinvertebrate taxa include Tubificidae, Chironomidae, ephemerid and baetid mayflies, various snails, and prosobranch and sphaeriid mollusks. Larvae of the dipteran family Chironomidae can be the dominant invertebrates in some floodplain habitats, as exemplified by the invertebrate biota of the Yolo Bypass (Benigno and Sommer 2008). Zooplankton and many macroinvertebrates are often associated with floating and emergent plants (Power et al. 1995) and densities may be greater along floodplain margins and littoral habitats than in open waters (Welcomme 1979).

Zooplankton

Similar to phytoplankton, zooplankton numbers are generally low in flowing waters but can become quite high in floodplain water bodies and backwaters. For example, in the Danube River, zooplankton density was found to be 15 times greater in a backwater and 30 times greater in an isolated floodplain water body than in the main river (Welcomme 1979). Similarly, Grosholz and Gallo (2006) found that zooplankton biomass was 10–100 times greater within floodplain habitats of the Cosumnes River than within the main river channel. Most zooplankton consume algal or detrital carbon.

Zooplankton generally cannot reproduce in waters with flow velocities >0.4 m/s (Rzoska 1978). In waters that allow reproduction, zooplankton population density can initially increase with hydrological residence time, in large part tracking the increasing algal productivity (Keckeis et al. 2003). With further increases in residence time, zooplankton may reach a peak and then begin to decline, due to a declining food base and increased fish predation (Baranyi et al. 2002; Keckeis et al. 2003; Grosholz and Gallo 2006). Zooplankton can be displaced from floodplain habitats by high-velocity water during floods but can reestablish quickly after floods, both from colonization from upstream areas and by emergence from diapause stages in the sediment, triggered by environmental cues (Moghraby 1977; Boulton and Lloyd 1992). Rotifers are characterized by rapid colonization and reproduction and so can initially dominate the zooplankton taxa after inundation (Keckeis et al. 2003).

Due to the long residence time and low water velocity of much of the Cosumnes River floodplain, zooplankton densities attained levels one to two orders of magnitude greater than in the main river channel. Zooplankton biomass generally peaked 10–25 days after flood events (Grosholz and Gallo 2006). This pattern of population growth is likely due to both increasing reproductive success as water velocities declined and increasing food resources, especially phytoplankton, with increasing residence time (Ahearn et al. 2006). On the Cosumnes floodplain, fish predation exerted strong influence on zooplankton populations: both zooplankton populations and the size of zooplankton declined as the abundance of larval and juvenile fish, which feed on zooplankton, increased. Later in the season, zooplankton populations again increased as fish predation focused more on aquatic insects (Grosholz and Gallo 2006).

Zooplankton can consume both algal and detrital carbon, but algal carbon has significantly greater food quality for zooplankton. The availability of algal carbon is therefore the most important regulator of zooplankton productivity, even where detrital carbon dominates the carbon budget (Sobczak et al. 2002; Jassby et al. 2003). Remnant Central Valley floodplains export water that provides high concentrations of high-food-quality algal carbon for zooplankton feeding, particularly during draining following flooding with long residence time (Ahearn et al. 2006; Schemel et al. 2004).

Aquatic Macroinvertebrates

Aquatic invertebrate populations can be maintained in seasonally inundated floodplain habitats through migration (e.g., decapod crustaceans), dormancy (e.g., some mollusks and insects), and recolonization (many aquatic insects). Macroinvertebrates within river-floodplain systems are distributed along both lateral (Malmqvist 2002) and longitudinal gradients (Reese and Batzer 2007). Along both gradients, the floodplain hydroperiod (e.g., frequency, duration, and season of inundation) strongly influences the distribution of invertebrate species and communities. In general, sites with infrequent and short-duration flooding are dominated by resident terrestrial invertebrates or by rapidly developing aquatic invertebrate species, while sites subjected to frequent, long-duration flooding are dominated by species characteristic of wetlands, often with desiccation-resistant stages (Reese and Batzer 2007).

Floodplains can support considerably higher invertebrate productivity than adjacent river channels (Gladden and Smock 1990; Benke 2001). In two forested streams in Virginia, Smock et al (1992) reported that 67–95% of total invertebrate production was derived from the floodplain. Various habitat features within the floodplain are important to invertebrates, such as floating and emergent vegetation, which can provide cover from predators (Welcomme 1979). Floating algal mats can provide habitat for some invertebrates. The mats, warmed by the sun, provide conditions for rapid growth and development of invertebrates, such as chironomid midges, and provide shelter from predators. Power (1995) described floating algal mats as "food-rich, sun-warmed floating incubators for invertebrates." Large wood can be a particularly important habitat feature for invertebrates, both in floodplains and in river channels, because wood provides habitat for burrowing invertebrates, shelter from predation, and a stable substrate for attachment and feeding. By trapping organic matter, wood can also boost food availability (Smock et al. 1992; Benke 2001; Benke and Wallace 2003; Winemiller 2004).

Some invertebrates persist in resting stages in floodplain soils during dry periods and may contribute a large proportion of the invertebrates during initial floodplain inundation (Boulton and Lloyd 1992; Benigno and Sommer 2008). Boulton and Lloyd (1992) took dry soil samples from a floodplain along the Murray-Darling River (Australia) and experimentally wetted them in a laboratory to examine invertebrate emergence. The soil samples were taken from portions of the floodplain with varying inundation recurrence frequencies (annually, 1 in 7, 1 in 11, and 1 in 22 years). They found that the dry floodplain soils acted as a "seed bank" from which invertebrates emerged soon after wetting. Even a soil that had not been inundated in 14 years produced an invertebrate biomass exceeding that reported from swamps in the area. The sediment that had been annually flooded produced the greatest biomass of invertebrates: cladocerans and rotifers emerged within 2 days and ostracods emerged after 2 weeks. Only protozoans emerged from the sediments that were rarely flooded. They suggest that the seed bank of invertebrates can provide important food resources for fish and waterfowl soon after inundation (until other sources can contribute) and that reduction of flood frequency can endanger this resource. While floodplain soils can act as a seed bank, with the potential to boost productivity during initial inundation, this productivity may be lost to tilling and cropping of the land during the dry season.

Terrestrial invertebrates can also be an important part of floodplain food webs (Welcomme 1979). For example, earthworms can be an important food resource for fish such as adult Sacramento splittail as they move onto newly inundated floodplains (Moyle et al. 2004).

Because adult stages of many aquatic invertebrate species leave the water, floodplain invertebrate productivity can provide a significant food subsidy to terrestrial organisms including predatory invertebrates, birds, and bats (Malmqvist 2002). For example, Nakano and Murakami (2001), in a nonfloodplain study, found that invertebrates from a stream provided 26% of the annual energy demand for forest birds with half of the bird species deriving 50–90% of their energy from aquatic invertebrates and one species, the winter wren (*Troglodytes troglodytes*), deriving 98% of its energy from aquatic invertebrate species. In a review of energy fluxes from rivers into terrestrial food webs, Ballinger and Lake (2006) cite studies that report that insectivorous birds are attracted to inundated floodplains to feed on emerging chironomids. However, they conclude by noting that despite the fact that floodplains are known to be highly productive and thus have the potential to be major sources of energy for adjacent terrestrial food webs, "there is a dearth of data on the value of flooding on lowland rivers to terrestrial food webs and we are not aware of any research addressing this knowledge gap" (p.23).

CONCLUSIONS

Floodplain food webs are supported by carbon from a variety of sources, with recent research suggesting that algal carbon, from both phytoplankton and periphyton, is the most important source for primary consumers, even in floodplains where the pool of available carbon is dominated by detritus. Algal growth rates tend to be greatest when flooding corresponds to periods with warm temperatures and high sunlight. Under these favorable conditions, algal productivity reaches its maximum within flooded areas with high residence time of the water. This primary productivity supports food webs encompassing diverse consumers, including various vertebrate species that are high priorities for management, such as fish and waterfowl (chapter 8). Thus, one of the key factors determining the abundance of floodplain-dependent vertebrate species is the overall productivity of floodplain food webs. Management strategies focused on increasing production of fish or other vertebrates can be guided by an understanding of the conditions that favor floodplain food-web productivity.

EIGHT

Fishes and Other Vertebrates

THE IMPORTANCE OF FLOODPLAINS to fishes was largely unappreciated until Welcomme (1979) and Goulding (1980) reviewed floodplain fisheries and fish ecology. While fishery managers and scientists had widely recognized that fishes were found on floodplains in abundance, even in temperate regions, they generally believed that floodplain use had, if anything, negative effects on fish populations, due to stranding as water receded. In this chapter, we discuss (a) general concepts of floodplain fish ecology (b) classification systems for floodplain fishes, (c) fish stranding on floodplains, (d) use of floodplains by Pacific salmon, as a well-studied example of fish use of floodplains in temperate rivers, and (e) floodplain food webs and fish. We then briefly discuss the importance of floodplains to other vertebrates and conclude with an example of a floodplain food web, integrating the topics of chapters 5–8.

FLOODPLAIN FISH ECOLOGY

Although the importance of floodplains to fishes was initially misunderstood, floodplains are now recognized as extremely important to the abundance and persistence of numerous fish species in river systems with climates ranging from tropical (Welcomme 1979; Agostinho et al. 2004) to arid (Modde et al. 2001) to temperate (Risotto and Turner 1985; Beechie et al. 1994; Pander et al. 2015). For example, in California's Central Valley, only recently has it been recognized that a number of declining fish species have a strong dependence on floodplains. These species include the endemic Sacramento splittail, *Pogonichthys macrolepidotus* (Sommer et al. 1997; Moyle et al. 2004), and Chinook salmon, *Oncorhynchus tshawytscha* (Sommer, Nobriga et al. 2001). The latter is one of the most important harvested species in the state.

Beyond the main river channel, floodplain systems have two primary habitat types that are important to fishes: permanent floodplain water bodies and seasonal floodplains. The permanent water bodies include oxbow lakes, ponds, marshes, sloughs, and secondary river channels that support fish year-round. Seasonal floodplains are the floodplain surfaces that are inundated on a seasonal basis but are dry for much of the year. Much of this chapter focuses on those seasonally flooded habitats because they can be extremely important as spawning and rearing areas for many species of fish and because of their high productivity (chapters 6). However, in many river systems, oxbow lakes are key to supporting high fish production and diversity (Buijse et al. 2002).

Fish species use floodplains in diverse ways, ranging from those that only appear on floodplains "by accident" to obligate floodplain spawners (species that require specific features of a floodplain for spawning) to year-round residents in permanent waters. The predictability (frequency), seasonality, and duration of flooding all influence the extent to which the fish fauna of a given river system will have developed adaptations to using seasonal floodplains (King et al. 2003). Where flooding is relatively frequent and predictable in terms of season, many fish species will show adaptations to take advantage of floodplain habitat and resources, such as floodplain spawning. Where flooding is unpredictable, fish species are much less likely to have evolved adaptations to take advantage of flood pulses, although floodplain access can still be highly beneficial, especially if oxbow lakes are a key habitat (Winemiller 2004).

The most productive freshwater fisheries in the world, such as those in the Mekong and Amazon Rivers, occur in tropical floodplain rivers that retain a natural flood pulse that is predictable and inundates extensive areas for long durations (Welcomme 1979). The Mekong River's fishery provides protein to tens of millions of people and is primarily supported by productivity derived within floodplain lakes and other habitats during long-duration monsoonal flooding (Coates et al. 2003). Rivers with intact floodplains have significantly higher production of fish than rivers or reservoirs that lack a dynamic flood pulse, a phenomenon characterized as the "flood-pulse advantage" (Bayley 1991, 1995). Illustrating linkages between floodplain inundation and fish production, Welcomme (1979) reported that fish production in the Danube River was directly proportional to the extent and duration of inundation of floodplains. Dutterer et al. (2013) reported similar findings for the Apalachicola River in Florida, United States. Risotto and Turner (1985) found that the amount of bottomland hardwood forest, which they used as a proxy for floodplain, was a significant predictor of fish biomass produced by various sections of the Mississippi River and its tributaries. Similarly, Alford and Walker (2013) demonstrated the importance of forest-covered floodplains to fisheries in a distributary of the Mississippi River.

Floodplain habitats provide numerous benefits to fish species and fish use floodplains for a range of behaviors including mating, spawning, rearing, feeding, and avoidance of adverse riverine conditions. For example, during periods of early-season flooding, when river water is relatively cold, floodplain water generally has lower velocities, is less turbid, and can be warmer than water in the river (Sommer, Nobriga et al. 2001). In the lower Mississippi River, blue catfish (*Ictalurus furcatus*) of all sizes grew larger when they had access to extensive floodplains for foraging. The more rapid growth was likely caused in part by warmer water temperatures on the floodplain; during the time catfish used the floodplain, water temperatures were between 16°C and 20°C, which tended to be several degrees warmer than the adjacent river (Schramm and Eggleton 2006). Further, fish may be less vulnerable to predators, such as birds and piscivorous fish, when dispersed on extensive, vegetated floodplains than when concentrated in river channels (Sommer, Nobriga et al. 2001). However, during draining phases, fish can be concentrated in pools and lakes and become increasingly vulnerable to predation (Moyle et al. 2007) as discussed in the section on fish stranding on floodplains.

Floodplains contribute to the productivity of fish populations both because fish gain access to terrestrial resources, including insects, seeds, fruits, and leaves (Chick et al. 2003; Winemiller 2004), and because high primary production by algae on floodplains (Ahearn et al. 2006) supports abundant zooplankton and aquatic invertebrates, often at densities considerably higher than in the adjacent river (chapters 7). Temperate floodplains are most productive when flooding coincides with warmer temperatures (e.g., spring to summer), and fish gain the most benefit from feeding on floodplains during these warmer floods. In the upper Mississippi, a system where flooding often coincides with cold temperatures, species with behaviors that allowed them to take advantage of floodplain habitat, such as largemouth bass (*Micropterus salmoides*) and bluegill (*Lepomis macrochirus*), showed increased growth during a year with an unusual warm-weather flood, while a species not known to use floodplains (white bass, *Morone chrysops*) did not show increased growth (Gutreuter et al. 1999).

The abundant food resources on floodplains can provide important nutrition to fish prior to spawning (Moyle et al. 2004) and for larval and juvenile fish after hatching (Ross and Baker 1983; Sommer, Nobriga et al. 2001; Agostinho et al. 2004; Grosholz and Gallo 2006). In the Central Valley, abundant food resources on floodplains support faster growth for juvenile Chinook salmon compared to adjacent main stem river habitats (Sommer, Nobriga et al. 2001; Jeffres et al. 2008). Because floodplains provide high-resource environments, many fish species time their reproduction so that their larvae and juveniles can rear on floodplains (Welcomme 1979). In tropical rivers, many of the most abundant and economically important fish species spawn only in floodplain habitats and have spawning periods that coincide with predictable annual flooding (Welcomme 1979; Hogan et al. 2004).

Because floodplains can provide important habitat for both spawning and rearing, the duration and extent of inundation can be particularly important for regulating year-class size of fish species that spawn on floodplains (Gomes and Agostinho 1997; Sommer et al. 1997; Madsen and Shine 2000; Dutterer et al. 2013; Pander et al. 2015). For fishes that spawn on seasonal floodplains, the minimum duration of inundation must be sufficient for spawning and hatching of embryos and for larval or juvenile fishes to reach a stage where they can leave the floodplain as it drains. Longer duration than this minimum can yield further benefits because the juveniles can take advantage of the high-resource habitats to grow large enough to actively forage and avoid predators once they leave the floodplain (Ribeiro et al. 2004; Magana 2013). The combination of abundant food, reduced velocities, warmer temperatures, and sanctuary from predators can make seasonal floodplains highly desirable habitat for juvenile fishes (Holland 1986; Paller 1987). In fact, survival and growth of floodplain-adapted larval and juvenile fishes generally increases on floodplains compared to adjacent rivers (Ribeiro et al. 2004; Jeffres et al. 2008).

Floodplain habitats are strongly influenced by hydrological characteristics (seasonality, frequency, magnitude, and duration of flooding) and require connectivity with the river. Therefore, fish populations that depend on floodplain habitats are vulnerable to flow alterations and separation of floodplains from rivers by levees. For example, many fish species in the Missouri River system, United States, spawn when water temperature is between 15°C and 25°C. Historically, floodplain inundation mostly coincided with water temperatures in that range. Currently, much of the floodplain has been disconnected by levees and, due to flow regulation, remaining floodplains are rarely inundated when water temperatures are within the range for spawning. In part due to these changes in floodplain hydrology and connectivity, the Missouri River commercial fish harvest declined 80% from the 1880s to the 1990s (Galat et al. 1998).

King et al. (2003) developed a conceptual model that describes the relative importance of seasonal floodplains to riverine fish faunas. The model holds that fish faunas are more likely to contain species with strategies for seasonal floodplain use when (a) the annual flood pulses are predictable in timing, (b) flooding regularly occurs during periods with favorable water temperatures, (c) annual flooding lasts for extended periods (months), and (d) extensive areas are flooded. Many of the large tropical rivers of Africa, South America, and Asia have flooding patterns that fulfill all of these criteria and, therefore, these rivers contain many fishes adapted for using floodplains for spawning, rearing, and foraging (Welcomme 1979; Hogan et al. 2004). At the opposite end of the floodplain-use spectrum is the Murray-Darling system, Australia's largest river. Flooding in the Murray-Darling is highly erratic in frequency and magnitude, and temporal patterns of inundation are largely decoupled from temporal patterns for water temperature. Consequently, no native fish species appear to be specifically adapted to using seasonal floodplains, although many species will use them for foraging and rearing on an opportunistic basis (see chapter 11; King et al. 2003).

A similar situation exists on floodplains of the Brazos River, Texas (Winemiller et al. 2000; Zeug et al. 2005; Zeug and Winemiller 2008b). The river is largely unleveed and has flows only partially regulated by flood-control dams. Flood events tend to be short (a few days to 2–3 weeks) and unpredictable in timing. As a result, there are no fishes adapted specifically for use of floodplains per se, but there is a fairly rich fish fauna present in oxbow lakes scattered across the floodplain. The lakes are connected to the river only during periods of inundation. More recently formed lakes tend to be deeper and closer to the river, while older lakes tend to be shallower and farther from the river, and these may periodically dry up. The oxbow lakes are generally highly productive and support a diversity of fishes. Not surprisingly, the fish fauna of each lake is dynamic over time with considerable variation among lakes, related to depth and the idiosyncrasies of colonization events. Flood events generally homogenize the faunas of the lakes, while intervening isolation, and occasional drying, promotes divergence, particularly if there are long intervals between flood events that allow connection of a given lake to the river. Pond-type nest building fishes (especially centrarchids such as bass [*Micropterus* spp.] and sunfishes [*Lepomis* spp.]) tend to become dominant under long isolation. Although oxbow lakes may connect only infrequently to the river, they apparently are a major source of recruits to river fish populations, especially for large long-lived species such as gar (*Lepisosteus* spp.) and buffalo (*Ictiobus* spp.). A similar pattern seems to exist for the Danube River in Europe (Buijse et al. 2002).

The floodplains of California's Central Valley represent an intermediate level of fish use of floodplains. In this region, the timing of runoff and flooding, from winter rainstorms and springtime snowmelt, is consistent and so the seasonality of flooding is predictable. However, due to high interannual variability in the amount of precipitation, the extent and duration of flooding is not predictable or consistent (see chapter 12; Feyrer et al. 2006). Thus, only a few species appear to be floodplain dependent, such as Sacramento splittail. Most native fishes and many alien fishes, however, use floodplains opportunistically. Many species—such as juveniles of Chinook salmon—have behaviors that allow them to take advantage of floodplain habitats, including entering floodplains when conditions are favorable and leaving floodplains before conditions become unfavorable and before they could become stranded. For these species, rearing and foraging on floodplains increases growth and survival for those individuals that use floodplain habitats (Sommer, Nobriga et al. 2001; Sommer, Harrell, Kurth et al. 2004; Sommer et al. 2005).

CLASSIFYING FISH USE OF FLOODPLAINS

Several systems have been developed for classifying floodplain fishes based on how they use floodplain habitats. Most simply, Ross and Baker (1983) classified fish species as either *flood-quiescent* (did not use the floodplain) or *flood-exploitative* (used the floodplain) although they noted it is often difficult to classify fishes into one or the other categories.

Winemiller (1989, 1996) classified fishes that use floodplains by three life-history strategies (periodic, opportunistic, and equilibrium) and applied this classification to floodplain use by fishes in the highly aseasonal and erratic Brazos River, Texas, United States (Winemiller et al. 2000). Górski et al. (2011) used these same classes to describe patterns of floodplain use on the Volga River, Russia, a regulated river but one that still retains a fairly predictable pattern of floodplain inundation. *Periodic strategists* were species with large body size, delayed maturity, and high fecundity, such as common bream (*Abramis bramis*) and ide (*Leuciscus idus*), that live primarily in permanent water bodies. These fishes were especially successful at spawning on floodplains that were inundated on a regular and predictable basis. If the seasonal flooding did not occur, they often did not spawn. *Opportunistic strategists*, such as sunbleak (*Leucaspius delineatus*) and gibel carp (*Carassius gibelio*), were smaller species with early maturation and long spawning periods that can take advantage of flooding when it occurs—they were typically species that are first to colonize new areas—but they do not require floodplains (e.g., for spawning). *Equilibrium strategists* tended to live in river channels and floodplain lakes but did not depend on seasonally inundated floodplains; they were typically fishes with delayed maturity, extensive parental care, and strong habitat preferences.

A widely used classification system, developed for European rivers, divides floodplain fishes into three categories: (a) species with a strong dependence on river habitats, preferring areas with flowing water (*rheophilic fishes*); (b) species that live in areas with low flows, mainly in backwaters and floodplain lakes (*limnophilic fishes*); and (c) species that occur in both broad habitat types (*eurytopic fishes*; Pander et al. 2015; Grift et al. 2003). Rheophilic species can often be floodplain dependent, requiring floodplain habitats for rearing of young. These categories have parallels to the classification of tropical floodplain fishes in Southeast Asia as white, black, and gray fish (Welcomme 1979; Junk and Bayley 2008).

- White fish are rheophilic fishes, mostly fairly large, that make extensive river migrations but usually make only limited use of the floodplain.
- Black fish are limnophilic fishes that mostly live on the floodplain, so they tolerate high temperatures and low dissolved oxygen; they tend to leave the floodplain only when the water is too low.
- Gray fish are eurytopic fishes that mainly live in the rivers but move on to floodplains during flooding.

Moyle et al. (2007) developed a classification of guilds of floodplain fishes based on studies of the Cosumnes River floodplain and other Central Valley floodplains. The Cosumnes floodplain is highly seasonal, reflecting the region's Mediterranean climate. During winter and spring, flood waters typically spread over large areas for 1–5 months, while, during summer and fall, the floodplain surface and even parts of the Cosumnes River channel become dry. Permanent water persists, however, in the many sloughs that drain the floodplain. The sloughs are weakly tidal and supported large populations of warm-water fishes, mostly alien species.

Moyle et al. (2007) classified fishes found on the seasonal floodplain and the intersecting sloughs into six user groups: (a) floodplain spawners, (b) river spawners, (c) floodplain foragers, (d) floodplain pond fishes, (e) inadvertent floodplain users, or (f) floodplain nonusers (table 8.1).

Floodplain spawners are eurytopic fishes that generally have a periodic life-history strategy and that use seasonal floodplains for spawning and juvenile rearing, generally corresponding to periodic strategists (sensu Winemiller 1996). Floodplain spawners can be further divided into *obligate* floodplain spawners and *opportunistic* floodplain spawners. For obligate floodplain spawners, reproductive behaviors and movements are strongly keyed to hydrological conditions associated with flooding, especially rising water levels. They may move onto the floodplain as inundation is beginning or

TABLE 8.1
Fishes associated with the seasonal floodplain of the Cosumnes River, California

User groups are explained in the text. Nonusers were not collected during a five-year study but were found in connected sloughs and rivers. For comparison with other classifications, the table also includes the column "RLE," which classifies fish as rheophilic (R), limnophilic (L), or eurytopic (E) (sensu Grift et al. 2003) And the column "POE" from Winemiller (1989, 1996), who classifies fish on floodplains as periodic (P), opportunistic (O), or equilibrium (E). Feeding guilds are identified as detritivores (D), zooplanktivores (Z), invertivores (I), piscivores (P), and omnivores (O).

Species	User group	RLE	POE	Feeding guild
Pacific lamprey, *Lampetra tridentata*	Inadvertent	R	E	D/P
American shad, *Alosa sapidissima**	Inadvertent	R	E	Z
Threadfin shad, *Dorosoma petenense**	Inadvertent	L	O	Z
Hitch, *Lavinia exilicauda*	River spawner	E	P	I
Sacramento blackfish, *Orthodon microlepidotus*	FP spawner	L	P	D
Sacramento splittail, *Pogonichthys macrolepidotus*	FP spawner	E	P	I/D
Sacramento pikeminnow, *Ptychocheilus grandis*	River spawner	R	E	P
Golden shiner, *Notemigonus chrysoleucas**	Forager	L	O	I
Fathead minnow, *Pimephales promelas**	Inadvertent	L	O	D
Goldfish, *Carassius auratus**	FP spawner	L	P	D
Common carp, *Cyprinus carpio**	FP spawner	E	P	I/D
Sacramento sucker, *Catostomus occidentalis*	River spawner	R	E	I/D
Brown bullhead, *Ameiurus nebulosus**	Nonuser	L	E	O
Black bullhead, *A. melas**	Inadvertent	L	E	O
White catfish, *A. catus**	Inadvertent	E	E	O
Channel catfish, *Ictalurus punctatus**	Nonuser	E	E	P
Chinook salmon, *Oncorhynchus tshawytscha*	River spawner	R	E	I
Rainbow trout, *O. mykiss*	Inadvertent	R	E	I
Wakasagi, *Hypomesus nipponensis**	Inadvertent	L	O	Z
Mississippi silverside, *Menidia audens**	Pond	L	O	I/Z
Western mosquitofish, *Gambusia affinis**	Pond	L	O	I/Z
Prickly sculpin, *Cottus asper*	River spawner	R	O	I
Tule perch, *Hysterocarpus traski*	Nonuser	E	E	I
Bluegill, *Lepomis macrochirus**	Forager	L	E	I
Redear sunfish, *L. microlophus**	Forager	L	E	I
Green sunfish, *L. cyanellus**	Nonuser	E	E	I
Warmouth, *L. gulosus**	Nonuser	E	E	I
Black crappie, *Pomoxis nigromaculatus**	Forager	E	E	I/P
Largemouth bass, *Micropterus salmoides**	Forager	L	E	P
Redeye bass, *M. coosae**	Nonuser	R	E	P
Spotted bass, *M. punctulatus**	Nonuser	E	E	P
Bigscale logperch, *Percina macrolepida*	River spawner	R	O	I

SOURCE: Based on Moyle et al. (2007)
*Alien species.

wait until flooding is well underway. The adults spawn on substrates, such as inundated plants, characteristic of floodplains. The embryos adhere to the substrate, hatch, and then rear for a few weeks in the complex cover provided by flooded vegetation. Once the juveniles are free-swimming, they stay on the floodplain as long as inundation continues and food is abundant, and they generally respond to floodplain draining by actively seeking out flowing channels to exit the floodplain. A good example of an obligate floodplain spawner is the Sacramento splittail (Moyle et al. 2004), which has year class strength that is highly correlated with the number of days of flooding of seasonal floodplains in its range (Sommer et al. 1997). The periodic strategists of Górski et al. (2011) are mostly in this category.

Opportunistic floodplain spawners use floodplains in similar ways as obligate spawners (spawning and juvenile rearing) but do not require floodplain habitats for spawning. However, their most successful spawning years typically occur during periods of extensive flooding (King et al. 2003). The best known examples of opportunistic spawners are common carp and goldfish, which are found worldwide in floodplain systems (Moyle et al. 2007). Many fishes native to the Mississippi-Missouri River system are opportunistic floodplain spawners, using floodplains for reproduction when inundation coincides with appropriate spawning season and water temperatures (Raibley et al. 1997; Galat et al. 1998). Likewise, many fishes that used a newly restored Danube River floodplain appear to be opportunistic spawners (Pander et al. 2015; chapter 11), as do most fishes using the floodplains of the Apalachicola River, Florida, United States (Dutterer et al. 2013). Generally, obligate floodplain spawners are found in river systems with predictable flood regimes, whereas opportunistic floodplain spawners are found in river systems where the extent and timing of floodplain are less predictable.

River spawners are rheophilic fishes that spawn in rivers upstream or adjacent to floodplains and have larvae and juveniles that can use seasonal floodplains for rearing (see box 8.1). These species may not require floodplains for persistence because juveniles also can rear on river-edge habitats. However, spawning upstream of floodplains may provide important benefits for these species because juvenile survival and growth, and ultimately abundance, can increase when juveniles can access abundant food resources and diverse habitats on floodplains. River spawner species vary in the degree to which they appear to have adaptations for floodplain rearing. For example, juvenile Pacific salmon (*Oncorhynchus* spp.) generally exit floodplains during draining, suggesting an

---

BOX 8.1 · Larval Fishes on a California Floodplain

In addition to fish that spawn on floodplains, many fishes that spawn in rivers have larvae that are passively carried to the floodplain by flows and can take advantage of floodplain habitats. Crain et al. (2004) noted a regular succession of larval fishes entering a restored California floodplain. The earliest to arrive were larvae of three native fish species and two alien species, entering as larvae during the high spring flows, when inundation was most extensive and water temperatures were cool (Crain et al. 2004). These larvae were typically most abundant in flooded areas close to the river, near levee breaches, and these same species were generally observed leaving the floodplain later as actively swimming juveniles (Moyle et al. 2007). As the season progressed, with reduced inundation and warmer temperatures, five alien taxa dominated the larval fish fauna, predominantly centrarchids that could spawn in the permanent sloughs that intersected the floodplain. A similar pattern of fish larvae succession was found in the floodplain of the Apalachicola River (Florida, United States), with larvae appearing in a progression of family taxonomic groups. The pattern was presumably related to water temperature, which increased during the flooding period, and temperature requirements for spawning (Walsh et al. 2009).

adaptation to variable floodplain hydrology. Conversely, juvenile Sacramento suckers (*Catostomus occidentalis*) grow faster within river habitats than in floodplain habitats (Ribeiro et al. 2004), indicating that not all river spawning fish may benefit from rearing on floodplains.

Floodplain foragers are eurytopic and limnophilic fishes that opportunistically use seasonal floodplains for feeding. They have opportunistic or equilibrium life-history strategies (sensu Winemiller 1996) and typically live in lakes, ponds, and sloughs within the floodplain. In the Cosumnes River, floodplain foragers tend to enter the flooded areas mostly late in the flood cycle when the water is warmer and food, especially insects and small fish, is abundant. In the Cosumnes, these floodplain foragers are mostly alien species native to the Mississippi River system, such as golden shiner, largemouth bass, and bluegill and redear sunfish (table 8.1). These species also use seasonal floodplains for spawning and rearing along their native rivers and may grow faster in floodplain habitats than in rivers (Gutreuter et al. 1999). Because these species evolved with opportunistic use of floodplains, they generally avoid stranding as the flood waters recede by moving back into lentic or lotic habitats. In years with prolonged flooding, allowing warmer water temperatures, adult floodplain foragers often spawn on the floodplain, along with floodplain pond fishes.

Floodplain pond fishes are short-lived limnophilic fishes with opportunistic life histories (sensu Winemiller 1996). They reside in permanent floodplain water bodies and become abundant, through rapid growth and reproduction, in shallow, seasonal floodplain habitats. These species share some characteristics with floodplain foragers, but pond fishes reproduce in most years and, through rapid reproduction, come to dominate shallow habitats. Floodplain pond fishes are often stranded in large numbers in temporary habitats as they dry and can attract flocks of piscivorous birds as a result. Examples from California include two alien species, Mississippi silverside and western mosquitofish (table 8.1).

Inadvertent floodplain users may compose a high proportion of species collected from seasonal floodplains but a low proportion of total fish numbers or biomass. During flood events, these species are transported onto floodplains from various habitats, including lakes, throughout the upstream watershed and show no specific adaptations for floodplain use. Larvae, juveniles, and adults can pass through the floodplain or become stranded during draining. In addition, there are some species that are almost never found on seasonal floodplains (*floodplain nonusers*) but are consistently found in deep waters of floodplain lakes and sloughs (secondary channels with permanent water). They probably would be better called floodplain avoiders. The equilibrium strategists of Winemiller (1996) may fall into either of these categories.

Because seasonal floodplains and associated permanent water bodies have high primary and secondary productivity and fish use these habitats to forage, floodplain fishes can also be assigned trophic guilds. The guilds typically reflect the dominant dietary items of adult fish when they are on the floodplains. Guilds usually include the following groups: (a) phytoplanktivores, (b) algivores, (c) detritivores, (d) zooplanktivores, (e) benthic invertivores (consuming invertebrates living on substrates, including aquatic plants), (f) piscivores, and (g) omnivores. Piscivores eat both fish and large invertebrates such as crayfish. Omnivores, such as bullhead catfishes, eat almost anything they can grab, from algae to invertebrates to fish, dead or alive. The separation of the seven groups is rarely clean because most fishes are fairly opportunistic in their feeding and change diets as they grow larger. Feeding guilds of a California floodplain assemblage are shown in table 8.1.

In short, there are many ways to categorize floodplain fishes based on how they respond to dynamic conditions and how they use floodplains as habitat. On seasonal floodplains, the assemblages of fishes with different life histories, diets, and ways of using the habitat are constantly in flux, both from year to year and from month to month during a flood event.

STRANDING ON FLOODPLAINS

A major concern for projects that reconnect rivers and floodplains is the risk that important fishes, such as endangered or commercially important species, may become stranded as flood waters recede. While consideration of stranding risk and careful design are certainly warranted, fishes native to a particular region often exhibit life history and behavioral adaptations to local flood regimes that reduce stranding risk. For example, fishes adapted to flooding regimes will begin to exit the floodplain based on a variety of cues, including decreasing inflow and/or depth, or increasing water temperature and/or clarity. Moyle et al. (2007) observed that many native species dominated samples early in the flood season but became rare later in the flood season because they left the floodplain as it drained and as water temperature warmed. These same species simultaneously appeared in the river and sloughs adjacent to the floodplains. Further, they found that the majority of fish stranded in isolated ponds after the floodplain drained were alien pond species.

However, particularly rapid and early disconnection between river and floodplain can lead to high levels of stranding. Rapid disconnection and draining can occur in some systems based on natural river hydrology patterns or can be the result of management, such as an upstream dam rapidly reducing reservoir releases. For example, in 2001 the Cosumnes floodplain became naturally disconnected from the river relatively early in the flood season and more splittail were stranded that year than in other years. Even in this year, splittail represented only 5% of the stranded fish and the majority of juvenile splittail successfully emigrated from the floodplain (Moyle et al. 2007). Where floodplains are flooded erratically for only short periods of time, stranding may occur with large flood events. In the Brazos River, Texas, United States, for example, large-scale flood events are infrequent and unpredictable in timing. When large floods do happen, "catastrophic mortality" of fish can occur as the floodplain dries up (Winemiller et al. 2000). In such situations, survivors are mostly fish that were able to move into oxbow lakes.

For many native fishes, use of floodplains likely represents a bet-hedging strategy in which different habitats provide a range of benefits over time. Advantages of floodplain use during "good" years likely outweigh the risks of stranding in "bad" years. Human-made structures, such as gravel pits, berms, and water-control structures, that interrupt natural drainage pose the greatest risk for stranding. These risks can be avoided or mitigated if restoration projects are designed to include effective drainage, including exit channels with minimal obstacles for emigration. For example, a levee setback on the Bear River (California, United States) included a contoured swale intended to drain a corner of the reconnected floodplain back into the river to minimize stranding (Williams et al. 2009). While stranding of individuals may still occur, most fish species will benefit at the population scale from promoting access to floodplains.

PACIFIC SALMON AND FLOODPLAINS: A NEW PARADIGM

The use of floodplains by fish in temperate rivers has gained increasing attention. In terms of how they use floodplains, Pacific salmon (*Oncorhynchus* sp.) of the western United States have received perhaps more attention than any other fishes in recent years. In this time, the general understanding of the relationship between floodplains and salmon has swung dramatically. There once was concern that floodplains represented a real threat for stranding, while now floodplains are seen as likely providing critically important habitat to maintain robust populations.

Salmon populations are in decline throughout much of their range along the Pacific coast of North America (Nehlsen et al. 1991; Nehlsen 1997), including those of the Central Valley of California (Williams 2006; Katz et al. 2013).

Until recently, researchers had not examined whether widespread loss of floodplain habitat contributed to salmonid declines, primarily because most floodplain habitats were disconnected from rivers prior to scientific study of salmon populations. For example, in their extensive review of salmonid declines in the Pacific Northwest, the National Research Council (1996) devoted a single paragraph to salmonid use of floodplain habitats. However, attention devoted to this issue, both through fieldwork (Sommer, Nobriga et al. 2001; Hall and Wissmar 2004) and historical research (Pollock et al. 2004), increased dramatically in subsequent years. Montgomery (2003) in his book *King of Fish*, which focused on the same region as the NRC review, recommended restoration of floodplain habitats as a high-priority action to restore salmonid populations.

Many early studies that describe salmonid use of floodplain habitats do not specifically use the word "floodplain," but instead use terms such as "off-channel habitat," which typically refer to secondary channels or other areas that are seasonally flooded. For example, Bustard and Narver (1975) reported that the overwinter survival rate of juvenile coho salmon (*O. kisutch*) was twice as high within beaver ponds as in the main river. Brown and Hartman (1988) found that juvenile coho had greater overwinter survival in off-channel habitats (ephemeral swamp sites) compared to the main channel. Due to the higher survival rates, off-channel habitats were responsible for a disproportionate amount of coho productivity relative to their area. The importance of off-channel habitat increased in years with very high flows, indicating these habitats served as refuges during floods. Swales et al. (1985) also concluded that juvenile coho salmon preferred off-channel sites as overwintering habitat, including side channels, back channels, off-channel ponds, and other low-velocity off-channel areas. These floodplain habitats were warmer, had lower water velocity, and higher cover compared to main channel habitats, with fewer predators and an abundant food supply. Growth and survival rates were higher for fish within off-channel habitats compared to those within the main channel.

Research in California has confirmed the value of floodplain habitats for salmon, showing faster growth of juvenile Chinook salmon due to warmer water and more plentiful food supplies (Sommer, Nobriga et al. 2001; Jeffres et al. 2008; Limm and Marchetti 2009; chapters 12 and 13). Beyond California, Roegner et al. (2010) also reported that Chinook and coho salmon juveniles reared for months within newly reconnected tidal freshwater marshes created by breaching dikes in the Columbia River estuary, and Teel et al. (2009) reported that juvenile Chinook salmon reared in seasonal floodplain wetlands of the lower Willamette River.

Through river regulation and diking, floodplains have been greatly modified and disconnected from rivers throughout the Pacific Northwest (Oregon, Washington, United States) and California (Montgomery 2003). This loss of floodplain habitat likely has had population-scale effects on salmon runs. For example, Pollock et al. (2004) estimated that the loss of beaver ponds and other floodplain water bodies has reduced the potential winter rearing population of coho smolts in the Stillaguamish River basin in Washington by almost 90%. Similarly, Beechie et al. (1994) concluded that almost 60% of winter and 32% of summer rearing capacity in the Skagit River basin (Washington, United States) was historically provided by side channels and distributary sloughs on the floodplain and river delta. Most of this habitat has been lost due to channelization and diking for flood control. Although forestry impacts on salmon populations have received a great deal of attention, Beechie et al. (1994) concluded that, in the Skagit River basin, loss of habitat due to flood control and drainage dwarfs the loss of habitat due to forestry activities.

Similarly, loss of habitats provided by beaver activity may have been extremely important as a cause of salmonid declines. Loss of this habitat due to widespread trapping of beavers occurred even before the commercial harvest of salmon began (Pollock et al. 2004). Salmon populations prior to their commercial harvest have often been used as the historical baseline. However, the preharvest baseline may have been an underestimate because salmon populations may have already been depleted due to the loss of beaver-created habitat. Thus, historical salmon populations may have been even larger than originally estimated from early harvest levels.

In short, seasonal floodplain habitats are important for Pacific salmon populations. Floodplain rearing offers numerous benefits for juvenile salmonids, including lower water velocities and predator densities and warmer, clearer, and more productive water. Studies from diverse river systems demonstrate population-scale benefits of floodplain restoration. For example, in the Skagit basin, Beechie et al. (1994) found that both seasonal and permanent floodplain habitats, such as sloughs and side channels, had much higher per-unit-area production capacities than main-stem or tributary channel habitats. Kareiva et al. (2000) examined the relative benefits of various restoration strategies for Snake River spring/summer Chinook salmon populations. They concluded that improving survivorship during the first year, when the fish use floodplains and the estuary, had the strongest benefit for overall populations. Floodplain rearing potentially improves survival and growth during early life stages, which can translate into greater oceanic survival (Ward and Slaney 1988). Ogston et al. (2014) report that boosting production of salmon smolts through floodplain restoration can be cost competitive with producing salmon smolts in a hatchery (approximately $1/smolt). They found that 16 ha of floodplain restoration in the upper Chilliwack River watershed (British Columbia, Canada) was responsible for approximately one-third of the annual production coho salmon smolts from the watershed. These studies suggest that floodplain restoration should occupy a major role in salmon recovery strategies.

Finally, it is worth noting the reciprocal nature of benefits of floodplain forests to salmon (see box 6.1). Ocean nutrients brought in by salmon are released when salmon die and decay. These nutrients are transferred to the forest though predators and scavengers, as well as by high flows. The result is bigger trees, which provide the complex habitat juvenile salmon need (Stokes 2014), as well as changed composition of floodplain vegetation (Hocking and Reynolds 2012). This relationship was also noted on a regulated California stream, although the main plant beneficiaries of the salmon nutrients were grapes in vineyards (Merz and Moyle 2006).

FISHES AND FOOD WEBS

Inundated floodplains can have enormous primary and secondary production (chapters 7), a seasonal phenomenon that is akin to ocean upwelling, the flush of grass on the African savannah following rain, or the arrival of Pacific salmon that die to fertilize their spawning streams. These events are often viewed as culturally or economically important because they lead to large populations of harvestable (or viewable) vertebrates, especially fish.

In rivers with extensive floodplains, fishes and other vertebrates typically take advantage of the high productivity of floodplains to feed and grow rapidly. This "flood pulse advantage" (Bayley 1991) happens through five interrelated pathways:

1. Larval and juvenile fish that feed directly on floodplain invertebrates and other food sources; these are fishes that arrive on floodplains either by being spawned there or by being carried in by flood waters.
2. Predators that consume juvenile fish as they leave the floodplain to move to other habitats (the trophic relay hypothesis of Kneib 1997).
3. Fish living in the adjacent rivers that feed on the productivity flushed from rivers including invertebrates and the algae and detritus that can contribute to riverine food webs (outwelling).
4. Adult fish that move onto the seasonal floodplain to forage, usually from permanent

waters located on or adjacent to the floodplain (ponds, river channels).
5. Consumption of plants, invertebrates, and fish on floodplains by birds and mammals, including consumption of fish stranded in floodplain pools.

The abundance of larval and juvenile fish (together known as "young of year" or YOY) on many seasonal floodplains is testament to the productivity of these habitats. While diverse fishes take advantage of inundated floodplains, most are typically the progeny of floodplain or riverine spawners. As discussed in chapter 7, the ability of temperate floodplains to support small fish depends mainly on in situ primary production which in turn is a function of timing, extent, and duration of flooding (but see Zeug and Winemiller 2008a). From the perspective of juvenile fishes, the ideal flood event on a temperate floodplain starts early enough in the season to "prime" flooded areas by mobilizing or importing nutrients to start primary production, followed by an extended duration of flooding that allows abundant invertebrate populations to develop, tracking the flush of detrital mobilization and algal productivity. The amount of primary and secondary production depends in large part on the overlap in timing between flooding and increasing spring photoperiod and temperatures. In many temperate systems, flooding tends to coincide with snowmelt, with water that is too cold and arrives too early to have high productivity available to YOY fishes. Where flooding is driven more by rain, such as in southeastern United States, exact timing, length, and extent of flooding are highly variable over the winter-spring rainy season, but the floodplain forests are still highly productive of aquatic invertebrates (Benke et al. 2000). In other systems, the timing of flooding may be highly erratic, so is essentially unpredictable for spawning and rearing of fishes (Schramm and Eggleton 2006). In these systems, many species may benefit from opportunistic use of floodplains, but few species actually depend on flood events for persistence (Humphries et al. 1999; Winemiller 2004).

Juvenile fish on seasonal floodplains feed primarily on invertebrates produced on the floodplain, either through rapid reproduction of zooplankton washed onto the floodplain or on invertebrates emerging from floodplain soils. For example, on the Yolo Bypass, Sommer, Nobriga et al. (2001) found juvenile Chinook salmon fed largely on emerging chironomid midges, while Katz (2015) found them to be feeding largely on zooplankton (*Daphnia magna*).

Young of year fish that move off floodplains as water levels drop are likely important seasonal prey for adult fish in the receiving waters such as river channels, lakes, sloughs, and estuaries. Kneib (1997) described energy transfer mechanisms within tidal marshes that may provide a model for how floodplains transfer energy to other aquatic habitats. Kneib proposed that the movement of small fish out of the marshes represents a major source of energy transfer to the aquatic habitats connected to the marshes and that, through this energy transfer, the marshes helped support fisheries. In a similar fashion, floodplains can export large numbers of small fish to other connected aquatic habitats and transport floodplain productivity to consumers in those habitats. For example, when seasonally flooded duck-hunting ponds were drained, O'Rear (unpublished data, 2012–2014, UC Davis) found that large (25–45 cm) striped bass (*Morone saxatilis*) congregated by the mouths of the pond outflows where they fed on abundant fish (mainly three-spine stickleback, *Gasterosteus aculeatus*) being flushed out, fish that would otherwise be smaller than their typical prey.

One of the hypothesized advantages of rearing on a seasonal floodplain is that predation pressure on YOY fish from large piscivorous fish is reduced across the shallow, extensive, and often well-vegetated habitats. While adult fish regularly move on to floodplains for spawning, movement of large piscivores to feed on floodplains is less well documented. For example, during 5 years of sampling a California floodplain, Moyle et al. (2007) found very few potentially predatory adult fish on the floodplain during flood events. Nevertheless, although predation pressure from piscivorous fish may be reduced, YOY fish rearing on seasonal floodplains can be vulnerable to predation from birds. Piscivorous birds can consume large quantities of floodplain fish, particularly if they become stranded.

In short, seasonal floodplains can greatly enhance fish production in river systems by providing a nursery area for early life-history stages, although the importance of this role will vary from river to river and from year to year in individual rivers. The more predictable the flood regime is on a river, the more likely floodplains will contribute primary and secondary production to support food webs that sustain large populations of adult fish. A wide variety of fishes can use floodplains successfully for reproduction and rearing; fishes adapted to the local flow regime are less likely to become stranded as water recedes. Fish, especially YOY, can be important in food webs connected to temperate seasonal floodplains. They may consume much of the secondary production (Grosholz and Gallo 2006) and are themselves important prey for other fishes and for birds and mammals. In the next section, we discuss these other vertebrates and how they use floodplains and then conclude with an example of a floodplain food web, from primary producers to piscivores.

OTHER VERTEBRATES

Worldwide, floodplains provide important features for a great diversity of reptile, amphibian, bird, and mammal species (Welcomme 1979). A key feature of floodplains that promotes this diversity is their dynamic nature, where floods of different magnitudes create a shifting mosaic of habitats that favor different species (see chapters 5). For example, Brawn et al. (2001) note that, due to the diversity of habitats created by dynamic flooding processes "terrestrial bird communities of floodplains are some of the richest in the world" (p.260). Floodplains also have strong interactions with neighboring upland habitats, often providing foraging areas for upland species or refuges from predation. The loss or simplification of the floodplain habitat mosaic has led to the decline or loss of many floodplain-dependent animals.

Floodplain habitats often support (or supported) rare and declining vertebrate species. For example, three birds species that are now extinct once depended on floodplain forests of the southeastern United States for their existence: ivory-billed woodpecker (*Campephilus principalis*), Carolina parakeet (*Conuropsis carolinensis*), and Bachman's Warbler (*Vermivora bachmanii*; Brawn et al. 2001). Floodplains, including riparian woodlands, within California's Central Valley support 20 species that are listed on either the Federal (United States) or State (CA) Endangered Species list (see table 13.1). These include mammals (e.g., the riparian woodrat), birds (least Bell's vireo and Swainson's hawk), amphibians (California red-legged frog), reptiles (giant garter snake, western pond turtle), as well as fish (Chinook salmon). Despite their great reduction in extent, floodplains are still extremely important habitat for more abundant wildlife. Birds are especially notable floodplain users because so many species are migratory or can otherwise move around to take advantage of the seasonally available and shifting habitats of floodplains. Their abundance is often a good measure of the health of floodplain ecosystems in a region. Here, we review in more detail the use of floodplain by four major groups of terrestrial vertebrates: migratory waterbirds, tropical migrant birds, resident riparian birds, and mammals.

Migratory Waterbirds

One of the most spectacular uses of floodplains worldwide is by migratory waterfowl that overwinter on floodplains or stop on floodplains during migrations to take advantage of abundant food. These birds include pelicans, various

> **BOX 8.2 · Pop-up Wetlands for Waterfowl**
>
> To increase the availability of shallow wetland habitat for migratory waterbirds in California's Central Valley, the Nature Conservancy developed a program called BirdReturns. The program relies on citizen science through a database called eBird that allows birdwatchers to report bird occurrences using their smartphones. These crowdsourced data allow managers to predict when migratory waterbirds will be arriving in the Central Valley. Based on that timing, they implement a reverse auction to identify farmers who are willing to "rent" their fields and receive compensation to allow them to be flooded for a period of time. Managed inundation of the rented fields creates "pop-up wetlands," providing shallow wetland habitat during the time that migratory waterbirds will benefit most from the flooding. BirdReturns has produced more than 16,000 ha of temporary wetland habitat between 2014 and 2016 (McColl et al. 2016) and was hailed in the *New York Times* as an illustration of reconciliation ecology (Robbins 2014).

waterfowl (ducks, geese, swans), wading birds (e.g., cranes), and shorebirds (e.g., sandpipers, plovers, and curlews). Indeed, it can be argued that locations of predictably flooded—and productive—floodplains play a major role in defining the paths of major flyways traveled by migratory birds. For example, millions of birds essentially track the Mississippi River on their annual migrations down the Mississippi Flyway, taking advantage of abundant food on floodplain wetlands along the way and the vast marshes of the Mississippi River Delta. Similarly, Central Valley floodplains and seasonal wetlands are critically important for waterbirds using the Pacific Flyway.

For ducks and geese, important floodplain habitats include wetlands, ponds, and lakes that retain water long after flood waters have receded from seasonal floodplains. For wading birds, the temporary habitats on seasonal floodplains, such as mudflats and beaches created by receding waters, are important foraging areas. Today, much of the area of such habitats can be found in intensely managed waterfowl refuges. The refuges are typically located on disconnected floodplains (behind levees) that rarely flood by natural processes, so they rely on water delivered to the refuges by canals and held in diked ponds and fields. In North America and Europe, these substitute floodplains provide much of the habitat used by large migratory populations of waterfowl along some flyways.

Similarly, agricultural fields can be flooded either as part of agricultural operations (e.g., rice fields; van Groenigen et al. 2003; Fleskes et al. 2005) or intentionally to benefit birds and other wildlife (see box 8.2).

Tropical Migrant Birds

Tropical migrants, mainly passerines ("perching birds"), spend winter in the tropics and return to more northern regions to reproduce in the summer, taking advantage of the flush of insects and other food often present during spring and early summer. While these birds can take advantage of the diversity of habitats provided by floodplains, floodplain/riparian forests are especially important habitat for migrant birds both as stopover habitat during migration and for breeding. These forests and the resources they provide—structure for nesting and cover, sources of food—are more predictable than the resources provided by seasonal floodplains, which vary between years depending on precipitation and runoff, particularly in regions where flooding has high interannual variability. Over long time periods, floodplain forests depend on flooding for regeneration, so tropical migrants depend on dynamic floodplain processes over longer timescales. In the floodplain forests of the Altamaha River (Georgia, United States), Hodges and Krementz (1996) found tropical migrants and resident birds were diverse (48 species) and that the abundance of the tropical migrant species increased with the width of the floodplain forest corridor along the river. In the Southeastern United States, prothonotary warblers (*Protonotaria citrea*) generally build their nests in inundated floodplain forests, potentially because this reduces predation from snakes (Petit 1999). In Wyoming, the diversity and abundance of songbirds in riparian forests was correlated with height of willows (Olechnowski and Debinski 2008).

Resident Riparian Birds

Resident passerine birds tend to be associated with riparian habitats in much the same way as tropical migrants. Woodpeckers of various species, for example, typically need large old trees in which to excavate nests and to forage for insects. Edge habitats—such as open fields or water adjacent to forests—seem to be particularly important for maintaining populations of resident birds and species diversity. Parkinson et al. (2002) found that heterogeneous habitat patches on the undammed Ovens River (Australia) promoted diversity of bird species and that birds associated with woodlands also used the seasonally flooded open habitats for foraging; the natural flood regime thus promoted bird habitat over both longer timescales (e.g., the development of diverse habitat patches through flood-driven geomorphic processes) and during short-term periods of inundation. Similarly, the little owl (*Athene noctua*) favors restored floodplain habitats along the Rhine River in Europe because while it breeds in trees, it forages in open floodplains on shrews, mice, and invertebrates (Schipper et al. 2012).

Mammals

Mammals are primarily associated with permanent wetlands and waters on floodplains, although the postflood flush of vegetation on seasonal floodplains provides opportunities for foraging by large grazers, such as elk (*Cervus elaphus*). In California's Central Valley, the tule elk (*C. e. nannodes*) received its name because of its association with the tules (*Schoenoplectus* spp.) and other dense vegetation on floodplain marshes, although this association was apparently largely the result of hunting pressure, which forced them to take refuge in the marshes (Moyle et al. 2014). On some African floodplains, such as the Okavango Delta, elephants and hippopotamus play a major role in creating channels through which flood waters flow (box 8.3; Mosepele et al. 2009). In North America, the principal mammal shaping floodplains is beaver (*Castor canadensis*) whose dams can increase retention time of water on large floodplains and create the principal floodplains on smaller streams. Beaver dams exert a major influence on vegetation and nutrients in the systems they flood (Naiman et al.1988). As discussed previously, beaver ponds can provide important rearing habitat for salmonids and the widespread loss of beavers in California and the Pacific Northwest may have precipitated a decline in salmon populations even before commercial fishing began.

Small rodents also can affect vegetation on floodplains, for example, by browsing seedlings of riparian trees. However, their population size, and thus their impact on vegetation dynamics, is sensitive to the magnitude and frequency of flooding; areas with annual flooding tend to have smaller populations of rodents than those with infrequent flooding (Andersen et al. 2000; Jacob 2003; Golet et al. 2013). In areas with frequent floods, even the rapid reproductive cycles of rodents may have a hard time recovering from flood-induced population crashes. Burrowing mammals such as pocket gophers (*Thomomys*) and moles (Talpidae) or nonclimbing rodents such voles (*Microtus*) often cannot find refuge areas from floods, while species that climb bushes or small trees for refuge (e.g., deer mice, *Peromyscus*) become more vulnerable to avian predators when they do so. In a floodplain along the Waal River, a tributary to the Rhine River (the Netherlands), Wijnhoven et al. (2006) found that at least a year was required for populations of mice, moles, and shrews to return to their former

> **BOX 8.3 · Okavango: An African Floodplain Ecosystem**
>
> The Okavango Delta is one of the largest wetland systems in the world, encompassing 800–1600 km², and swelling with floods up to 28,000 km² in wet years (Mosepele et al. 2009). This immense floodplain is the terminus of a large river system, the Okavango, which originates in tropical Angola and flows through the deserts of Namibia before reaching its subtropical delta in Botswana. The delta is notable for its large biomass of herbivorous mammals (up to 12 metric tons per km²), including elephants and hippopotami. It is also notable for how those massive animals—along with some tiny ones—serve as ecosystem engineers that modify floodplain structure and topography (Mosepele et al. 2009).
>
> Hippos move daily between deeper water and feeding areas on land and, because they follow consistent pathways, they create and maintain vegetation-free channels through the floodplain. These channels influence how water flows through the floodplain and can serve as points for avulsion when sedimentation of main channels leads to channel switching. Vegetation-free flow paths are due to animals, as are the well-vegetated high areas that break up the flow across the floodplain. The islands that dot the Okavango Delta, which support much of the floodplain's woody vegetation, began as termite mounds. Termites colonize peripheral or isolated dry areas during dry years and construct high (2–4 m) mounds of clay. When a mound is flooded, it collapses but persists as a patch of higher ground, providing favorable conditions for recolonization by termites during the next dry cycle. This repeated process ultimately creates an island around which the water flows. The islands provide infrequently flooded, vegetated habitat for large mammals and birds, which help distribute seeds and promote further colonization of the islands by woody species.
>
> Temperate floodplains lack termite mounds and hippos, but they have (or had) their own ecosystem engineers, such as beavers and bison (both are discussed in the text of this chapter). Unlike the Okavango—which maintains a full staff of "engineers"—temperate floodplains are largely missing theirs, and understanding how they affected floodplain topography, vegetation, and processes requires historical reconstruction.
>
> A temperate floodplain that bore considerable resemblance to the Okavango was the Tulare Lake system in the southern Central Valley of California. The Tulare system consisted of rivers that flowed from the Sierra Nevada into deltas and a terminal lake surrounded by vast seasonally inundated marshlands (the Tulare basin is "partially endorheic" because during wet years it overflowed into the San Joaquin River basin to the north). The lake, marshes, and floodplain forests hosted dense populations of fish and waterfowl, as well as ecosystem engineers such as tule elk and beaver. The system also supported large indigenous human populations, as the Okavango still does. The Tulare system, and its wildlife and engineers, is now gone, drained for agriculture—a fate that the Okavango can hopefully avoid.

numbers after a flood event and that populations appeared to spread slowly from refuges. A similar pattern was found for a restored floodplain in California, along the Sacramento River (Golet et al. 2013). In a Chinese floodplain (Dongting Lake), it took at least 3 years after flooding for populations of Norway rats (*Rattus norvegicus*) and house mice (*Mus musculus*) to achieve population levels similar to those in surrounding agricultural land (Zhang et al. 2007).

On the restored floodplain along the Sacramento River, all small mammals species showed significant declines after a flood event but some species, such as California vole (*M. californicus*), recovered faster than others (Golet et al. 2013). This study also showed that (a) alien rodents (e.g., house mice) had no advantage over native species in surviving flooding and (b) riparian/floodplain restoration sites were not an important source of rodent pests to adjoining agricultural fields. In fact, Golet et al.

(2013) argue that reduced populations of potentially harmful rodents in floodplain areas are an ecosystem service provided by flooding (and floodplains that can be flooded regularly). This service of rodent population control can be observed during a flood event, when flocks of herons, egrets, gulls, and other predatory birds aggregate in the front of a flood that is just starting to flow through fields, picking off the displaced rodents and insects.

Ungulates, such as deer and elk, can affect riparian regeneration and reproduction. For example, deer herbivory suppressed regeneration of riparian vegetation along small streams in northern California (Opperman and Merenlender 2000). In Yellowstone National Park, Kay (1995, 1997) and Kay and Chadde (1992) reported significant effects of elk herbivory on riparian willows; willows within areas that excluded ungulates were nearly 10 times taller than those exposed to herbivory, and the grazed willows had no seed production. Beschta and Ripple (2006) documented widespread decline in willow forests due to elk herbivory along the Gallatin River in the northwestern part of Yellowstone, attributing the increased browsing pressure to the loss of wolves as their main predator. They documented that the decline of riparian forests resulted in channel widening and incision, decreasing the hydrologic connectivity between the Gallatin River and its floodplain. The loss of wolves from the Olympic Peninsula (Washington, United States), and consequent increase in elk herbivory on willows and cottonwoods, was also associated with declining floodplain forests and river channel widening and a dramatic increase in the amount of channel with a braided morphology, an example of a trophic cascade—loss of wolves leading to increased browsing leading to reduced recruitment of riparian trees—culminating in a change to river and floodplain geomorphology (Beschta and Ripple 2006).

In a grand unintended experiment on factors regulating riparian vegetation, wolves were reintroduced to Yellowstone National Park. The reintroduction caused a change in behavior in their primary prey—elk. Rather than congregating along valley-bottom floodplains, as they did in a wolf-free environment, the elk dispersed more widely across the landscape (Ripple and Beschta 2003). As a result of this wolf-induced change in elk behavior, riparian vegetation—including willows, cottonwood, and aspen—recovered, a trophic cascade touched off by the intentional reintroduction of an apex predator (Ripple and Beschta 2012). These recovered trophic cascades can extend further than just wolves, ungulates, and riparian vegetation. In Banff National Park (Canada), Hebblewhite et al. (2005) compared a portion of a valley with wolves with a portion of the valley with minimal wolf presence. The area with wolves had an order of magnitude fewer elk and, consequently, higher density of riparian vegetation and also significantly greater diversity and abundance of riparian songbirds.

Here we have emphasized that wolves may have been a missing player that, once returned, restored a previous dynamic between ungulates and floodplain forests. However, there is another missing player in western North American that may also have had a considerable influence on the extent and composition of floodplain forests: American bison (*Bison bison*). North America's largest ungulate, bison once roamed central North America in herds of tens of millions (Shaw 1995; Lott 2002) but were near extinction by the late nineteenth century. These massive animals with massive aggregations likely had significant influences on riparian forests through browsing and trampling and by changing floodplain morphology and structure through excavation of wallows and clearing trails through vegetation (Cordes et al. 1997; Butler 2006). While bison no longer congregate in their once-massive herds, it is worth contemplating how different our riparian ecosystems in the Great Plains and northern woodlands have become in the absence of bison.

FOOD WEBS

This chapter has focused on floodplain vertebrates that are generally the most prized by people, in terms of economic, recreational, cultural, spiritual, and aesthetic values. These vertebrates are therefore also the most likely to be protected by regulations that govern harvest levels or protection of endangered species. Because of these regulatory and economic values, these species are often intertwined with floodplain-management actions, whether it is restoration to enhance habitat or decisions about land use and how it affects species viability. Understanding the ecological processes that govern these species' population dynamics is therefore critical to effective management. Many fishes and other vertebrates are embedded within floodplain food webs. In short, the vitality of the species people care about depends on the overall productivity of floodplain food webs. We therefore conclude this chapter with a synthesis of food-web processes during a flood event, focusing on floodplains of the Central Valley (California, United States), the system that is the focus of the extended case study in chapters 12–14.

Food webs within inundated floodplains of the Central Valley can be characterized as having two basic stages: a "first responder" stage that initiates rapidly upon inundation (figure 8.1A) and a "long-duration" stage that develops during long-duration flood events (figure 8.1B). The carbon source of the "first responder" food web includes terrestrial vegetation and leaf litter and phytoplankton with rapid growth and adapted to faster velocity and more turbid waters. Primary consumers include aquatic invertebrates carried to the floodplain by the flood waters (e.g., from upstream river or wetland habitats) or those that emerge rapidly from floodplain wetlands or soils, such as chironomids (Sommer, Harrell, Solger et al. 2004; Benigno and Sommer 2008). Zooplankton also enter food webs when they are carried to the floodplain from lakes, ponds, and ditches (Katz 2015). Terrestrial insects also get caught up in flood waters and enter floodplain food webs. This "first responder" food web can provide a food base for fish, such as juvenile Chinook salmon, as soon as they enter the floodplain and before food webs based on autochthonous productivity have developed (the "long-duration" food web described below; Sommer, Harrell, Solger et al. 2004).

A more complex food web develops with longer-duration flooding and is based on autochthonous algal productivity (figure 8.1b). During long-duration inundation, biogeochemical processes liberate inorganic nutrients that, in addition to nutrients imported from the river, can support the growth of phytoplankton and periphyton. Within high residence time habitats, phytoplankton densities can increase to levels that support high productivity of zooplankton and invertebrates (Grosholz and Gallo 2006). Aquatic macrophytes grow and, along with inundated terrestrial vegetation, provide a substrate for periphyton. The productivity of algae, zooplankton, and aquatic insects supports a range of fish species that then are preyed upon by piscivorous fish. All fish species are consumed by piscivorous birds, such as herons, egrets, kingfishers, cormorants, and pelicans, and by mammalian predators such as mink and otters.

CONCLUSIONS

Temperate floodplains, despite being diminished and degraded, still support a high diversity and abundance of wild vertebrates, especially fishes and birds. Much of the value of floodplains for vertebrates is based on the productivity of floodplain food webs. Due to variability in flooding, this productivity can vary greatly between and within years. Generally, species adapted to take advantage of floodplains are found in systems that are flooded with some predictability in season and frequency. Thus, floodplains with predictable, frequent, and long-duration flooding will often support many fishes that are adapted to take advantage of floodplain resources. The flyways of migratory birds generally include a series of dependable

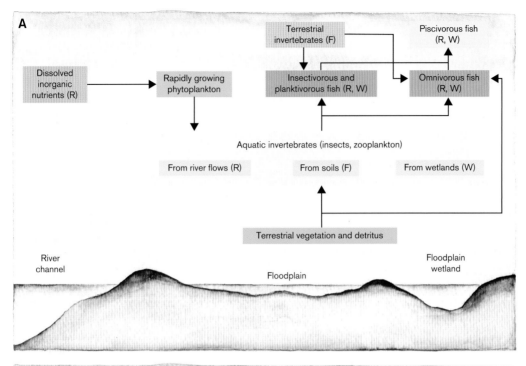
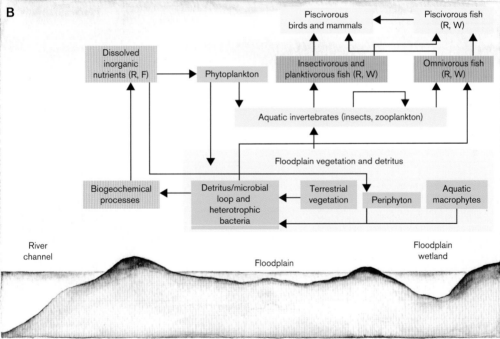

FIGURE 8.1 Food webs on a Central Valley (California) floodplain. (A) The "rapid responder" food web. (B) The food web that develops during long residence time flood events. Letters indicate origin of specific components: F = floodplain (soils, vegetation); W = wetland; R = river. Green box = primary productivity and detritus; blue box = primary consumers; indigo box = secondary consumers; yellow box = tertiary consumers.

floodplain habitats as stopover points. For resident and tropical migrant songbirds, as well as many small mammals, floodplain forests and grasslands often provide important habitat and, to persist over the long term, these habitats generally require dynamic flooding processes. In summary, production of fish, migratory birds, and other vertebrates is among the most important ecosystem services provided by functioning floodplains.

NINE

Ecosystem Services and Floodplain Reconciliation

TRADITIONAL MANAGEMENT OF floodplains has emphasized prevention of flooding and conversion of floodplains to uses such as agriculture, housing, and industry that are incompatible with periodic inundation. This management emphasis has had enormous economic benefits through urbanization and industrialization of flat lands and development of intensive agriculture on the rich alluvial soils. While promoting economic development, the widespread conversion of temperate river floodplains has resulted in two overarching problems: (1) continued risk from flooding of developed land and (2) loss of ecosystem services due to land-cover change and hydrological disconnection of rivers from their floodplains. Solving the first problem is the subject of the next chapter (chapter 10). In this chapter, we briefly review ecosystem services that floodplains support. We then describe how hydrological disconnection leads to the loss of these services and how alien species contribute to the difficulty of managing floodplain ecosystems. Finally, we describe reconciliation ecology as an approach to restoring ecosystem services to floodplains.

ECOSYSTEM SERVICES

Dynamic floodplains support productive ecosystems that generate a broad range of benefits for people (Gren et al. 1995; Barnett et al. 2016). These benefits can be characterized as ecosystem services. In their review of the value of the world's ecosystem services, Costanza et al. (1997) found that floodplains were the second-ranked ecosystem type, behind estuaries, in terms of their per-hectare value to society. Despite representing <2% of Earth's terrestrial land surface area, floodplains are estimated to provide approximately 25% of all nonmarine ecosystem service benefits (Costanza et al. 1997). Working at a finer scale using data from Illinois (United States), Sheaffer et al. (2002) estimated costs that would be required to replace goods and services provided by functioning floodplains. They concluded that each hectare of floodplain has a replacement cost of approximately US$150,000. Among the wide range of ecosystem services, the ability of floodplains to contribute to flood-risk reduction, such as by attenuating flood peaks, is often ranked as the most valuable (Costanza et al. 1997; Sheaffer et al. 2002), and chapter 10 focuses on that service. Traditional structural approaches to flood management do not take advantage of this ecosystem service and in fact reduce the ability of floodplains to absorb flooding (Pinter et al. 2006). Structural approaches also reduce the ability of floodplains to provide the ecosystem services described in the following sections. Some of the best documented services include sediment and nutrient reduction, carbon sequestration, groundwater recharge, fisheries, recreation, and support for biodiversity.

Sediment and Nutrient Reduction

As flood water spreads out on floodplains, it slows down and loses its ability to transport sediment. As a result, most sediment drops out, with the coarsest sediment being deposited close to the river (chapter 3). The finer sediments are deposited across the broader floodplain. With recurrent deposition, floodplains develop deep, fertile soils, which historically supported productive forests (Brinson 1990; Yarie et al. 1998; chapter 6). Floodplains are justifiably famous for these soils, the resource that makes them so attractive for agriculture. These soils benefit crops that require annual inundation (flood-recession agriculture) and also those that require prevention of flooding by levees. Even crops that suffer from flooding in the short term benefit from the long-term legacy of past flooding, the fertility and depth of floodplain soils.

Because nutrients such as phosphorous are largely adsorbed to sediment particles, floodplain deposition can result in improved water quality in downstream water bodies, including estuaries and near-shore marine habitats (Noe and Hupp 2005). In addition, biogeochemical processes within floodplain wetlands, such as denitrification, can reduce nitrogen concentrations in river water (chapter 4). Consequently, floodplain reconnection and restoration have been recommended as strategies to reduce nutrient pollution to large marine systems such as the Chesapeake Bay (Noe and Hupp 2005) and the Gulf of Mexico (Mitsch et al. 2001).

Carbon Sequestration

Sediment deposition over centuries of flooding creates deep, fertile soil that supports dense, floodplain forests (Brinson 1990; Yarie et al. 1998). These forests sequester carbon within rapidly growing trees. Other forms of vegetation, such as tules (bulrush) and cattails, also store carbon, as do floodplain soils (Zehetner et al. 2009; Bernal and Mitsch 2012). Maraseni and Mitchell (2016) found that riparian forests in Australia contained up to 293 tons of carbon per hectare within standing and dead trees and shrubs, and so they suggested that reforestation and management of degraded riparian areas offer significant carbon sequestration benefits. Based on current carbon market values, Eisenstein and Mozingo (2013) estimated an approximate annual value of $100 per acre for floodplain carbon storage for California.

Groundwater Recharge

During flooding, a portion of flood waters percolates into the shallow groundwater. For example, Valett et al. (2005) reported that approximately half the flood water entering an experimental semiarid floodplain in New Mexico (United States) contributed to groundwater recharge. The Yolo Bypass (California) is frequently inundated for long periods of time, contributing to recharge of the groundwater in shallow aquifers. This recharge contributed to a valuable "groundwater bank" during drought (Jercich 1997). In arid, semiarid, or Mediterranean climate regions, flooding can contribute to ecosystem productivity by recharging groundwater and thus increasing the water available during the growing season; this water can be particularly valuable if flooding occurs during or just before a period of drought (Robertson et al. 2001). Groundwater recharge is often one of the most important ecosystem services provided by floodplains (Schuyt 2005), supporting much of the agriculture in parts of Africa (Acharya and Barbier 2002, 2003; Opperman et al. 2013).

Fisheries

River systems that have active floodplains support the most productive freshwater fisheries in the world, especially in the tropics, feeding hundreds of millions of people in Africa, Asia, and Latin America (UNEP 2010; Opperman et al. 2013; McIntyre et al. 2016). Floodplains support river fisheries by boosting total productivity (chapter 7) and by providing favorable habitat conditions for fish (chapter 8; Galat et al. 1998). Productive river-floodplain fisheries generally come from systems with flood pulses that are predictable, frequent (annual), and of long duration. River systems with these hydrological characteristics typically have fishes that are adapted to take advantage of floodplain resources and habitat (King et al. 2003; Winemiller 2004). In many temperate floodplains, fisheries are supported by the creation of complex habitat features on floodplains, including permanent water bodies such as oxbow lakes, and these features are created and maintained by regular flooding (see chapters 3, 8, and 11).

Illustrating linkages between floodplain inundation and fish productivity, Welcomme (1979) reported that the productivity of fisheries in the Danube River was directly proportional to the extent and duration of inundation of floodplains. Likewise, Risotto and Turner (1985) found that the amount of bottomland hardwood forest, which they used as a proxy for floodplain, was a significant predictor of the amount of fish biomass produced by the lower Mississippi River and its tributaries. Rivers that exhibit annual flood pulses onto extensive floodplains have significantly higher productivity of fish per unit area than water bodies that lack a dynamic flood pulse, including regulated rivers or reservoirs, a phenomenon characterized as the "flood-pulse advantage" (Bayley 1991, 1995). Eisenstein and Mozingo (2013) reviewed how floodplains in California's Central Valley provided benefits to the salmon fishery. For example, Southwick et al. (2009) estimated that if reconnected floodplain could restore just 1% of the historic salmon fishery, the per-acre fishery value provided by the floodplain would be approximately $950, which is fairly close to other estimates of the per-acre fisheries values of floodplains.

Recreation

Increasingly, floodplains provide "open space" in regions dedicated to intensive human use. As open space, floodplains can support activities during periods of inundation (e.g., bird-watching, boating, duck hunting) and when not inundated (hiking and camping). They support habitats with diverse fish and wildlife, which can have high recreational value, often because of their accessibility (chapter 8). For example, Eisenstein and Mozingo (2013) estimated that the potential value of hunting, fishing, bird-watching, and hiking on Central Valley floodplain lands may generate several hundred dollars per acre in recreational value annually. When flooded in winter, temperate floodplains often support huge flocks of migratory ducks, geese, and shorebirds. These overwintering birds not only generate recreational use through hunting and bird-watching on the major flyways, but also, in their Arctic breeding grounds, support indigenous hunters and are important components of wetland ecosystems.

Biodiversity

There are few species of animals and plants that are strictly confined to floodplains, but there are many species that are most abundant on floodplains and their associated wetland and forests (chapter 8). In the western United States, birds such as yellow-billed cuckoo, willow flycatcher, and least Bell's vireo primarily nest in riparian/floodplain trees and shrubs, and their decline in abundance is generally tied to the loss of riparian habitat. In the United States, the federal Endangered Species Act (ESA) provides strong

TABLE 9.1

Ecosystem service values of restoring agricultural lands to bottomland hardwood forest wetlands in the Lower Mississippi Alluvial Valley, compared to the net income from agriculture ($/ha/year).

The "social welfare value" column reflects economists' estimates of the value to society from the restored ecosystem services. The "+" indicates services that were not quantified but that likely have positive value. The "private market value" column indicates the actual monetized value of that service, with one column for currently available markets and one column for values in prospective markets (from Murray et al. 2009).

Ecosystem services	Social welfare value ($)	Private market value	
		Current ($)	Potential ($)
Greenhouse gas mitigation	162–213	59	419
Nitrogen mitigation	1268	0	$34
Wildlife recreation	16	15	15
Flood attenuation & other services	+	+	+
Total	1446–1497+	74+	1068+
Agricultural net income		368	

protections for endangered species and can provide a regulatory driver for the protection of floodplain habitats that support listed species (Van Cleve 2012). Similarly, the European Union has issued directives to protect biodiversity and important habitats such as floodplains. Thus, the habitat requirements of endangered species can be a strong driver for floodplain restoration and protection. For example, "habitat banks" or "mitigation banks" can be established on floodplains (Bunn et al. 2014). Landowners, such as a developer, can pay into a habitat bank as a requirement for receiving a license to modify the habitat of endangered species elsewhere; these funds can be used to support conservation of floodplains.

Other Ecosystem Services

Each floodplain provides its own distinctive set of ecosystem services. Some, for example, may be sources of native insects that pollinate nearby orchards and other crops. Others may sequester heavy metals and other toxins, such as mercury. Some may be sites that support seasonal duck-hunting clubs. And some may be valued primarily for aesthetics, because they are open vistas in intensely used human landscapes.

Overall, these multiple values illustrate that although a single ecosystem service may not provide as much economic value as agriculture, multiple services combined can achieve more comparable values. Eisenstein and Mozingo (2013) describe a set of ecosystem services for Central Valley floodplains that can, in aggregate, provide significant value. Opperman et al. (2010) provide a similar comparison of agricultural revenue to that of "stacked" ecosystem services for Mollicy Farms, a reconnected floodplain on the Ouachita River, Louisiana, United States (table 9.1).

While the above discussion focuses on how stacked ecosystem services from natural floodplains could compete with agriculture in terms of economic value, a complementary perspective is that various forms of profitable agriculture can provide some of the ecosystem services of floodplains. Certain types of agriculture are compatible with periodic flooding, such as flood-recession agriculture (common along tropical rivers) and annual row crops in flood bypasses where the season of flooding does not overlap the growing season (see the case study on the Yolo Bypass in chapter 12). These agricultural areas can provide benefits to fish and

> **BOX 9.1** · A Love Affair with a Fish: Floodplains, Aquaculture, and Sustainable Food
>
> Dan Barber, a prominent chef from New York, has probably introduced more people than anyone else to the concept that food production systems on floodplains can also provide broad benefits to fish and wildlife. His TED talk, *How I fell in love with a fish*, has been viewed nearly 2 million times. While he doesn't actually use the word "floodplain"—instead referring to a river and its wetlands—his talk paints a vivid portrait of a healthy floodplain system that happens to also be an aquaculture operation. Barber's talk features the Veta La Palma (VLP) aquaculture business, located in the floodplain and estuary of the Guadalquivir River, near Seville, Spain. He sequentially describes the components of a functioning floodplain ecosystem that could have come from this book's table of contents, describing VLP's hydrology, its biogeochemical effects on water quality, and its productivity of phytoplankton, zooplankton, invertebrates, and, of course, fish. He also focuses on the birds that prey on those fish and how, surprisingly, the owners view the abundance of piscivorous birds as a great indicator of the health of their operation. VLP covers 28,000 ha within the Doñana National Park that had previously been drained for pasture. When the owners of VLP acquired the property in 1982, they implemented operations consistent with the management objectives of the reserve. They reconnected the property to the Guadalquivir River, creating a mix of habitats including brackish and freshwater marshes along with some rice production and pasture. Unlike intensive aquaculture operations, VLP does not feed its fish. Instead, the fish consume invertebrates that flow from the estuary into the brackish fishponds. VLP provides habitat for 250 species of birds, compared to approximately 50 species prior to the floodplain reconnection (Svadlenak-Gomez 2010) and is particularly valuable as habitat for birds during dry years when natural flooding is limited elsewhere in the Reserve. Márquez-Ferrando et al. (2014) report that populations of black-tailed godwits (*Limosa limosa*) have increased in the Doñana region due to an increase in habitat—mostly flooded rice and fish farms, and Toral and Figuerola (2010) suggest that widespread conversion of other agriculture to seasonally flooded rice fields in Europe after World War II was responsible for increased populations of many waterbirds, particularly those that feed in wetlands during migration. This continental trend, along with the specific example of VLP, shows the potential for both reconciled floodplains and novel ecosystems to support wildlife in intensively managed landscapes. Barber concludes his talk by noting that, in addition to environmental benefits, VLP provides an "ecological model" that holds the "recipe for the future of good food," by which he means both sustainable and delicious.

wildlife during periods of flooding. Rice fields are often intentionally flooded at times of the year that benefit migratory wading birds and waterfowl (chapter 8; Toral and Figuerola 2010). Agriculture—or in some cases aquaculture (box 9.1)—that sustain ecosystem services provide interesting examples of novel ecosystems and lessons for the sustainability of food production systems.

LOSS OF FLOODPLAIN ECOSYSTEM SERVICES

Ecosystem services—such as fisheries, clean water, and nutrient-rich soil—have drawn people to floodplains for millennia. River valleys have served as important places for settlement, with the first civilizations arising along the Nile, the Yangtze, Ganges, and the Tigris and Euphrates Rivers. Regions that developed industry and intensive agriculture tended to apply flood-control approaches that disconnected floodplains from rivers, thus allowing land uses that were largely incompatible with flooding. This conversion of floodplains to other uses is widespread along temperate rivers (Tockner and Stanford, 2002).

In this section, we review the processes that disconnect floodplains from their rivers and facilitate land-use conversion. As discussed, disconnecting floodplains from rivers has produced significant economic benefits, including globally important agricultural production in floodplains of rivers such as the Sacramento, San Joaquin, Mississippi, Murray-Darling, Rhine, and Danube. However, disconnection has also led to dramatic declines in floodplain ecosystem services (Richter et al. 2010).

Two primary processes disconnect rivers and floodplains: (1) structures that physically prevent flood waters from flowing out of the channel and onto the floodplain, mainly levees and floodwalls, and (2) flow regulation from dams and diversions that change the hydrograph, including reducing peak flood flows. Note that, in addition to influencing hydrological connectivity, infrastructure described in this section (levees and dams) also disrupts lateral and longitudinal connectivity for organisms, sediment, and organic material.

Levees and Floodwalls

As described in the next chapter, levees are a central feature of flood management. They are essentially walls or berms that are intended to prevent river flow from flooding land behind the levees. Levees were built along the Yangtze and Yellow Rivers in China more than 4000 years ago; the Rhine, Mississippi, and Sacramento Rivers were first leveed in the eighteenth and nineteenth centuries (Sayers et al. 2013). Levees can be built to protect existing development, such as agriculture and residential and commercial buildings, and/or to facilitate future investment. Levees have been built along portions of most major temperate rivers, particularly in lowland river valleys. For example, in the United States, there are approximately 23,000 km of levees managed by the US Army Corps of Engineers in addition to approximately 160,000 km of nonfederal levees (National Committee on Levee Safety 2009). Large portions of major cities have been developed in floodplains. Many cities in the United States have 20–40% of their developed area within floodplains, including Charleston, South Carolina; Fargo, North Dakota; Dallas, Texas; Omaha, Nebraska; St. Louis, Missouri; and San Jose, California (Freitag et al. 2009).

Dams and Diversions

Dams and diversions impound water and can alter the flow of rivers. The number and placement of dams within a river system, along with how they are operated, can greatly affect riverine processes including the functioning of floodplains (McCluney et al. 2014). Some dams are intended to store water and regulate the hydrograph in order to shift river flow toward periods when it can provide a desired function, including power generation, urban water supply, irrigation, flood management, and recreation. This storage and regulation function can occur at timescales varying from hours to years. Thus, a hydropower dam may be operated to generate electricity only a few hours a day during periods of peak demand—causing dramatic within-day fluctuations in flow—while many large reservoirs are capable of storing water during wet years and releasing it during dry years.

Dams can have single purposes, such as hydropower or flood control, or multiple purposes. Flood-control dams and multipurpose dams with major flood-control responsibilities are operated specifically to store flood waters and reduce downstream flood peaks (figure 9.1). Dams operated for other purposes can also dramatically reduce flood peaks. For example, dams on the San Joaquin River, United States, such as Friant Dam, are operated primarily for water supply and irrigation. Friant Dam is capable of storing moderate flood flows, though it spills during large flood events. Nevertheless, in most years, the dam's storage of water for release during periods of high demand for irrigation results in reduced flood pulse magnitudes and extent of inundation.

Levees and dams—either independently or, in many river systems, managed in coordination—can dramatically reduce the extent of floodplain subject to inundation. As discussed in the next chapter, these structures often prevent

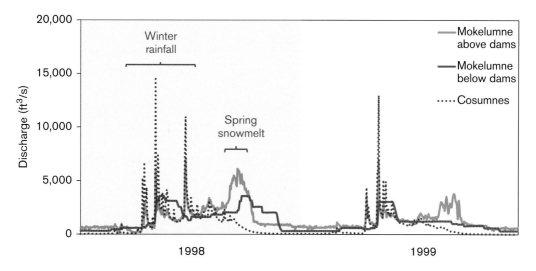

FIGURE 9.1 Discharge of two adjacent rivers in California's Central Valley with roughly similar sized basins. The Cosumnes River is essentially unregulated and displays natural flood peaks (red dotted line). The Mokelumne River has a greater proportion of its basin above the elevation that receives snow during the winter and is regulated by two large dams that provide flood control, hydropower and water-supply storage. The Mokelumne's hydrograph below the lower dam (purple line) shows the influence of dam operations, including the reduction in winter peak flows and the capture of the spring snowmelt pulse, which is clearly visible in the Mokelumne's hydrograph above the dams (green line) (USGS 2017).

most floods from reaching a river's floodplain. Although every system is vulnerable to infrequent floods that can overwhelm flood-control infrastructure, flood risk in most managed rivers has been sufficiently reduced so that floodplain land can be converted to other uses. This hydrological disconnection catalyzes land-use changes such as forest clearing, wetland draining, crop planting, and building construction.

Other Impacts on Floodplain Ecosystems and Their Services

Floodplains are also affected by infrastructure and management that alter their sediment supply. Dams, by capturing bed load and a portion of suspended load in their reservoirs, interrupt the transport of sediment from areas of erosion (e.g., mountains) to areas of deposition (e.g., floodplains and deltas). For example, capture of sediment by reservoirs in the Mississippi basin—particularly behind a series of large, multipurpose dams on the sediment-rich Missouri—has resulted in a loss of approximately half of the system's total load of suspended sediment. This reduction in sediment contributes to the high rate of loss of coastal wetlands in the Mississippi River Delta (Wohl 2011). Due to the capture of bed load by reservoirs, rivers downstream of dams can become incised, reducing the frequency of connectivity between river and floodplain (Kondolf 1997).

Sand and gravel are both valuable as construction materials, and these resources are often mined from river channels and floodplains. This removal of sand and gravel affects sediment supply and morphology of the mined area as well as downstream channels, floodplains, and deltas. Brunier et al. (2014) attribute incision of channels in the Mekong Delta to the large-scale mining of sand upstream. Local impacts from mining include the conversion of floodplain forest or agriculture into large pits. Pits close to the main channel can become "captured" during high flows and become incorporated as part of the channel. A captured pit can cause incision both upstream, by the propagation of headcutting from the nick point of the edge of the pit, and downstream, as the pit captures sediment and reduces the supply available downstream. Incision contributes to the lowering of the alluvial aquifer and floodplain water table (Kondolf 1997).

Groundwater pumping can further negatively impact floodplain ecosystems through decreasing connectivity between rivers and floodplains. By lowering water tables, groundwater pumping can alter vegetation composition on floodplains. For example, on Idaho's Big Lost River (United States), groundwater pumping contributed to channel dewatering and lowered water tables, leading to mortality of narrow-leaf cottonwoods (*Populus angustifolia*) and sandbar willows (*Salix exigua*; Rood et al. 2003).

Overall, most temperate rivers flow through regions with high population densities, intensive agriculture, and other land uses incompatible with flooding. These temperate rivers have therefore been subject to massive changes through flood-management infrastructure (dams and levees) and consequent land-use conversion. For example, Mississippi River floodplain forests below the Ohio River have declined by 80% from their historic extent (Llewellyn et al. 1996) and less than 10% of historic floodplains remain hydrologically connected to rivers in California's Central Valley (Bakker 1972, Barbour et al. 1991; The Bay Institute 1998). On the lower Missouri River, the surface area of floodplain wetlands declined by 40% between the 1890s and 1980s due to channelization and levees (Galat et al. 1997). The Danube River of Europe has lost over 80% of its floodplain habitats (Günther-Diringer and Weller 1999). Similar losses have been recorded for large rivers across the temperate world, including elsewhere in Europe and in China.

Not surprisingly, the massive decline of connected floodplains along temperate rivers has resulted in the loss of much of the ecosystem services they once provided, such as fisheries (Galat et al. 1998) and freshwater and riparian biodiversity. Freshwater species are endangered at higher rates than either terrestrial or marine species (WWF 2014), in large part because of hydrological disconnection and changes in flow from water-management infrastructure (Richter et al. 1997; Ricciardi and Rasmussen 1999). The people who live along these rivers—and society more broadly—generally are not aware of what is missing from the landscape, a phenomenon described as the *shifting baseline syndrome* (Papworth et al. 2009).

ALIEN SPECIES ON FLOODPLAINS

Changes to land and water management in river-floodplain systems often facilitate invasions by alien (nonnative) plants and animals. Alien species are now pervasive on most modern temperate floodplains, even those considered to be "natural." For example, common carp (*Cyprinus carpio*), native to Eurasia, are well adapted to floodplain ecosystems and now have a global distribution. Common carp often have large populations in temperate river-floodplain systems in the United States and in Australia. In the Mississippi River and some of its major tributaries, such as the Illinois River, invasion of several species of Asian carp has had a dramatic effect on the fish fauna. Three species of carp from China—grass carp (*Ctenopharyngodon idella*) and particularly silver carp (*Hypophthalmichthys molitrix*) and bighead carp (*H. nobilis*)—have become well established in much of the Mississippi River basin and, in some places, represent the majority of fish biomass in river and floodplain habitats (e.g., 63% of fish biomass in the main channel of the Illinois River [Roth et al. 2012]) and may be having a negative effect on native species through competition (Irons et al. 2007). Likewise, altered flow regimes, including diminished flood pulses, have contributed to increases in alien fish and riparian trees such as tamarisk in rivers in the western United States (Stromberg et al. 2007). Temperate river systems feature increasingly homogenized faunas, especially in the United States and Europe, due to the spread of alien species with broad distributions and intentional introductions of species that do well in modified ecosystems (Moyle and Mount 2007; Hermoso et al. 2011). This faunal homogenization also reflects the homogenized flow regimes generated by dams (Poff et al. 2007) that have decreased river-floodplain connectivity.

Increasingly, temperate floodplains have alien species integrated into their fish fauna. Floodplain lakes and ponds in both Europe and North America, for example, often support alien species such as common carp, largemouth bass (*Micropterus salmoides*), mosquitofish (*Gambusia* spp.) and ictalurid catfishes (*Ameiurus* spp.). These fishes are adapted to using floodplain or lake habitats in their native range and they are also tolerant of adverse conditions, such as high temperatures. Thus, they are well adapted for using altered habitats where they have been introduced. Lasne et al. (2007) found that nonnative species were most abundant in floodplain lakes with little connectivity to the Loire River, France. Where river-floodplain connectivity was high, alien fishes were less common. Conversely, where nonnatives have become widespread in river systems, connectivity through flooding can threaten remnant populations of native species that have persisted in isolated floodplain water bodies. For example, nonnative species are common throughout the Willamette river-floodplain system (Oregon, United States) and the Oregon chub (*Oregonichthys crameri*), a small fish endemic to the Willamette, tends to be low in abundance or absent where nonnative species are abundant. Chubs are generally found in floodplain water bodies that lack alien species and these tend to be habitats with low frequency of connectivity to the river. In this system, increased connectivity can actually lead to greater dispersion of nonnatives and a decline in the endemic chub (Scheerer 2002). Successful introductions of chub to numerous isolated habitats in the basin allowed the species to be removed from the endangered species list. Similarly, recommendations for restoring Sacramento perch (*Archoplites interruptus*) to their original floodplain habitats include some prevention of connectivity between rivers and floodplains in order to exclude nonnative fish (Crain and Moyle 2011). Reintroduction to floodplain habitats could involve raising perch in floodplain ponds that are disconnected from the overall river system and periodically allowing connection to allow adult perch to disperse, followed by removal of all the alien fishes that would have colonized the rearing ponds during the flood (Crain and Moyle 2011).

Alien aquatic invertebrates can also be significant parts of floodplain biota. Paillex et al. (2009) found that four species of aliens (*Corbicula fluminea, Potamopyrgus antipodarum, Crangonyx pseudogracilis,* and *Dikerogammarus villosus*) quickly invaded new floodplain channels on a restored floodplain on the Rhone River, France. The aliens subsequently appeared to exclude native invertebrates that originally formed a fairly diverse assemblage in the area.

Likewise, floodplain plant assemblages can often become dominated by alien weeds, especially grasses such as giant reed (*Arundo donax*, native to Eurasia) and shrubs such as tamarisk (*Tamarix ramosissima*). On a newly colonized area of a New Zealand floodplain, Peltzer et al. (2009) found that the most abundant alien plants were forbs and grasses, while the most abundant native plants were shrubs and grasses. In later succession, 90% of the plant biomass was made up of two shrub species, one native and one nonnative. The most unexpected finding of this study, however, was that relatively low biomass alien plant species caused significant changes to the microbial and invertebrate communities in the soil.

These examples underscore the prevalence of alien species on temperate floodplains and the management challenge they pose. Reduction or eradication of alien species will require management actions tailored to the species and context. In some cases, increasing connectivity or restoring the natural flow regime can boost native species over aliens, while in others it may have the opposite effect. Increasingly, floodplain ecosystems of large temperate rivers in highly manipulated landscapes are likely to be novel ecosystems (Hobbs et al. 2013), with mixtures of native and alien species, requiring a reconciliation approach to long-term management.

PROTECTION, RESTORATION, AND RECONCILIATION

The considerable decline of ecosystem services from floodplains has inspired interest in the conservation of remaining healthy floodplains and the development of management approaches that allow degraded floodplains to provide services. Protection, restoration, and reconciliation are three general approaches to improving floodplains to support ecosystem services. Protection of existing functioning floodplains is usually just a starting point because most temperate floodplains are highly altered and many are heavily invaded by alien species. Maximizing the ability of protected floodplains to sustain ecosystem services will generally require further actions that fall under the general concepts of restoration and reconciliation.

Restoration of floodplains tends to focus on reinitiating physical processes, such as reconnection with rivers, lowering of floodplain elevations, vegetation management, and increasing flood flows in regulated rivers. Reconciliation attempts to increase ecosystem services within floodplains that are managed for multiple purposes and are often novel ecosystems.

This section provides an introduction to floodplain protection, restoration, and reconciliation, while chapter 10 focuses on linking these actions with flood-risk management. Chapter 11 features case studies of protection, restoration, and reconciliation. The third part of the book, the extended case study on floodplains in California's Central Valley, also features various ways to increase ecosystem services from floodplains.

PROTECTION

Functioning floodplains can be protected through regulatory and policy approaches, including zoning and regulatory protection for wetlands (Brown et al. 1997). Existing floodplains can also be protected by real-estate transactions or agreements funded by government agencies or private entities. The European Union has mechanisms to protect floodplains under the European Water Framework (e.g., Ebro River; chapter 11). The United States has funding programs that can be used to protect floodplains, such as the Conservation Reserve Enhancement Program (CREP) of the Department of Agriculture (USDA). As part of the American Recovery and Reinvestment Act, the USDA had additional funds to acquire floodplain easements and received applications for >10 times the area of land (192,000 ha) than they were able to enroll (14,400 ha; USDA 2009). Nonprofit organizations, including land trusts, can protect existing floodplains through easements and purchases. For example, along the Savannah River (Georgia/South Carolina, United States), the Nature Conservancy (TNC) is acquiring title or easements on land to ensure that it remains in low-intensity land uses compatible with periodic inundation (Opperman et al. 2010). Protection alone often cannot accomplish conservation objectives; active management or restoration may be required.

RESTORATION

Restoration actions include reinitiating ecosystem processes, such as increasing connectivity between rivers and floodplains, modifying flow regimes, and lowering floodplain elevations.

Restoring Connectivity

As described throughout this book, hydrological connectivity drives floodplain processes and the disconnection of floodplains and rivers is generally the underlying cause of loss and degradation of floodplains. Thus, floodplain restoration often requires floodplain reconnection. Reconnection can occur through physical reconnection (e.g., removing barriers or adding controlled connections) or hydrological reconnection (e.g., the release of environmental flows from a dam upstream of a floodplain), or a combination of the two.

Physical reconnection can occur through cutting breaches through a levee, such as was implemented on the Cosumnes River in California (chapter 12) and at Mollicy Farms in Louisiana, where levee breaches reconnected 6400 ha of floodplain to the Ouachita River (Opperman et al. 2010). Levees can be set back further from the river to reconnect portions of the floodplains, as has happened on the Sacramento, Bear, and Feather Rivers in California (Williams et al. 2009; Opperman et al. 2010).

Modifying Flows

Dam operations can be modified to restore characteristics of the flood pulse to increase river-floodplain connectivity downstream and promote floodplain benefits. Reservoirs can release "managed floods" that are designed to replicate various characteristics of historical flood patterns, such as magnitude, season, duration, or rate of change (Postel and Richter 2003). For example, the US Army Corps of Engineers released an experimental flood from Alamo Dam on the Bill Williams River (Arizona, United States) and a multistakeholder team monitored the ecosystem responses. The hydrology and geomorphic processes promoted by the flood favored the regeneration and growth of native riparian willows (*Salix* spp.) over alien tamarisk and also removed beaver dams that had converted much of the river's lotic habitat to lentic habitat (Shafroth et al. 2010). On the Truckee River (Nevada, United States), dam operators mimicked the gradual recession of the snowmelt hydrograph to promote regeneration of willow-cottonwood floodplain forest (Rood et al. 2003). Several dams in Africa have released managed floods to partially restore downstream floodplain ecosystems and livelihoods that had declined following reduced river-floodplain connectivity (Duvail and Hamerlynck 2003; Loth 2004). A broad range of methods are available to evaluate managed flow releases from dams and to recommend flow characteristics (Tharme 2003; Harman and Stewardson 2005; Beilfuss 2010; Esselman and Opperman 2010).

In some cases, floodplain restoration may require both physical reconnection and hydrological reconnection. For floodplains along rivers with natural or near-natural flow regimes, physical reconnection of river and floodplain can potentially restore important processes. For floodplains along rivers with regulated flow regimes, restoration of these processes may also require changes to flow management. For example, levee setbacks along regulated rivers may contribute to flood-management goals but allow only minimal re-creation of historic floodplain processes and habitats if the upstream dam does not provide flow releases that mimic the most ecologically valuable types of flood events, such as long-duration, low-magnitude spring floods (Williams et al. 2009; Opperman et al. 2010). Conversely, managed flow programs intended to reinitiate flood pulses may need to address downstream physical connectivity, such as implementing levee breaches or building structures that allow managed connectivity between river and floodplain (Loth 2004).

Dam reoperation can also be used for floodplain restoration on rivers with navigation dams. Unlike the environmental flows described above, in these rivers the dam reoperation is intended to influence river and floodplain habitat upstream, rather than downstream. For example, on the Mississippi River, navigation dams can be operated to provide lower water levels upstream during the summer growing season, exposing substrate and creating areas of shallow water. This practice mimics natural seasonal variability which promotes growth of wetland plants and improves habitat for fish and waterfowl (Kenow et al. 2015).

As described in chapter 2, floodplains become inundated by water from a range of sources and direct connection with the river may not be required to restore some components of floodplain hydrology. On the Illinois River (Illinois, United States), TNC reinitiated inundation of a floodplain at the Emiquon site by turning off pumps that had previously drained the floodplain of water derived from local sources, such as precipitation, tributary inflows, and high water table (Lemke et al. 2014). A gate to allow managed connections with the Illinois River was completed in 2016.

Lowering Elevation

In certain situations, increasing the hydrological connectivity between river and floodplain can occur through the lowering of floodplain surface elevations. However, due to the cost of moving earth to change topography, these projects are generally limited in scale, such as a small swale constructed to provide fish habitat in a levee setback area on the Bear River (California; Williams et al. 2009) or are only done as part of flood-risk-management projects, such as the lowering of floodplain surface elevations along some rivers in the Netherlands to increase conveyance capacity (see chapter 11).

Passive versus Active Restoration

Reconnection and restoration of floodplain hydrology can often promote geomorphic and ecological processes that rebuild floodplain habitats and restore species and communities. The restoration of processes can be described as "passive restoration" in contrast to "active restoration" such as planting trees on a floodplain. Active restoration and planting are common on floodplain restoration projects, such as those done for regulatory mitigation. However, passive restoration often produces conditions that are more comparable to historic floodplain habitats, at lower cost. For example, at the Cosumnes River Preserve, TNC planted approximately 500 acres with typical floodplain tree species such as cottonwood and Valley Oak. Monitoring of the preserve revealed that restoration of process—levee breaches that allowed reconnection between the unregulated river and floodplain—promoted more rapid restoration of riparian forests. Based on these results, TNC now emphasizes passive or process-based restoration (Reiner 1996) at the Cosumnes River Preserve. However, on the nearby Sacramento River the flow regime is altered due to reservoir regulation, and floodplain sites that are physically connected to the river are not hydrologically connected with flows that promote forest regeneration. At these sites, TNC still engages in active restoration with planting of riparian seedlings (Opperman et al. 2010). Further, in some places herbivory by ungulates, such as elk or deer, may be sufficiently intense to retard or prevent natural regeneration and so exclusion of herbivores may be required (Opperman and Merenlender 2000).

Overall, restoration of floodplain structure, function, and ecosystem services generally requires restoration of various dynamic processes. These processes, in turn, require both river-floodplain connectivity and key components of the natural flow regime. However, while restoring processes is necessary, it is not necessarily sufficient. Many temperate floodplain sites are quite small and thus require active management.

Restoring floodplain ecosystem services requires people to confront numerous challenges: temperate floodplains exist in a context dramatically different than historical conditions, are laden with a legacy of built infrastructure and other developments, support many alien species, and have to serve multiple functions. Overcoming these challenges while maintaining or enhancing the desired ecosystem services of floodplains—including native species habitat—requires flexible vision and new approaches. Below we describe how reconciliation ecology provides a pragmatic and effective framework for managing floodplains for multiple benefits, including ecosystem services.

RECONCILIATION

In chapter 1, we offered floodplain reconciliation (Rosenzweig 2003)—rather than just floodplain protection or restoration—as a theme for this book. A reconciled floodplain is one in which people are part of the ecosystem, as active managers and participants. They work with a range of processes and tools—both "natural" and not—to achieve well-defined goals for the floodplain and for what it should produce. These goals depend on management needs and societal values and will vary from place to place and may change over time. For some floodplains, the conservation of native biodiversity and natural processes will be the top priority, while in other places flood management will be the top priority.

In much of the world, temperate floodplains are embedded within complex cultural and economic landscapes, with values shaped by diverse uses and expectations. At one extreme, Felipe-Lucia et al. (2014) describe the floodplain of the Piedra River, Spain, as an "agroecosystem," in which natural habitat, best represented by riparian forests, is extremely limited (less than 3% of the area) and is just one ecosystem service that must compete with others such as food production. The Piedra floodplain (19 km²) has a long history of intensive farming and the Piedra River is dammed and its flow highly regulated to provide water for irrigation and municipal use. Within these constraints, the likelihood of increasing the extent of natural floodplain habitat is low. However, a reconciliation approach to the Piedra floodplain focuses on maximizing the value to native birds and wildlife of both the limited natural riparian habitat and the extensive agricultural areas (an approach supported by the European Water Framework Directive; see case study of the nearby Ebro River in chapter 11). Felipe-Lucia et al. (2104) state "we . . . suggest that for the Piedra River, and similar floodplain agroecosystems, a mosaic of habitats comprising productive crops, poplar groves, fruit groves, and restored riparian habitats would increase the supply of [ecosystem services] and the resilience of the floodplain ecosystem."

The reconciliation approach on the Piedra River floodplain evolved over centuries of change and management. A much newer example of a reconciled floodplain is emerging on another Mediterranean-climate river 9000 km away from the Piedra. The Dos Rios Ranch (Stanislaus County, California) was historically planted with a variety of crops but, due to its location at the confluence of the San Joaquin and Tuolumne Rivers, had a history of flooding during wet years. The 650 ha property was acquired by River Partners, a nonprofit organization, with the goal of restoring the farmland to floodplain forest (River Partners 2015; figure 9.2). In addition to reforestation, the limited remnant patches of riparian forest currently on the ranch will continue to be managed for habitat, such as nesting for the rare Swainson's hawk (*Buteo swainsoni*). One of the authors (PBM) visited the site in June 2015, during the fourth year of a historic drought in California. The two rivers were barely flowing; in fact, the San Joaquin was actually flowing *backward*, toward a pump operated by a local irrigation district. Fish observed in the river revealed nothing but alien species, such as largemouth bass, western mosquitofish, and common carp; generalist species common on floodplains of other parts of the world. In the short term, River Partners is focused on restoring an 80 ha site at the confluence, currently still disconnected from the river by levees. To minimize costs, they planted trees (14 species) and shrubs (8 species) in rows, randomly interspersed. This plantation-style planting allows for efficient watering of the saplings by a drip irrigation system, necessary due to a lowered water table on the floodplain. Herbicides are used on occasion to control competing weeds. When the plants are well established, with deep roots, the levees will be breached so that flooding can occur on the site. The flooding will come from combination of natural hydrology and flow releases from upstream dams.

The Dos Rios project provides an illustrative example of reconciliation ecology at work. The project is deeply imbedded in a broader floodplain that is intensely farmed with high-value crops. As if mimicking its surroundings, the restoration of riparian forest is using an

FIGURE 9.2 Restoring a floodplain forest at Dos Rios (California, United States), June 2015, using an agricultural model. These plantings are about 2 years old (photo by Peter Moyle).

agricultural model, with irrigated rows of plantings. The project was funded from a broad range of sources, including those focused on fish and wildlife restoration, flood management, and mitigation for impacts from dams and water management. The restoration of some dynamic hydrologic processes will occur due to a broader flow-management program on the San Joaquin River, a process underway as result of a citizen lawsuit and court order (San Joaquin River Restoration Program 2015).

The ultimate trajectory of the Dos Rios project will be establishment of a novel ecosystem, one managed by people and supporting a mixture of native and alien species. With continued management, the planted forest will remain largely free of alien trees, although the understory of forbs and grasses will be dominated by nonnative species. One goal of the project is to provide flooded forest for use by rearing by juvenile Chinook salmon outmigrating down the San Joaquin and Tuolumne Rivers. However, juvenile salmon will only be present if there is active management upstream, including salmon hatcheries. Although most of the other fish species on the site will continue to be nonnative, they will be prey for native piscivorous birds. In its complicated mix of native and nonnative species, natural processes, and managed interventions, the Dos Rios is a model of a reconciled floodplain ecosystem—a place where managers are not trying to turn back the clock to some imagined pristine ecosystem, but instead create a place where active management will achieve specific goals, in this case an increase in habitat for rare native species.

While Dos Rios represents reconciliation as focused management of a relatively small area to promote biodiversity and ecosystem services, the Yolo Bypass, also in California, provides a compelling illustration of another dimension of reconciliation. As described below and in subsequent chapters, reconciliation can also gener-

> **BOX 9.2** · Floods Are Beautiful
>
> While some myths and legends portray floods as punishment from a wrathful god, cultures that depend on floodplain productivity celebrate floods in their art and song. The Barotse people, who live on the floodplains of the Zambezi River in western Zambia, hold an annual festival that celebrates the return of the flood. The festival—the Kuomboka (literally, the "moving out of the water")—is marked by drumming, and the benefits of flooding are vividly celebrated in song and poem:
>
>> It is flood time in Bulozi. The floodplain is clothed in the water garment.
>> Everywhere there is water! There is brightness! There are sparkles!
>> Waves marry with the sun's glory
>> Birds fly over the floods slowly; they are drunken with cool air.
>> They watch a scene which comes but once a year
>> Floods are beautiful.
>
> SOURCE: NAMAFE (2004)

ate significant ecosystem services on very large areas that were not originally intended to support those services.

CONCLUSIONS

Floodplains are among the most valuable ecosystem types on the planet in terms of providing services, including regulation of floods, fisheries, groundwater recharge, sequestration of nutrients and carbon, habitat for rare biota, and recreation. Particularly along tropical rivers, food-production systems that depend on the natural flood pulse and floodplain productivity (fisheries and flood-recession agriculture) feed hundreds of millions of people and provide one of the most important and tangible examples of how floodplain ecosystem services are intertwined with the well-being and culture of people (box 9.2).

Due to widespread conversion of temperate floodplains, many of these services have been diminished or lost. Management agencies are now giving increasing attention to floodplain restoration, motivated to protect or restore habitat for endangered species and/or commercially valuable species (including fish and waterfowl) and for other services. However, restoration of temperate floodplains occurs in a complicated context, including altered hydrology, a legacy of infrastructure and other development, widespread nonnative species, and multiple, often competing, demands on land and water. Due to this complicated context, we suggest that reconciliation ecology offers a pragmatic and effective conceptual model for securing the most benefits out of floodplains. A reconciled floodplain is one that accommodates people as part of river-floodplain systems, which are managed for specific goals that include nature conservation as part of a broader mix of economically and socially important activities.

Reconciliation projects generally require active management to continue to provide natural values. One reason for this is that reconciliation sites are often small in area, given that many floodplains are already extensively occupied by economically valuable land uses. However, integrating natural processes into flood-risk management is one way that a reconciled floodplain can become large in area. For example, the Yolo Bypass (chapters 12 and 13) is roughly three orders of magnitude larger than the confluence site of Dos Rios described above. The Yolo Bypass is a critical part of the flood-management system of the Sacramento Valley, but it also provides the best remaining lowland floodplain habitat in that valley. The bypass can be viewed as a vast reconciled floodplain whose value as natural habitat arose almost by accident but can now be managed to promote multiple benefits. Reconciliation and management for multiple benefits are the dominant themes of the remaining chapters of this book.

TEN

Floodplains as Green Infrastructure

RIVER FLOODPLAINS PRESENT two seemingly conflicting challenges to sustainable management of the world's rivers: the need for actions to reduce flood risk to people and economies and the need to support biodiversity, fisheries, and other ecosystem services that often require flooding (chapter 9). This chapter focuses on flood-management approaches that strive to reconcile these two challenges. Under traditional water management, these objectives—reducing flood risk and maintaining floodplain ecosystems and services—have generally been incompatible. As we will show, reconciliation of these objectives is not only possible, but also essential. Strategies that reduce flood risk for people can also promote high-value ecosystem services, while restoration projects can contribute to regional flood-management strategies. We first summarize traditional structural approaches to reducing risks from floods and then discuss new challenges that these traditional approaches do not address well. The subsequent sections then make the case for developing green infrastructure by looking at it first as a general approach, then at methods for creating it, and finally at the multiple benefits that can result.

TRADITIONAL APPROACHES TO FLOOD-RISK REDUCTION

Flooding is the most damaging natural disaster in the world, with average annual losses exceeding $40 billion (Cooley 2006). These damages have been increasing through time, and flood risk will continue to rise due to increasing numbers of people living in flood-prone areas. In addition, climate change is contributing to more intense storms even as flood-management infrastructure is aging. This infrastructure—including dams, levees, and floodwalls—has been the traditional response to managing flood risk. Here we provide a general overview of traditional management of floodplains and flood risk. However, a full review of these subjects is beyond the scope of this book. For a broader review of flood management, with an emphasis on new approaches, see Freitag et al. (2009).

Rivers have always attracted people because they provide abundant benefits such as fisheries, transportation, drinking water, and waste removal. For much of their length, most large rivers are located within valleys that provide flat, fertile land for farming, which are mostly floodplains. The cornucopia of river benefits has always been intertwined with the risk of periodic flooding. In fact, flooding is directly responsible for many of the advantages of rivers, including fisheries and fertile soil from deposition of nutrient-rich sediment. Originally, settlements dealt with flooding by having residents move to higher ground during floods or by restricting buildings to the very highest ground. But as civilizations grew more dense, large populations with permanent buildings became vulnerable to floods. It is because of this intertwined fertility and danger that the Yellow River is known as both China's "cradle" and its "sorrow" (box 10.1). In stories, folk tales, and songs, the Mississippi serves as a backdrop not only for romantic adventure, but also for inexorably rising floods (e.g., the blues song "When the Levee Breaks" by Kansas Joe McCoy and Memphis Minnie (1929), and William Faulkner's (1939) novel *The Wild Palms*, both set against the backdrop of loss and disruption of the 1927 flood; see chapter 11).

The 1931 floods in China, along the Yellow, Yangtze, and Huai Rivers, are estimated to be the deadliest natural disaster ever; official statistics reported 145,000 deaths, but external estimates placed the death toll at nearly 4 million (Kundzewicz and Takeuchi 1999). Today, flooding remains the most damaging type of natural disaster. Since 1900, nearly 7 million people have been killed in floods and 3 billion people have been displaced or suffered economic losses. While flooding is still a major source of mortality, the frequency of flood events causing massive mortalities is decreasing, due to improved warning systems and improved delivery of medical care, food, and water to flood refugees. Most (90%) of the 7 million fatalities occurred in just five individual flood events in China, all between 1900 and 1960. However, the number of people negatively affected by floods continues to increase and has been rising steadily since the 1970s,

---

> **BOX 10.1** · Giant Floods
>
> The earliest civilizations emerged along the floodplains of great rivers such as the Euphrates, Nile, and Yangtze. These civilizations both reaped the bounty of floods—the nutrient-rich deposits of Nile floods were the lifeblood of Egyptian dynasties—and occasionally suffered from damaging floods. But what about *really* big floods, the earth-inundating, Biblical floods found in the legends of many cultures?
>
> There is ample evidence of truly epic river floods, albeit they were regional, not global. These floods were larger than what can be produced by rainfall and runoff but, rather, were the product of some geologic event, such as the failure of a natural dam (Montgomery 2016). The "channeled scablands" of eastern Washington state (United States), which cover over 13,000 km², were created by floods that surged from the breaking of giant ice dams. The dams formed repeatedly as the continental glaciers retreated, 10,000–13,000 years ago, creating huge lakes that covered large areas of Washington, Idaho, and Montana. The dams would periodically burst due to increased pressure from impounded water, rapidly draining the lake and sending immense walls of water roaring across the landscape (McDonald and Busacca 1988). Similar ice dam effects, though generally not so large, have been recorded elsewhere in the world.
>
> Working in the Yellow River basin (China), Wu et al. (2016) found geologic evidence for a legendary giant flood, the Great Flood of Emperor Yu, that occurred about 4000 years ago. A massive landslide had dammed the Yellow River and created a huge lake in the Jishi Gorge, where the Yellow emerges from the Tibetan Plateau. Catastrophic failure of this dam, perhaps due to an earthquake, unleashed a flood that caused an avulsion of the Yellow River in the floodplain region downstream, leading to a period of channel instability and frequent flooding. Legends claim that Yu organized the people to dredge and contain the river, allowing floodplain agriculture—and civilization—to emerge from the ruins. For this accomplishment, Yu received the "divine mandate" and founded the Xia Dynasty, the earliest recorded Chinese dynasty and regarded as the beginning of Chinese civilization.

with currently approximately 100 million affected annually. In recent years, annual flood damages worldwide have been $10–40 billion annually, with an average of 10,000 fatalities (Kundzewicz and Menzel 2005; Cooley 2006). In the United States alone, flood damages have continued to rise over the past century. Costs have risen from less than US$1 billion annually in the 1930s to over US$6 billion currently (in constant dollars; Pielke et al. 2002).

These rising damages have occurred even as massive investments have been made in flood-control infrastructure including the following:

1. *Flood-control dams and reservoirs.* Dams can be operated to reduce downstream flood levels. By maintaining empty reservoir space behind the dam, a flood-control dam and reservoir can capture incoming flood waters and release a lower flow rate downstream for an extended period. A dam can store flood water, and thus reduce the downstream volume and height of the flood, until the reservoir is full. After the reservoir is full, the dam "spills" water and the rate of water leaving the dam is the same as the rate of water entering the reservoir. Thus, the ability to manage a flood is a function of the empty storage space in the reservoir relative to the size of the flows the upstream watershed produces. However, the need to maintain empty storage space can compete with other reservoir purposes that benefit from storing water, such as water supply and hydropower.

2. *Levees.* Levees are essentially walls constructed along one or both sides of a river. These walls prevent rising flood waters from inundating land behind them. The Chinese began constructing levees along the Yangtze and Yellow Rivers over 4000 years ago and levees were constructed along major rivers in the United States and Europe, such as the Rhine and Mississippi, in the eighteenth and nineteenth centuries (Sayers et al. 2015). Levees can range from simple berms of earth pushed up along a river to highly engineered "super levees" intended to protect extremely dense populations and high-value real estate, such as 300 m wide levees along the Ara and Sumida Rivers in Tokyo, Japan (C40 and CDC 2015). It is estimated that there are more than 180,000 km of levee in the United States alone (National Committee on Levee Safety 2009). An area protected by levees can be flooded if the levee is overtopped when a flood exceeds levee height or if it fails. Levee failure can occur from multiple causes, such as poor construction, earthquakes, erosion as water cuts into it, burrowing animals, and also by water seeping under or "piping" through the levee (figure 10.1).

3. *Floodwalls.* Floodwalls serve a similar function as levees, functioning as barriers to prevent flood water entering the area behind the structure, but they are vertical and constructed of material such as concrete. Levees, on the other hand, are generally made of soil and are sloping on both sides. Thus, floodwalls take up much less space than levees but are much more expensive to build. They are used to protect very dense or valuable real estate, such as the downtown of a city.

4. *Channel modifications.* Flood-management projects often modify river channels. Modifications can include "armoring" of banks to prevent rivers from changing course (e.g., meandering) and to reduce the erosion of stream banks (Florsheim et al. 2008). Other modifications can include dredging and straightening of channels to increase the speed at which a flood travels down the channel. Channel straightening and armoring can lead to channel incision.

Developed countries have made massive investments in flood-management infrastructure and this infrastructure has certainly saved lives and reduced damages in many places. However, flood damages continue to rise,

FIGURE 10.1 Levees along the Walla Walla River, Oregon, United States, failed as high flows reestablished meanders, point bars, pools, and riffles (photo taken on January 30, 1965; US Army Corps of Engineers 1971).

suggesting the need for more innovative and resilient approaches (Freitag et al. 2009; Opperman et al. 2009; Sayers et al. 2015). Further, structural approaches necessarily break the connection between rivers and floodplains and eliminate or diminish the biophysical and ecological processes described in previous chapters that are responsible for a range of important ecosystem services from floodplains. While structures will continue to play an important role in preventing flood waters from inundating cities, towns, and crop lands, flood managers are increasingly looking to new approaches that balance the use of structures with a wider variety of approaches, especially "nonstructural" flood-management approaches. These approaches use a broad range of tools, including zoning that avoids development in flood-prone areas, elevating or flood-proofing structures, pricing insurance according to actual risk, forecasting and evacuation systems, and using natural features to manage flood waters ("green infrastructure"). Nonstructural approaches allow for flood-risk management that can maintain some, or most, of the beneficial connections between rivers and floodplains (Freitag et al. 2009). Nonstructural approaches are not necessarily new tools; many have been in use for the past 100 years.

CHALLENGES FOR TRADITIONAL FLOOD-RISK REDUCTION

Flood management based on engineered structures ("flood control") has saved lives and permitted greater use of floodplains. But this path has also resulted in many unintended negative consequences, including decline of freshwater ecosystems, the loss of fisheries, and other ecosystem services (chapter 9).

Further, dams and levees create a false sense of security. Believing that all flood risk has been eliminated, people move into and make investments within areas where structures have eliminated regular flooding (Mount 1995). However, dams and levees are generally not designed to stop *all* floods and, further, structures fail, even during floods below the level they were designed to contain (figure 10.1). For example, levee breaks along rivers of California's Central Valley have occurred during 25% of the years since 1900; failures have occurred even during relatively small floods (Florsheim and Dettinger 2007). Flooding from failure of a structure can be rapid and catastrophic and thus very dangerous and costly (Tobin 1995). Because this hidden risk remains after flood-management structures are built, the Association of State Floodplain Managers (2007) recommends that levees not be used to facilitate new development. However, because development has increased on land that is imperfectly protected by structures, flood damages in the United States have continued to rise over the past century (Pielke et al. 2002). Levees may even increase flood risk elsewhere. By constricting the floodplain and reducing floodplain storage, levees can increase flood stages, and thus risk, in areas downstream (Pinter et al. 2006).

Overall, structural approaches to flood management face three major new challenges: (a) increasing risks from population growth, climate change, and aging infrastructure; (b) increased demand for ecosystem services; and (c) conflicts with other water-management objectives.

Rising Risk from Population Growth, Climate Change, and Aging Infrastructure

As described above, flooding is already one of the most damaging types of natural disaster worldwide. Several trends indicate flood risk will continue to increase along temperate rivers—and along coasts and rivers in general—especially as more people live in flood-prone areas. The problem is exacerbated by upstream land-use changes, such as increase in impervious surfaces in cities and rapid field drainage, which continue to alter runoff and downstream flood dynamics while development continues in places vulnerable to flooding (Pinter 2005). Guneralp et al. (2015) project that, globally, half of all urban expansion between 2000 and 2030, representing 500,000 km², will occur in flood-prone areas; in 2030, approximately 40% of urban areas will be in flood-prone areas. Further, forecasts indicate that climate change will lead to greater flood magnitudes and frequencies. Evaporation rates increase with air temperature and warmer air can hold more water vapor, increasing the intensity of heavy precipitation events (Kundzewicz et al. 2008). Knox (1993) reported that a period with somewhat higher amounts of annual precipitation (10–20%) in the upper Mississippi valley had significantly greater flood magnitudes, with flood levels that are today considered "500 year events" occurring several times a century. Similarly, Kwadijk and Middelkoop (1994) reported that relatively small changes in annual precipitation in the Rhine Basin (10–20%) could increase the frequency of floods capable of inundating most floodplains along the Rhine River in the Netherlands.

Beyond historical analysis of past climate and modeling of future climate, Easterling et al. (2000) reported that greater and more intense precipitation has already been observed in parts of world. For example, the frequency of intense rainfall events in the Midwestern United States has approximately doubled over the past half century (Saunders et al. 2012). Scientists have already found a climate "fingerprint" on a rise in precipitation or flood frequency and magnitude (Min et al. 2011; Pall et al. 2011). Based on both increased flood frequency and magnitude, and growing populations in flood-prone areas, a 2011 study by the US Federal Emergency Management Agency (FEMA) predicted that the land area vulnerable to a "100 year flood" will increase by 45% (Lehmann 2011).

On many temperate rivers, flood-management infrastructure was built decades or centuries ago and is now degrading. In the United States, for example, many dams require extensive rehabilitation, with an estimated 5 year price tag of US$12.5 billion, including 1743 high-hazard dams in need of repair. Levee maintenance is chronically underfunded, requiring an estimated 5 year investment of

US$50 billion (American Society of Civil Engineers 2009).

Due to all of these trends—changing land use, population increases in flood-prone areas, a warmer climate generating bigger floods, and aging infrastructure—flood risks along temperate rivers are greater than commonly perceived and are growing.

Increased Demand for Ecosystem Services

The decline and endangerment of freshwater species is a strong indicator of how structural modifications to river systems cause significant harm to other values and benefits that people derive from rivers, collectively called ecosystem services (chapter 9). Due to widespread floodplain disconnection and flow alteration, freshwater fish populations have declined dramatically in many rivers. For example, the Missouri River commercial fish harvest declined 80% from the 1880s to the 1990s due to flow regulation and levee construction (Galat et al. 1998). The loss of the flood pulse due to flow regulation from dams, and consequent decline in floodplain extent and function, has had negative impacts on millions of river-dependent people worldwide (Richter et al. 2010), such as those who farm along rivers such as the Tana in Senegal and Niger in Africa (Opperman et al. 2013).

In developed countries, social values and environmental legislation are challenging river managers to meet traditional objectives, such as flood management or hydropower, while also restoring river ecosystems and their environmental, cultural, and recreational values (Warner et al. 2011). For example, the US Endangered Species Act of 1973 requires that those charged with flood management, including the US Army Corps of Engineers and the FEMA, consider the needs of endangered aquatic species in flood-control policies and projects. This is reflected in a 2008 decision by the US National Marine Fisheries Service that FEMA's flood insurance program was having a negative effect on habitat for endangered salmon because it was linked to development of floodplain habitats (Freitag et al. 2009; Le 2012).

In later-developing countries, river managers face the challenge of increasing economic values from rivers, such as hydropower or flood-risk reduction, without compromising other values. In these countries, large rural populations may still depend directly on the productivity of river-floodplain ecosystems (Richter et al. 2010). Therefore, finding water-management solutions that allow rivers to continue to provide the broadest range of benefits are urgently needed.

Conflicts with Other Water-Management Objectives

Reservoirs that provide flood control often also have other purposes, such as water supply, hydropower, and recreation. These diverse uses often compete with each other. Flood control is maximized by maintaining low reservoir levels during periods of flood risk, whereas hydropower generation is maximized by high reservoir levels. Water supply risk can increase when reservoirs are lowered for flood control, especially if the basin that depends on the water enters a period of drought after reservoir drawdown. These conflicts may be exacerbated by climate change, which is forecasted to increase the unpredictability of precipitation and runoff and thus may increase both flood and drought risk in many parts of the world (Kundzewicz et al. 2008). In short, improving flood control through reservoir operation can come at the expense of other reservoir benefits, especially water supply and hydropower production. Both are coming under increasing demand due to population growth and climate change.

USING GREEN INFRASTRUCTURE IN FLOOD-RISK REDUCTION

The need to balance reducing flood risks with diverse ecosystem benefits has led flood managers in many countries to consider a broader range of responses to flood risk beyond strict reliance on engineered infrastructure. Major floods in Europe in the 1990s that overwhelmed engineered defenses led flood managers to think more broadly about how to address future risk (Nienhuis and Leuven 2001). The governments of France, the Netherlands, and Germany developed a "Flood Action Plan" for the Rhine River, which aims to reduce flood risks and damages by 25% by 2020 with an emphasis on nonstructural flood reduction measures, including the concept of "Room for the Rivers" (Kundzewicz and Menzel 2005; chapter 11). These plans essentially call for better integration of "green" and "gray" infrastructure. Gray infrastructure refers to engineered structures such as dams and levees, whereas green infrastructure refers to natural features, including forests, wetlands, and floodplains, that can replace or complement services provided by engineered infrastructure (UNEP-DHI Partnership 2014).

The integration of green infrastructure into flood management is a key thread of a larger trend of river management. The management of rivers and floodplains has been evolving, including a diversification of approaches and a refinement of language used to describe the discipline: the term "flood control" has gradually been replaced with terms such as "flood-risk management" or "flood-risk reduction." Flood control suggests mastery over rivers and the capacity to stop flooding, whereas the latter terms acknowledge that risks can never be completely eliminated, although they can be managed and reduced (Williams 1994). While flood control relies primarily on structures that attempt to determine where water should go, flood-risk management draws upon a broader range of tools, interweaving structural with nonstructural approaches, striving in some places to keep floods away from people and, in other places, to keep people away from floods (Larson and Plasencia 2001; Larson et al. 2003; Freitag et al. 2009).

Integrating green infrastructure into flood management promotes a *diversified portfolio approach*, which can increase resiliency of a flood-management system (Aerts et al. 2008; Dawson et al. 2011). The 2011 flood on the Mississippi River demonstrated the value of this diversified approach. The reconnection to the river of portions of the historical floodplain during the highest flood stages was critically important to the performance of the overall system (Opperman et al. 2013; chapter 11). Below we summarize three emerging principles for flood-risk management that strive to reduce risk while maintaining a broader range of values from rivers, drawn from practice and policy recommendations made over the past few decades (Galloway 1994; Blackwell et al. 2006; Galloway et al. 2007; Freitag 2009; Opperman et al. 2009; Sayers et al. 2015; Whelchel and Beck 2016):

1. *Work with, not against, natural processes.* Green infrastructure interventions rely on understanding how natural systems produce, store, and convey flood waters, and working with those processes. Traditional flood-management methods often seek to contain or reverse natural processes, such as confining flood waters between narrow levees and modifying channels to maximize the speed that floods travel downstream. Working against natural processes requires continuous vigilance (e.g., against levee failures) and maintenance (e.g., repair of erosion sites). Working with natural processes can allow nature to do some of the work and reduce long-term maintenance. Further, these natural processes are responsible for other diverse benefits and resources from river-floodplain ecosystems, and so flood-management methods that *promote* natural processes can produce a much broader array of benefits than can methods that *prevent* natural processes.

2. *Plan and implement flood management with a river-basin perspective.* Flood management is most effective when implemented at the scale of the entire river basin, as illustrated

by recent management of floods on the Mississippi River (Sayers et al. 2015; chapter 11). Flooding at a given site along a river is due to flood waters that are generated throughout the upstream watershed and that have moved along the entire upstream channel network. Thus, condition of the whole river basin contributes to flood risk at a given point. For example, is the land upstream mostly forested or mostly paved? Do agricultural fields have tile drainage? Are streams flanked by wetlands and floodplains or have they been channelized and straightened to maximize the speed of drainage? Flood-management projects that do not consider these river-basin conditions and processes are vulnerable to being overwhelmed by them. Not all flood projects can be implemented at the scale of an entire river basin, usually due to economic cost, institutional complexity, or political boundaries. However, project managers should at least ensure that they fully understand how individual infrastructure projects are influenced by basin-scale conditions.

3. *Deploy a diverse portfolio of methods.* Green infrastructure is not *the* solution for flood-risk management, but it should be a key part of an overall solution that draws upon diverse approaches (Muller et al. 2015; Palmer et al. 2015). Because green infrastructure methods produce diverse benefits and work with natural processes, flood managers are increasingly advocating their use. However, such solutions are not appropriate in all situations and their effectiveness varies with location and size of the river or watershed—a method that may reduce flood risk for a small watershed in New England will not necessarily be equally effective for the Mississippi River. Similar to an investment portfolio, a diversity of approaches to flood management can increase resiliency (Aerts et al. 2008; Dawson et al. 2011). A sustainable and resilient approach to flood-risk management will deploy diverse approaches throughout the river basin, with methods varying by local conditions while still reflecting the overall basin conditions. These approaches include a mix of structural tools, such as floodwalls to protect urban areas and levees to protect farmland, and nonstructural tools, such as early-warning systems, insurance incentives, and reliance as much as possible on green infrastructure. Large-scale examples of the latter include maintaining forests and wetlands to retain runoff and reserving agricultural floodplains to act as "relief valves" to convey water during large floods. In fact, many of the green infrastructure approaches described below are themselves a mix of structural and nonstructural approaches. For example, setting levees back from the river still requires a structure (the levee) and also incorporates natural floodplain surfaces to increase conveyance, a nonstructural approach.

USING GREEN INFRASTRUCTURE IN FLOOD MANAGEMENT: METHODS

Green infrastructure for flood management encompasses four major categories: (1) managing land to retain runoff and flood water, (2) large-scale preservation of floodplains, (3) setting back levees, and (4) creation of floodways and flood bypasses.

Managing Land to Retain Runoff and Flood Waters

Flood waters are generated by precipitation falling on the land, and becoming runoff at a sufficient rate to produce a marked rise in flows of streams and rivers (chapter 2). A key green infrastructure principle is that land can be managed so that it retains, or slows down, runoff as much as possible (Hey et al. 2004). In other words, to the extent possible, manage flood waters where they are generated, which is on the land surface, and in the soils, throughout the watershed. Specific methods to do this vary with the setting. Examples include the following:

- *Reforestation* can reduce peak flows produced from watersheds. Forests have numerous features that help retain water, beginning with the interception and storage of precipitation on the leafy canopy. Forests generally have a thick layer of organic matter on the ground, such as leaves and wood, and this layer serves as a sponge to hold water and also protects the soil from the direct impact of raindrops, which can increase erosion. Finally, forests tend to have deep soils that promote infiltration, allowing much of the precipitation to move into the shallow groundwater, a much slower path to the stream than via surface runoff. In contrast, a deforested or overgrazed watershed is characterized by a limited or absent organic layer and more compacted soils. Lacking the water-retaining features of the organic layer and permeable soils, much of the precipitation that hits the ground is rapidly converted to surface runoff. In fact, any feature that tends to retain and slow water down—including wetlands, meadows, wood in streams, or beaver dams—can reduce peak flows (Nyssen et al. 2011).

- *Agricultural best practices*, such as the addition of wetlands and small detention ponds, can reduce the magnitude of runoff produced from farm fields. Such features can be particularly important in croplands that feature tile drains. These features have added benefits of improving water quality by processing and removing excess nutrients from farm runoff (Fiener et al. 2005; Mitsch et al. 2006; Magner and Alexander 2008).

- *Slowing urban runoff* can reduce flood peaks, increase water quality, and have other benefits. Because cities are characterized by impervious surfaces, such as roads, parking lots and rooftops, a high proportion of rainfall swiftly becomes surface runoff, with a fivefold increase compared to an undeveloped watershed (Freitag et al. 2009). In urban settings, various methods can be used to retain storm water and prevent floods from overwhelming storm drain systems and causing urban flooding. These methods can range from those deployed at the scale of individual buildings, including porous pavement, "green" (vegetated) roofs, rain gardens, and rain barrels, to features that can attenuate runoff for larger areas, including grassy swales, wetlands, and detention basins (Freitag et al. 2009). Fortuitously, most methods for slowing runoff also help make cities greener, healthier, and cooler in summer while improving aesthetics and recreational value. Parks, greenways, "day-lighted" creeks, and urban gardens all contribute to a more vibrant city and also slow and retain storm water (UNEP-DHI Partnership 2014). In addition, on-site retention of urban storm water runoff reduces the amount of pollutants and sediment that enters streams and rivers.

The ability of these methods to reduce peak flood levels depends on size of the watershed and magnitude of the flood. Generally, techniques to retain water within forests and wetlands of a watershed are most effective at reducing moderate floods of short duration and within small watersheds (Bathurst et al. 2011). During very large or very long floods, the capacity of soils and wetlands to hold water is exceeded—they become full of water—and thereafter have limited or no further influence on the flood (Pitlick 1997). While broad patterns suggest that forested regions in general experience lower flood damages (Bradshaw et al. 2007), managing land to retain and slow flood waters is most effective in small watersheds and for reducing peaks of short-duration floods. Urban areas in particular are characterized by small watersheds that can be overwhelmed by intense, very short storm events, so these techniques show promise in addressing urban flood challenges.

> **BOX 10.2** · Acquiring Floodplains to Protect Boston
>
> In the 1970s, the US Army Corps of Engineers studied alternatives for reducing flood risk for Boston (Massachusetts, United States) from the Charles River and concluded that acquisition and protection of floodplain wetlands could provide similar protection, for approximately one-tenth the cost, as new levees and a flood-control reservoir. The Corps acquired 3440 ha in the Charles River floodplain for approximately $10 million (Postel 2005). The Corps reports that the "total annual cost" of the project is $477,000 and estimates that annual benefits are $722,000, with $125,000 of those benefits derived from recreational and environmental benefits (e.g., open space, habitat, recreation; US Army Corps of Engineers 1993).
>
> Based on the success of this project, in the 1990s the Corps studied the potential of natural valley storage to reduce flood risks in other Massachusetts rivers. On the Nashua River, the Corps identified 4800 acres of natural valley storage. If 30% of this storage were to be lost through encroachment and development on the floodplain, the Corps estimated that flood heights for the 100 year flood would rise by 1.7 feet downstream. Thus, protection of the full natural valley storage could prevent this rise in flood levels from occurring. The Corps cautions that natural valley storage is most effective at reducing flood peaks for floods of short duration, noting that during long-duration flooding, the storage capacity of floodplain wetlands becomes overwhelmed and inflow equals outflow.

Large-Scale Preservation of Floodplains

In addition to land management that aims to hold the rain where it falls, large-scale preservation of floodplains along a river can reduce flood levels and risk in some situations. The US Army Corps of Engineers has studied and implemented this approach in Massachusetts, focused on "natural valley storage" which they define as wetlands or floodplains along a river or stream that can temporarily store overflow from rising rivers and release it slowly back into the river, thus reducing the flood peak (US Army Corps of Engineers 1993; box 10.2).

Otter Creek and the town of Middlebury (Vermont, United States) provide a recent example of the value of large areas of connected floodplain to reduce flood risk for communities. In August of 2011, Hurricane Irene caused record flood levels throughout Vermont, including on Otter Creek in the town of Rutland with a peak discharge of 540 m^3/s. Middlebury lies approximately 50 km downstream of Rutledge but experienced a much lower peak, 200 m^3/s, nearly a week later. Flood damages were much lower in Middlebury than they were in Rutland. In between the two towns lies the "Otter Creek swamp complex," more than 7000 ha of floodplain forest and wetland which allowed the flood waters to spread out and attenuate the flood peak. Remarkably, the storage provided by Otter Creek's floodplain resulted in a gage with twice the drainage area (Middlebury at 1600 km^2) recording a peak discharge that was less than half the discharge recorded at the smaller upstream gage (Rutland at 795 km^2). Watson et al. (2016) estimate that the floodplains reduced damages in Middlebury from between US$500,000 and US$1,800,000 during Hurricane Irene.

Hydrologically connected floodplains can be maintained through policies, such as zoning or regulatory protection for wetlands, or through real estate transactions such as acquisition, easements, or government-funded programs, such as the Wetland Reserve Program. For example, to maintain potential for floodplains to be inundated along the Savannah River (Georgia/South Carolina, United States), the Nature Conservancy is working with willing landowners to acquire, or place under easements, floodplain lands to maintain them in low-intensity land uses that can be flooded (Opperman et al. 2010). In addition to acquisition or easements, emerging markets for ecosystem services, including carbon and nutrient sequestration, floodwater storage, and recreation, may be able to provide revenue to landowners that maintain floodplains connected to rivers (Opperman et al. 2009, 2010).

In addition to retaining and slowing flood water, maintaining hydrologically connected floodplains provides another important benefit: minimizing economic damages from floods. Land that remains in forest, wetland, or other flood-tolerant land uses is generally not vulnerable to damage during floods. The conversion and development of these lands, conversely, transitions them into land uses that become vulnerable to flood damages. Preventing floodplains from becoming developed, and thus vulnerable to future damages, can be a cost-effective approach for flood-risk reduction. This is particularly true when multiple benefits of floodplains are included, such as open space and improved water quality. Using models that incorporate potential flood risk, damage, and land values, communities can target the protection of those parcels that will provide the most reduction in future flood damages for the lowest cost of acquisition or easement (Kousky et al. 2011).

The benefits of maintaining floodplain in land uses compatible with flooding can be substantial. Brown et al. (1997) provide a clear illustration of this through a comparison of damages experienced by adjacent portions of Michigan and Ontario, including urban areas, during a series of storms and floods in the late summer of 1986. While floods in Ontario actually had greater magnitudes, total damage to property was only $500,000, while damages in Michigan were approximately $500 million. The factor most responsible for this dramatic difference in flood damages was that Ontario had far fewer structures in the floodplain due to a regulatory policy that discouraged such development (Brown et al. 1997). Biron et al. (2014) describe a process to map a "freedom space" for rivers that encompasses the area needed for river's geomorphic processes (e.g., meander migration) and inundation during large floods. Buffin-Bélanger et al. (2015) demonstrated that limiting development within the freedom space would produce net economic benefits from avoided damages, reduced maintenance costs, and ecosystem services.

Setting Levees Back from the River

Along many major rivers, such as portions of the lower Mississippi and California's Sacramento, levees were originally constructed very close to the edge of the river channel. This alignment of levees maximized the amount of land protected. By placing levees close to the channel, river managers tried to convert rivers into more effective conduits to rapidly flush flood water and sediment through the system.

However, constructing levees close to the channel creates a set of problems and challenges. By preventing flood waters from spreading, such levees greatly narrow the area available to transport floods. This constriction does work to rapidly flush flood waters through the system, but it also exposes levees to high-velocity water along their "wet" side (figure 10.2), increasing erosion and requiring high maintenance and repair costs. In much of the United States, the funds needed to repair levee systems are far greater than the amount available, with delayed maintenance and repair contributing to increased risk of failure (American Society of Civil Engineers 2009). Levees close to a river also reduce the extent of floodplain that is regularly connected to the river and can provide floodplain ecosystem services. Because of their vulnerability to erosion, levees close to the channel often require armoring to prevent erosion and meandering, further diminishing high-value habitats, such as riparian forest, found along river edges.

Setting levees close to the river, and armoring them, also causes other management challenges and environmental impacts. As described in chapter 4, meander migration is a fundamental process for many lowland rivers, and levees and bank armoring are often placed without consideration for the natural tendency for channels to migrate. Failure to understand how geomorphic processes interact with narrow levees can result in dangerous and expensive damage to structures. For example, a flood-

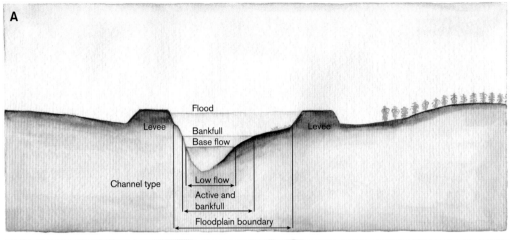

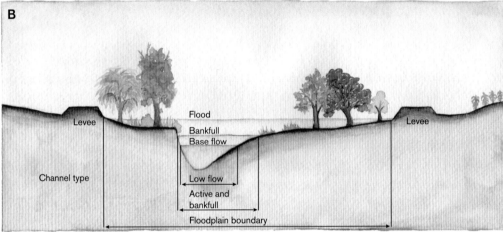

FIGURE 10.2 (A) Levees that are located relatively close to the channel. During a flood, the levees are exposed to high stages of fast-moving water, increasing the risk of erosion and the need for maintenance. There is limited area for river-floodplain habitats and processes between the levees. (B) Setback levees. For the same flood, the levees are exposed to lower water stages and velocities, reducing erosion risks and maintenance costs. The area that can support other floodplain benefits is greatly expanded.

management project on the Walla Walla River (Oregon) entailed channel straightening with levees built very close to the straightened channel. Not long after project completion, the river reestablished a meandering planform during a 1965 flood, resulting in multiple levee breaks (Waananen et al. 1971; figure 10.1).

Moving levees further back from the channel to create "setback levees" can help address these problems. Setback levees increase channel capacity for carrying flood waters, which is often the primary objective of moving or setting levees back from the channel. However, setting levees back has several other benefits. The area exposed to periodic inundation from the river increases, thus increasing the variety of benefits from river-floodplain connectivity, such as water-quality improvements. The expanded area on the "wet side" of the levee provides greater room for the channel to meander and create floodplain habitat features, such as wetlands and forests. While levees close to the channel are exposed to deep, high-velocity water during floods, setback levees are less frequently exposed to flood waters because of the increased channel capacity. Additionally, setback levees are less vulnerable to

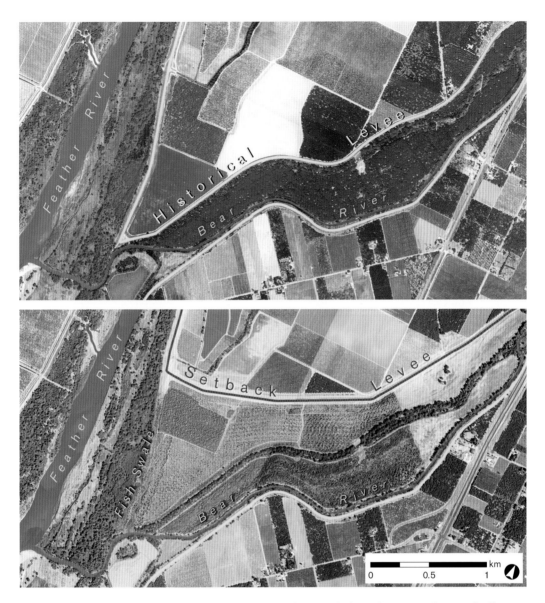

FIGURE 10.3 A setback levee project on the Bear River at its confluence with the Feather River in the Central Valley, California, United States. The top photo shows the original levees in yellow before the project. To increase conveyance and reduce backwater flooding, the north levee of the Bear and a section of levee along the Feather River were removed and a two-mile long setback levee was built (purple). The project lowers flood risk along the Bear and restored hundreds of acres of floodplain habitat. Note the "fish swale" in the lower photo, a feature intended to reduce stranding of fish and to support ecologically valuable long duration and frequent inundation (USGS 2005, 2012, 2014b).

erosion from flood waters because flow over floodplains is generally much shallower and slower than flow in river channels (Mount 1995; Dwyer et al. 1997; Freitag et al. 2009).

By increasing conveyance through a section of river, setback levees can relieve "bottleneck" points on a river where flood waters would tend to back up and potentially cause flooding. For example, the Bear River (Central Valley, California) once had narrow levees along the channel, including at its confluence with the Feather River. When both rivers were flooding, flood waters would back up into the Bear. Engineers determined that the best solution to reduce this backwater flooding was a levee setback project (figure 10.3). The setback levee is nearly 3 km long and restores 240 ha of floodplain habitat. The increased conveyance provided by the setback levee is projected to lower flood stages by 1 m during major floods, reducing flood risk along the lower Bear River (Williams et al. 2009).

When levee setbacks allow restoration of floodplain forests, flood managers must evaluate how restored vegetation will affect flood stage. Vegetation is a form of *hydraulic roughness*, or resistance to flow, and different types of land cover have very different values for hydraulic roughness. Bare ground has very low roughness, whereas a mature forest has high roughness. Because roughness slows down flood waters, it can increase flood elevations. Hydraulic models can be used to understand the effect of different vegetation types on roughness, and thus flood stage, and can guide where different vegetation types can be planted (Leyer et al. 2012). Baptist et al. (2004) developed a model to assess options for balancing objectives of floodplain management such as forest restoration and maintaining conveyance of water across reconnected floodplains along the Rhine River (the Netherlands). For their study area, they found that clearing vegetation from 15% of a floodplain over a 25–35 year return period would produce diverse floodplain plant communities, simulating effects of meander migration, and allow sufficient conveyance to maintain maintenance of safe flood stages. Hydraulic modeling also informed the mix of vegetation planned for the Bear River levee setback project. The lower portion of the reconnected floodplain was planted with floodplain forest species, while the upper portion will be maintained as grassland (Williams et al. 2009; figure 10.3). Greco and Larsen (2014), in a study of three areas of the Sacramento River flood control system, designed a multifunctional open channel that met objectives for both flood-risk reduction and conservation, including increasing riparian forest. They used optimization gaming scenarios to show that increased roughness (forest restoration) could be accommodated through expansion of the cross-sectional area of the flood channel.

The value of levee setbacks has been demonstrated for both the Mississippi and Sacramento Rivers (chapters 11 and 13, respectively). Both rivers experienced floods that overwhelmed levees built close to the channel, exposing the vulnerability of narrowly confining a river between levees, and both rivers have subsequently moved toward a more diverse flood-management system, with flood bypasses and some levees set back further from the river. A similar evolution is happening today in the Netherlands (chapter 11). However, many rivers throughout the world are characterized by levees that are quite close to the channel along much of their length in lowland valleys. Thus, there is great potential to reduce vulnerability and increase other benefits by planning or realigning levee systems such that levees are set back from river channels, to the extent possible.

Design of levee alignment along a river (e.g., distance from channel along various river reaches) requires assessment of multiple factors, including the value of floodplain land and land-use activities, the benefits of connected floodplains, costs of levee construction and maintenance over the long-term, and risk. Alternative levee alignments can be compared based on these factors during planning for the design—or redesign—of flood-management systems. The aftermath of a flood, during which levee repairs or replacement needs to be planned, approved, and funded, can be an opportune time for such comparisons of levee alignment (Jacobson et al. 2015).

Comparing alternative levee alignments can also be undertaken by management agencies during long-term planning for comprehensive flood management. For example, the California Department of Water Resources (CDWR) is developing a Central Valley Flood Protection Plan and, during the planning process, they are conducting a "Floodplain Restoration Opportunity Analysis" to identify the potential for levee alignments that promote floodplain restoration and contribute to flood-risk reduction goals (chapter 11). Larsen, Girvetz et al. (2006) compared levee alignments and identified setback distances that balanced flood management and ecosystem benefits, such as allowing meander migration. Larsen et al. (2007) then showed how long-term planning for levee alignments could both accommodate natural channel processes and protect built infrastructure. With a case study from the lower Illinois River, Guida et al. (2016) provide a hydrodynamic and economic model to compare maintaining levees and alternative options for setting levees back and reconnecting floodplains, in terms of their financial costs and benefits for flood management and habitat.

Floodways and Flood Bypasses: Large-Scale Storage and Conveyance of Flood Water

Major temperate lowland rivers, such as the Mississippi, Yangtze, and Rhine, were once flanked by vast floodplains that annually flooded and drained. While the extent of inundation varied among years, the river channels never carried the floods on their own, but rather the river and floodplain were an integrated system for moving water from the continental interiors to the ocean.

Attempts to control floods along these great rivers have been continuous, in order to allow land to be settled and farmed with greater certainty and security. The Chinese began building levees along the Yangtze and Yellow Rivers thousands of years ago, but floods continued to overwhelm the defenses, killing hundreds of thousands to millions of people. In the late nineteenth century, both the Mississippi and Sacramento Rivers became arenas in which engineers debated about how best to manage flood risk. Both rivers had advocates for a "levees-only" approach and both had advocates for an approach that would allow the river to continue to access portions of its floodplain during major floods. The "levees-only" side won the debate in both rivers, in large part because that approach maximized the amount of land that could be "reclaimed" from the river (Kelley 1989). However, great floods provided the final counterargument for both rivers because extensive and frequent levee failures exposed the vulnerability of the levees-only approach.

For both rivers, the US Army Corps of Engineers responded to levee failures by redesigning the flood-management system to include areas designated as "flood ways" or "flood bypasses." These are portions of the historic floodplain that receive flood waters and become inundated during major floods. The Yangtze River, too, has portions of the historic floodplain, known as Flood Detention Areas, which can be reconnected during floods, while the Meuse, Rhine, and Elbe Rivers in northern Europe have systems of polders that can be intentionally flooded to reduce river flood stage (Klijn et al., 2004; Förster et al. 2005). Along the Rhine River, the governments of the Netherlands, Germany, and France are pursuing a program called "Room for the River" which includes features that allow the flood waters to move into portions of the historic floodplains (chapter 11).

Flow from rivers into floodways are managed in a variety of ways, ranging from weirs that allow flow into the floodway once river stage exceeds the elevation of the weir (e.g., the Yolo Bypass on the Sacramento) to gates that can be opened by managers (e.g., the Jin Jiang Flood Detention Area on the Yangtze) to "fuse-plug" levees that are removed (e.g., by dynamite) in order to allow flood waters to enter the floodway, such as the Birds Point-New Madrid Floodway on the Mississippi River. These portions of the floodplain, managed to connect to the river during floods, are often very large. For example, the Birds Point-New Madrid Floodway encompasses 52,600 ha, while the Yolo Bypass spans 24,000 ha. Because floodways are only inundated during floods, they serve as "relief valves" for the system in two ways: conveyance and storage.

Conveyance

Floodways, or bypasses, increase the area available to convey flood waters safely through a particular reach of river. This is analogous to opening additional lanes at a bridge toll crossing during periods of intense traffic. For example, the Yolo Bypass conveys approximately 80% of the volume of major floods safely around the city of Sacramento. By increasing conveyance, strategically placed floodways can also reduce "backwater flooding," which is caused by the piling up of flood waters at and behind a bottleneck, such as where bluffs constrict the river. Vegetation increases roughness within a floodway, reducing its ability to convey flood waters. As a result, floodways are often managed for vegetation with low roughness, such as annual crops or wetlands.

Storage

Floodways can also store water and function similarly to a flood-control reservoir. While conveyance is analogous to adding lanes to a highway, floodways that provide storage can be viewed as a system of parking lots in a road network. During heavy periods of traffic (with an intense rush hour equating to a flood peak), management can direct many cars to remain in parking lots rather than joining the main flow of traffic (e.g., through policies such as congestion pricing). The parking lots release their cars slowly over time after the peak of traffic has passed so that the highway "downstream" of the parking lots experiences lower peak traffic. The analogous process in rivers with floodways (or flood control reservoirs) is known as "peak shaving," which reduces the height of the flood peak experienced at some downstream point. The Jianjiang Flood Detention Area along the Yangtze is intended to function in this manner with floodgates that can be opened as the flood is rising and has the capacity to hold 5 billion cubic meters of water, reducing the height of the peak against the levees that protect cities with millions of people. The Mississippi River and Tributaries Project has four large areas designated for "backwater flooding" that function to store water for a period of time to reduce downstream flood peaks.

Floodways provide conveyance and storage benefits that vary with location in a river basin, size of the floodway, and characteristics of the flood. Floodways also vary in the frequency with which they are used. Some floodways on the Mississippi have been used only a few times in 80 years, while the Yolo Bypass is inundated to some extent in most years.

Because floodways are only inundated during floods, they can be used for a variety of economic activities. Land use within a floodway varies with frequency of inundation (Opperman et al. 2009), as illustrated by two floodways in the Mississippi River and Tributaries Project (chapter 11). The New Madrid Floodway has been flooded only twice in nearly a century and so is intensely farmed and includes 200 homes. In contrast, the Bonnet Carré Spillway has been used 10 times in that period and it is uninhabited, with land managed for fishing, hunting, and recreation. Although the Yolo Bypass experiences some flooding nearly every year, much of it is in productive agriculture because, in California, the flood season (winter to early spring) has little overlap with the growing season (spring to fall). The agriculture in the Bypass is in annual crops that are generally not planted until after the period of flooding has ended.

Floodways can provide significant environmental benefits. The Mississippi River floodways in Louisiana are managed largely for natural vegetation and support abundant fish and wildlife (Alford and Walker 2011; Piazza 2014) and approximately a third of the Yolo Bypass is in wildlife refuges, including managed wetlands. Even the agricultural land within floodways can provide environmental benefits, particularly during periods of inundation. The Yolo Bypass is now recognized as the most valuable remaining floodplain habitat in the Central Valley and fishes and birds are abundant on farm fields during periods of both natural and managed flooding (chapter 13). The agricultural New Madrid Floodway also provided some benefits when it was inundated in 2011 (Rantala et al. 2016; chapter 11).

The floodways described above were planned and implemented 50–80 years ago, during a time when scientific understanding of river-floodplain systems was limited and society expected a much narrower range of benefits from rivers. Scientists now understand that connectivity between rivers and floodplains drives processes that produce a broad range of resources and benefits and, increasingly, societies expect these benefits—including fish and wildlife habitat, improved water quality, and recreation—from rivers. Thus, the design of future floodways can more fully reflect this evolution in knowledge and expectations and incorporate design features and operations that can produce a broader set of benefits. For example, a portion of each new floodway could specifically be managed for frequent, long-duration inundation, similar to the "fish wetland" in the Bear River setback levee (Williams et al. 2009).

Further, existing floodways can be modified or managed to increase their output of benefits beyond flood-risk management. For example, the Yolo Bypass provides important floodplain habitat, but it is not always inundated during the most important time of the year for native fish (early to midspring). River managers and conservation organizations have been exploring options for a control structure on the weir that shunts water from the Sacramento River to the Bypass. This control structure would allow for managed flooding of portions of the Bypass to maximize benefits to native fish while still avoiding negative impacts to agricultural management. Other options are also possible for the Sacramento River bypass systems. A study of this bypass system (Greco and Larsen 2014) showed that by expanding the flood channel footprint while increasing design roughness coefficients, the bypass system could effectively meet multifunctional objectives—flood risk reduction (greater conveyance and storage) and increased ecosystem benefits due to greater floodplain area. Another study found that lowering the elevation of the weir controlling flows from the Sacramento River to the Yolo Bypass could increase long-duration springtime flows that result from a few large storms that bring Central California most of its water (Florsheim and Dettinger 2015).

GREEN INFRASTRUCTURE: MULTIPLE BENEFITS

Gray infrastructure generally provides a single benefit, such as a levee that prevents flood waters from inundating farmland. Moreover, gray infrastructure often causes impacts to other benefits. For example, a levee system can move water quickly through a reach, increasing flood heights and velocity downstream. Further, the levee narrows the area inundated and restricts natural channel evolution, resulting in a loss of other floodplain services, such as water quality improvements and fish productivity. Conversely, green infrastructure solutions often provide a broad range of benefits beyond their primary purpose. For example, the Yolo Bypass was built in the 1930s, solely to reduce flood risk for Sacramento. Without specifically intending to do so, river managers had created a feature that would later be recognized as providing the best remaining floodplain habitat for migratory waterfowl in the Central Valley and one that also provides recreation, open space in a rapidly growing region, and a groundwater bank that proved important during drought (Jercich 1997; Sommer, Harrell et al. 2001). Similarly, levee setbacks not only provide "room for the river" to provide a broader range of ecosystem services, but also can provide area for open space and recreation. Green infrastructure features within a city double as parks, greenways, and recreational areas that make a city a more appealing place to live (UNEP-DHI Partnership 2014). These other services are often called "co-benefits" or "secondary benefits" of green infrastructure projects focused on flood-risk reduction. As described below, advances in quantifying these benefits may promote increased uptake of green infrastructure approaches.

Although green infrastructure approaches to flood management have numerous benefits, due to a variety of constraints they have been implemented far less frequently than traditional engineering approaches. First, green infrastructure approaches are much less known among engineers and decision makers, and this limits the extent to which they are even considered. Second, green infrastructure solutions, such as levee setbacks and flood bypasses, require more land than structural approaches such as building (or rebuilding) levees close to the channel. Because land is expensive, this can make green infrastructure solutions more expensive in the short run. Further, taking land out of agricultural production can have negative economic consequences for rural communities, and thus solutions that change floodplain land use can be controversial and require consideration of these economic impacts. Finally, the diverse services that river-floodplain systems provide—such as improving water quality and supporting fisheries—can be difficult to quantify or monetize (chapter 9). The inability to quantify these services results in benefit-cost analyses that don't fully consider the negative impacts from traditional approaches nor the secondary benefits resulting from green infrastructure approaches. Finally, even if these services could be quantified, mechanisms, such as markets, often do not exist to link the beneficiaries of the service with those whose land provides the service.

OVERCOMING CONSTRAINTS TO INFRASTRUCTURE INTEGRATION

Integrating green and gray infrastructure into one flood-management system is not easy. Here we present five potential strategies for overcoming constraints to integrating green infrastructure into management. The case studies in chapter 11 provide greater detail on some of these solutions.

1. Combine Multiple Sources of Funding

Although levee setbacks can lower flood risk and reduce long-term maintenance costs, acquiring the necessary land is usually expensive. But because levee setbacks can also create new ecologically functional floodplain, this flood-management tool can also draw on sources of funding for environmental restoration. Along the rivers that drain to Puget Sound (Washington, United States), a public-private partnership called Floodplains by Design has emerged not only to address flood-risk reduction, but also to promote multiple benefits from floodplains. Partners include the Washington Department of Ecology, The Nature Conservancy, and multiple city, county, and tribal governments. The partners identify areas of floodplain that if protected or reconnected, such as through a levee setback, will produce multiple benefits, including flood-risk management, ecosystem restoration (e.g., for salmon habitat), agricultural viability, open space, and water quality. By combining these objectives into their prioritization processes, the partners can identify areas that provide multiple benefits and can combine funding sources to achieve floodplain reconnection. Thus far, US$33 million from the state has been matched with US$80 million from other sources to support floodplain projects (Whelchel and Beck 2016). Tapping into restoration funding can potentially allow flood-risk projects to go forward that would otherwise be unable to secure sufficient funding, as illustrated by Hamilton City along the Sacramento River (Opperman et al. 2010).

2. Create Flood-Compatible Agriculture

Proposals to reconnect floodplains, for either ecological restoration or flood-management objectives (or both), can lead to concerns from rural communities that loss of farmland will threaten the vitality of the agricultural economy. However, reconnected land doesn't necessarily need to mean a loss of private land or a cessation of revenue to landowners. Reconnected floodplain lands can continue to generate revenue to landowners through markets for ecosystem services (described below) or through flood-compatible agriculture (Opperman et al. 2009).

Lands exposed to periodic flooding, such as the land on the river side of a levee setback or within a floodway, do not necessarily need to be used to recreate natural habitat. These areas can also be in traditional crops (e.g., intolerant of flooding) if they are in "relief valve" areas that will very rarely be flooded (such as the New Madrid Floodway) or if floods rarely or never occur during the growing season (such as the Yolo Bypass). Areas that are flooded more frequently during the growing season can be maintained with flood-tolerant forms of agricultural land use, including pasture, timber, or flood-tolerant crops. Potential flood-tolerant crops include feedstocks for cellulosic ethanol biofuels, such as willow, switchgrass, or diverse native prairie grasses (Tilman et al. 2006; Volk et al. 2004). Flood-compatible agriculture can thus allow green infrastructure solutions that do not take agricultural land out of production and can be consistent with vibrant agricultural economies and rural communities (Donath et al. 2004). For projects where hydraulic roughness is a concern, agriculture can provide a land cover with much lower hydraulic roughness than forest. The Yolo Bypass provides a clear example of a hydrologically connected floodplain that supports both profitable agriculture and significant ecological benefits (chapter 13).

3. Stack Multiple Ecosystem Services

While some portions of land within green infrastructure flood projects can be maintained in flood-compatible agriculture, other projects, or portions of land within a project, may be most appropriately managed as natural floodplain vegetation due to the frequency of inundation. Because floodplains provide a range of valuable ecosystem services (chapter 9), natural floodplain habitats can still potentially provide revenue to landowners. Some of these services have existing markets or other mechanisms that can provide revenue to floodplain landowners who provide the service. For example, the Yolo Bypass includes land that is leased to duck hunting clubs, and, during a drought, the high water table of the Bypass contributed to a groundwater bank (Jercich 1997; Sommer, Harrell et al. 2001). Habitat "banks" also exist for wetlands and endangered species, providing compensation for landowners that maintain or restore land providing these values. Other services, such as habitat and open space, can be promoted by public sources of funding, such as the Wetland Reserve Program. Carbon sequestration has potential markets for climate change mitigation (Joyce 2011) and markets may emerge for removing nutrients from rivers, for example, to reduce nutrient inputs that cause the "dead zone" in the Gulf of Mexico (Mitsch et al. 2001). Hey et al (2004) suggest a market in which agricultural landowners who created wetlands on their property (sited and designed to attenuate flood peaks), or levee districts that allow their land to flood, could sell flood-storage credits to downstream cities. The Sacramento Area Flood Control Agency has explored the potential for compensating farmers whose lands flood during large storms, and thus serves as "relief valves" that ease pressure on downstream developed areas (Opperman et al. 2009). Revenue from the services described above may allow agricultural landowners to continue to receive income from land in natural floodplain vegetation that is competitive with agriculture, particularly if revenue from multiple services can be stacked together. In other words, while carbon sequestration revenue may not be competitive with crops, that service plus recreation plus nutrient sequestration may be comparable, particularly for marginal agricultural land—such as land that is frequently too wet for successful cropping. With the Mollicy Farms project, the Nature Conservancy reconnected 18,000 acres of marginal farmland on the floodplain to the Ouachita River and is now implementing a research program to better understand the potential for stacked ecosystem services to provide comparable revenue to agriculture (Opperman et al. 2010).

4. Improve Valuation of Floodplains

Policies to evaluate the benefits and costs of alternate projects (green infrastructure versus traditional engineering) often focus on a narrow set of benefits and do not incorporate the diverse values provided by floodplains. Additionally, structural projects have long-term costs for maintenance, rehabilitation, and replacement that are often underestimated, while green infrastructure projects tend to have lower long-term maintenance costs. Because of limitations of science and policy, processes to select among flood-management options generally fail to capture the full value of benefits supported by green infrastructure projects and, similarly, do not reflect the full costs and impacts from traditional engineering projects. Improvements in the science of valuation and policies governing cost-benefit analysis will advance green infrastructure strategies, including the "stacking" of ecosystem services and combining multiple funding sources.

5. Increase Management Flexibility

As a component of flood-risk-management projects, floodplain reconciliation can also have value for water supply management objectives. Within multipurpose reservoirs, the need to leave reservation space to capture potential floods can compete with water storage for water supply and irrigation (Das et al. 2011). In California and the western United States, climate change will exacerbate this competition. California's large multipurpose reservoirs rely on refilling during the period of April to July as the snowpack melts in the mountains. During this period, managers can safely increase reservoir levels by filling the flood-storage space because there is low risk of the kind of precipitation event that can cause flooding. Dettinger et al. (2011) predict that, under a warming climate, reduced snowpack will lead to significant declines in April–July flows for rivers draining the Sierra Nevada. At the same time, more precipitation is predicted to come as winter rain, creating sudden peak flows that may be hard for reservoirs to accommodate. Conversely, leaving large portions of reservoirs empty can exacerbate water storage challenges when the system enters a period of drought. Large floodplains, downstream of reservoirs, that can accommodate unpredictable runoff events may allow reservoir managers to maintain higher reservoir levels rather than lowering them in preparation for storms that may never come. Because of the interrelationship between flood management and water supply, the 2005 California Water Plan Update called for integrated flood management as one of the tools for improving water supply.

CONCLUSIONS

Integrating green infrastructure strategies into flood-management systems has great potential to reduce flood risks for people while maintaining or restoring the diverse benefits that river-floodplain systems can provide. With climate change, population growth, and aging dams and levees contributing to rising flood risk, green infrastructure solutions will increasingly play a key role in community safety and resilience (Blackwell et al. 2006; Freitag et al. 2009; Opperman et al. 2009; UNEP-DHI Partnership 2014).

Green infrastructure cannot replace traditional infrastructure. Dams and levees will continue to play a key role in keeping people safe. However, within already developed river basins, green infrastructure projects increase the flexibility and resiliency of current water-management systems and should be given full consideration when flood-management systems are retrofitted, rehabilitated, rebuilt, or reoperated. Thus, in the wake of flood damages to levees, managers should consider alternate alignments of levees that include setting them back from the river. For example, following repeated flooding of a levee district in Iowa, the levee district, USDA, FEMA, and the Iowa Natural Heritage Foundation developed a plan that included not repairing the levees and converting the flood-prone area into a wildlife refuge. The farmers received flood easements for their land from the USDA's Wetland Reserve Program. Flood recovery and levee repair would have cost at least $4 million, while the easements and transition to a refuge cost $2 million. That land is no longer subject to harmful flooding, avoiding future repair and rehabilitation costs, and the area now provides flood storage, habitat, and recreation values (Freitag et al. 2009).

Within river basins that are just beginning to be developed for flood and water management, alternatives need to be explored that fully incorporate green infrastructure methods. This approach is more likely to allow these river basins to "get it right" the first time, rather than pursuing strictly structural approaches and later seeking to retrofit the system because of structural failures, the desire to restore broader river-floodplain benefits, or both.

Green infrastructure approaches, and flood-management approaches generally, are most effective when implemented as part of a diverse portfolio of tools that are deployed within a river-basin perspective (Sayler et al. 2015). This perspective allows green infrastructure approaches to be targeted to where they can be most effective. Specific approaches are more or less appropriate depending on the river system and the location within that river system to which they are applied (Morris et al. 2016). Natural valley storage may be an effective approach for reducing flood risk in a small watershed, but the utility of this approach typically diminishes within larger watersheds. In some locations levee setbacks may reduce flood risks for areas upstream or downstream (Williams et al. 2009), while in other locations, the primary benefit of levee setbacks may be flood damages avoided in the area reconnected to the river, such as previously farmed areas now in natural vegetation (Jacobson et al. 2015).

Through a river basin perspective, green infrastructure approaches can be effectively integrated with traditional flood management and with water management more broadly. As the examples in chapter 11 show, this integrated approach strives to create a mosaic of land uses in the floodplain that provide the most benefits for society.

ELEVEN

Case Studies of Floodplain Management and Reconciliation

NEW APPROACHES TO managing temperate floodplains have emerged across the world. Managers are recognizing that integrating functioning floodplains into flood-management strategies results in safer, more cost-effective, and more resilient systems. Further, these approaches can provide significant environmental benefits. Floodplains are increasingly recognized as regional centers of biodiversity that provide habitat for charismatic species of fish and wildlife, ranging from species that are economically important to those that are endangered. Increasingly, they are also being regarded as valuable open spaces, important for recreation and aesthetics. Floodplain-management projects and plans are as diverse as the rivers themselves. In this chapter, we provide a set of case studies of floodplain management, with a focus on management approaches that demonstrate that the reconciliation model can produce systems that provide multiple benefits.

CASE STUDY 1: ROOM FOR THE RIVER IN THE NETHERLANDS

The Netherlands' large-scale "Room for the River" program is an influential and often-cited effort for innovative management of floods and floodplains. In fact, the phrase "room for the river" has come to be synonymous with river management that features natural river processes including floodplain reconnection. An examination of this program sheds light on using green infrastructure for floodplain management and illustrates some of the challenges of reconciling flood management with improved ecological functions of floodplains.

A childhood story has introduced most people to the essence of the Netherlands' water-management challenges. The image of the little Dutch boy with his finger in a leaking dike is whimsical but telling. Most of the 42,000 km² surface area of the Netherlands, a country one-sixth the size of the United Kingdom, is prone to flooding. Historically, these flood-prone lands were alluvial floodplains and deltas built by the Rhine and Meuse Rivers. The

Rhine River is one of the longest in Europe (1320 km), important for navigation and home to 60 million residents in its basin. The river traverses nine countries before crossing the Netherlands and emptying into the sea.

For 2000 years or more, as in many other areas in Europe, the fertile alluvial lands along the Rhine and other rivers were used for agriculture and were separated from the floods that had created them. In addition to constructing river levees, the Dutch also built sea walls to allow farming on land that would otherwise be periodically flooded by seawater. Low-lying areas of land, called polders, were diked and then drained with ditches. These actions caused polder surfaces to subside and thus the polders effectively became protected "islands" of land that lay below sea level. Continual draining of polders required pumping, leading to the development of the iconic Dutch windmill. Currently, approximately 40% of the land in the Netherlands is below sea level and would be regularly inundated without seaward dikes. About 65% of the densely populated land is flood prone (van Stokkom et al. 2005; Klijn, pers. comm., Jan 4, 2016).

Beginning about 1000 years ago, extensive levee systems were built along major rivers including the Rhine, the Meuse (running from Belgium and France in the south), and the Scheldt (also from Belgium and France in the south). The extensive constriction of channels by levees, along with sediment deposition on floodplains between the levees, led to increasingly higher flood stages. In response, the Dutch have raised levee heights multiple times over the centuries. This management approach continued until its wisdom was called into question by paradigm-changing floods in 1993 and 1995, when over 250,000 people had to be evacuated due to threats of failing levees (Klijn et al. 2004).

The questions arising from these events catalyzed the "Room for the River" concept (figure 11.1). The change in thinking within the Netherlands has been described as a historical "switch from technological river management to ecological river management" (Nienhuis et al. 1998). One reason for this evolution in approach was recognition that the old methods, including dams, levees, channel stabilization, straightening, and other flood protection schemes, simply could not work indefinitely. The prospect of more floods like those in 1993 and 1995, coupled with predictions of sea-level rise, required the country to change its approach. Planners agreed that new management measures needed to be sufficiently flexible to accommodate higher flood levels expected in the future. Simply continuing to raise levees in response to rising flood levels was unsafe and costly. The Dutch government chose methods that accommodated predicted increased flows (e.g., due to climate change), while halting work on projects to raise and reinforce levees (van Stokkom et al. 2005). A plan was developed to increase the channel cross-sectional area available to convey floods, allowing more space for the rivers to spread into their former floodplains in some places—lowering the height of flood waters rather than raising the height of levees.

After planners identified hundreds of potential locations for interventions, 39 locations were selected based on an analysis of costs and benefits. In 2007, construction began on 34 of the projects (figure 11.1), focused on the three main branches of the Rhine River system (Klijn et al. 2013). Specific actions include the following:

a. Relocating levees away from the river, allowing more room for flood waters;
b. Creating side channels to carry high flows;
c. Lowering groins and training spurs;
d. Excavating rivers to increase channel capacity;
e. Removing or raising obstacles in the river (such as bridges) and/or lowering the floodplain surface to increase flow capacity;
f. "De-poldering" or allowing water to flow into areas formerly protected; with this approach, some farms and homes will be moved to adjacent high ground that was raised up above flood level.

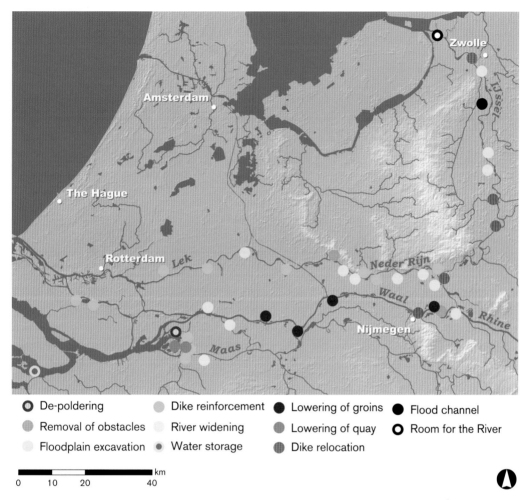

FIGURE 11.1 Sites of active Room for the River projects (after Zevenbergen et al. 2013; European Environment Agency 2016; Eurostat 2016).

Implementing management actions for the Room for the River program has not been easy. As with other floodplain reconciliation projects, appropriating land in densely populated areas is challenging and careful facilitation is needed to build cooperation across a range of interests. Further, the program must contend with basin-scale influences that are beyond local control. The social cost has been high, because homes and farms have been displaced. From the outset, it was decided that, although the overall plan was formulated and guided by a central government agency, the implementation of site-specific measures would be decentralized, with local or regional authorities or private parties taking the lead (Klijn et al. 2013).

The translation from scientific and government policy to local implementation has had successes and failures (Warner and van Buuren 2011). Immediately following the flooding of 1993 and 1995, public support was widespread because people had vivid memories of recent events. The plans were initially embraced by local citizens and an international agreement with other countries in the Rhine River basin was reached. But opposition arose as plans moved toward decisions about specific locations. Particularly challenging was the identification of land that could be converted from uses that required flood protection, such as crops and living space, to those that can be flooded, such as pasture or forest. Although the science and management community embraced the Room for the River concept, Dutch citizens, having grown up with polders protected by levees and sea-dikes, did not accept the change easily (Wolsink 2006). There was a shortage of public support, which was ameliorated with extensive outreach and other measures aimed at public communication and collaboration. A growing body of literature documents the social, political, and collaborative processes that were required to win, and maintain, enough support to move forward—without requiring another major flood to spur action (e.g., Wolsink 2006; Warner and van Buuren 2011; Rijke et al. 2012; Zevenbergen et al. 2013).

The comprehensive plan for both the Rhine and the Meuse Rivers grew out of these efforts (Hooijer et al. 2004) and emphasized four guiding principles for large river systems. These thoughtful principles apply to most large floodplain rivers in temperate climates:

1. Flood risk and potential damages are increasing.
2. In large systems, structural measures alone will not be sufficient for effective flood management, nor be sustainable over the long term.
3. Effective and sustainable flood-risk reduction can be achieved through adapting land use, giving the river more room.
4. Land use adapted for flooding should go hand in hand with ecological enhancement of floodplains.

In addition to these overarching principles, the Room for the River program provides insight into the social aspects of implementation. Outreach and communication are critical for project success, including seeking local cooperation early and throughout the process.

Room for the River provides a clear case of reconciliation on floodplains. This program will not return Dutch rivers and floodplains to their presettlement form because the landscape will continue to be heavily occupied and managed. However, by strategically reconnecting portions of the floodplain and allowing lands that once were defended vigorously to now flood, the Dutch will lower risks, have greater flexibility to address changing conditions, and increase the range of benefits from their floodplains, including improving conditions for native fishes and wildlife (box 11.1).

CASE STUDY 2: DANUBE RIVER— A RECONCILED FLOODPLAIN

The Danube River flows 2800 km from its headwaters in the Alps to its terminus in the Black Sea, making it the second longest river in Europe. The Danube's 805,000 km^2 basin spans 10 countries and encompasses a number of tributaries that are themselves large rivers, including the Isar, Prut, Tisza, Sava, Drava, and Morava Rivers. Historically, much of the Danube's length flowed through vast floodplains with many meandering channels (Günther-Diringer and Weller 1999). The floodplains contained diverse habitats and were centers of regional biodiversity. Like most European rivers, the Danube is now highly modified, primarily for flood control, navigation, and hydropower. Its floodplains have been largely converted to agricultural and urban uses, but forests do remain in some places. It has been estimated that over 80% of the historic floodplains have been lost, leaving 70 major wetland complexes along the river and its tributaries with some degree of protection (Günther-Diringer and Weller 1999). Most of the major changes to the river system, such as channel deepening and widening, occurred in the nineteenth and twentieth centuries, mainly to improve navigation. However, traces of the original floodplain channels, lakes, and other features are still present and these can be used to guide restoration. Most restoration projects are small relative to the amount of habitat lost and take advantage of local opportunities and infrastructure.

> **BOX 11.1** · Room for an Urban River
>
> Setting aside room for the river is not only applicable in rural settings where agricultural land can be fairly easily converted to floodplains, but can also work in cities. Nijmegen, the oldest city in the Netherlands, is involved in a $370 million project that will excavate a new channel for the River Waal, creating a large island and an urban river park. The existing levee will be set back 350 m, allowing more room for the river and increasing capacity of the channel and floodway. The project will provide greater, and more sustainable, flood-risk reduction than that provided by the current channel and levee alignment (Climatewire 2012).
>
> The new channel and island will provide water recreation opportunities and serve as a unique urban park (figure 11.2). With new bridges, water features, restaurants, and a walkable waterfront, the plan provides many amenities and recreation opportunities. The river, which used to run along the side of the city, and the newly created channel and island, will become a focal point for the city. Although about 50 residences will be relocated, the result will be a more sustainable urban living space. Zevenbergen et al. (2013, p.11) summarize the change, writing "in the future, instead of turning its back on the river, Nijmegen will embrace it."
>
>
>
> FIGURE 11.2 Conceptual image of levee setback plans for the Waal River in Nijmegen, the Netherlands. The original channel is on the left side of this image. The right side shows a conceptual image of the new island, new channel, new bridges, and anticipated development (from Zevenbergen et al. 2013).

One example of such a restoration project is on the Danube River about 68 km north of Munich, Germany. The area chosen for restoration was disconnected from the river in the 1970s, but still had significant floodplain forests remaining, as well as remnant channels, ponds, and oxbow lakes. Restoration options were constrained by needs to maintain channel form and flows for hydropower and navigation. Working within these constraints, the project, completed in 2010, created a set of floodplain habitats with a highly regulated and limited connection to the river rather than a large-scale "natural" flood connection to the river across the entire historical floodplain surface. Commenting on this approach, Stammel et al. (2012, p.58) noted that "the human impact (hydropower plants, silviculture) will persist for at least another 40 years. Therefore, the pristine situation, an unleashed river, could not be the general guiding principle."

The project directs flows through floodplain channels and reconnects floodplain lakes and channels to the main river. The project's three specific goals are to (1) create a floodplain channel or "river" (9 km long) that carries water from a hydropower dam through the floodplain, (2) create a narrow active floodplain along the channel (at least 25 m on each side) that is inundated under pulse flows from the dam (termed ecological flooding), and (3) create a drainage system in the downstream section of the project that lowers groundwater levels where they are artificially high due to water backed up by a downstream hydropower dam (Stammel et al. 2012). The restored channel mostly followed historic channels, but a new section had to be cut through a dry oxbow lake. In-channel fish habitat was improved by adding fallen trees, boulders, and gravel (Pander et al. 2015). The flow through the floodplain channel now ranges between 1.5 and 5 m³/s and is regulated to rise and fall based on flows of the Danube River, which has a mean annual discharge of 313 m³/s at the project site.

Because fish had the potential to respond most rapidly to the new habitat, populations were monitored both in the main river and in new channels and lakes before and after the new channel was connected to the main river (Pander et al. 2015). The initial fish fauna of the ponds and isolated channels consisted of 21 species. Twelve of the species and 74% of the individuals were eurytopic, relatively unspecialized in habitat requirements, while only 4 species (15%) were limnophilic and 5 (11%) were rheophilic (see chapter 8 for definitions). Within two months, the species number had increased to 38, representing about half the species pool in the main river. Few additional species arrived over the next 4 years. A majority of the first successful colonizers were eurytopic, although rheophilic and limnophilic fishes became more common as time progressed. Juveniles were the predominant life form for most of the new species, suggesting the importance of restored floodplain channels and ponds as nursery areas for river fishes. Pander et al. (2015) noted that some species were strongly associated with habitat characteristics such as large wood in new channels, indicating the importance of restoring not only hydrologic patterns, but complex habitat as well. In natural floodplains along the Danube River, fish biomass is typically higher by a factor of 10 or more on the floodplains than in the adjacent main river (Junk and Bayley 2008).

This project on the Danube illustrates reconciliation ecology at work. While more comprehensive restoration of processes could have been achieved by reopening the entire historical floodplain area to inundation, this was not feasible from an economic and social perspective. Nevertheless, the new habitats that were created clearly have benefits to river fish populations and presumably other aquatic organisms. Benefits to vegetation and associated birds and mammals will likely accrue primarily in areas close to new channels that are wetted periodically. The success of this project, which partially restores areas and processes, can help catalyze more expansive projects on rivers like the Danube, especially if rigorous monitoring programs inform management (Stammel et al. 2012).

CASE STUDY 3: EBRO RIVER—CAN FLOODPLAINS BE RESTORED ALONG A DEGRADED RIVER?

The Ebro (Ebre) River is the second largest river on the Iberian Peninsula, with a length of about 910 km and a watershed of 85,362 km². It flows from northeast Spain through Catalonia and into the Mediterranean Sea. The Ebro has a hydrograph typical of Mediterranean climates, with highest flows normally occurring November through March and very low flows during late summer. Most of its discharge derives from rain and snow falling on the Pyrenees. Characteristics of the historical flow regime have been partially reversed with elevated baseflows in summer to deliver irrigation water for agriculture and decreased flows in winter due to capture and storage by dams. However, large, damaging floods can still occur (e.g., 2015).

Two important natural areas along the Ebro depend on components of the natural flood regime: the middle reaches of Ebro River (350 km in length), which feature remnant floodplain forests and lakes, and the Ebro Delta, renowned for its diverse birds and fisheries. The floodplain of the middle Ebro originally covered 740 km^2 and was the largest floodplain on the Iberian Peninsula (Ollero 2010).

The watershed has been heavily settled for hundreds of years and is largely devoted to agriculture, industrial, and urban uses. To protect developed areas from flooding and to allow increased agriculture in the riparian zone, the river has been extensively dammed and lined with levees and other structures to control flows (Cabezas et al. 2009). At least 187 dams have been built in the watershed; together they store 57% of average annual runoff, although they spill during wet years (Batalla et al. 2004). The middle reach of the Ebro is undammed, but the limited floodplains that are still subject to flooding are highly modified. Ollero (2010) noted that in the past 50 years, 81 projects have extensively modified the channel. In addition to direct channel modifications, upstream reservoirs trap most of the sediment in the system, resulting in channel incision and armoring of the downstream riverbed (Batalla and Vericet 2013). Due to these changes, overbank flows are much less frequent than they were historically (Cabezas et al. 2009) and little sediment is deposited on the floodplains. The floodplain has been largely converted to agriculture, although remnant riparian habitat is protected. Due to lack of connection between the floodplain and dynamic river flows, remaining habitat patches are covered mostly with late successional vegetation, which is senescent in places (Gonzalez et al. 2010); early successional vegetation, characteristic of active floodplains, is largely absent (Cabezas et al. 2009). Isolated oxbow lakes are gradually filling with organic matter, so the aquatic invertebrate fauna in riparian areas is more characteristic of lakes and other permanent waters than of floodplain or riverine habitats. Species richness and abundance is highest in sites with higher frequency of flood connectivity to the river (Gallardo et al. 2008).

The river supports both native and alien fishes, including endemic species such as Ebro barbel (*Barbus graellsii*), Ebro nase (*Parachondrostoma miegii*), and Catalan chub (*Squalius laietanus*; Maceda-Veiga et al. 2010). In the main river, native species are in decline, while alien species, adapted to altered and polluted habitats, including reservoirs, dominate the fauna. The most recent invader is the wels catfish (*Silurus glanis*), a large predatory fish that appears to be having detrimental effects on populations of other fishes. In addition, the fish shows signs of being affected by diverse pollutants, from pesticides to heavy metals to industrial compounds. Residents are told not to eat fish from the river because of the high levels of contaminants in their tissues. The benthic invertebrate fauna of the river is dominated by polychaete worms, indicating severe pollution.

The Ebro River flows into the Mediterranean Sea through the Ebro Delta, developed over centuries by deposition of sediment during high flows. Today, the sediment supply is almost entirely captured by two large upstream dams, Mequinença and Ribarroja, which were built in the 1960s. Deprived of sediment-laden flood waters, the delta is no longer growing and is presumably shrinking, especially with rising sea level. While much of the delta is farmed for rice, it is also a high priority for conservation. Reserves in the delta include a Natural Park, a Special Protection Area for birds, and a Ramsar wetlands site (DeltaNet Project 2015).

Can a river-floodplain system this highly degraded be restored to a more natural condition that supports native plants and animals, albeit in a novel ecosystem? Based on the European Water Framework Directive, restoration of the middle reaches of the Ebro has been identified as a priority for enhancing biodiversity in Europe. Ollero (2010) suggested that changes to reservoir operations could restore some components of the natural flow regime that, coupled with strategic levee setbacks, could create a "Fluvial Territory." This area would feature a wide undefended corridor that would allow at least some room for the river to meander, recreating early successional riparian/floodplain habitats and reconnecting oxbow lakes. Modeling by Gómez et al. (2014) indicates that short flood-like pulse flows released from Mequinença and Ribarroja Dams would have substantial benefits for ecosystem services, potentially justifying the cost of the water released. Ollero (2010) suggested that establishing the Fluvial Territory would require acquisition of approximately 8000 ha of agricultural land. If the Fluvial Territory were sufficiently large to attenuate large floods, floodplain reconnection might even contribute to flood-risk reduction; large floods during February and March of 2015 highlighted the potential need for this.

There is growing realization that integrated water planning and management is needed in the Ebro (e.g., Omedas et al. 2012), but achieving this will require overcoming numerous challenges. The Ebro River therefore illustrates both the potential and the difficulty of reconciling floodplain uses in Mediterranean areas. It is widely recognized as a river system with many natural values that can be enhanced by modifying the flow regime and floodplain land use. However, the Ebro also illustrates the great challenges of implementing these measures in regions where water and land are both used intensively with economic activities that depend on current management regimes.

CASE STUDY 4: MANAGING FLOODS IN THE MISSISSIPPI RIVER

In 1927, the lower Mississippi River experienced a flood that is considered the most destructive river flood in the history of the United States: levees were breached or overtopped at 145 locations, 70,000 km² were inundated, and 700,000 people were displaced for weeks to months. Officially, hundreds of people died but more likely the number of deaths of rural residents was in the thousands (Barry 1998). In 2011, the Mississippi carried the greatest volume of flood waters ever recorded, exceeding even the historic flood of 1927, but the impacts of the flood—in terms of loss of life and damages—were dramatically different. The intervening changes in the flood-management system, responsible for the starkly different outcomes, illustrate the value of floodplain management as a method of flood-risk reduction.

The Mississippi is one of the largest temperate rivers in the world and, along with its tributaries, it drains over 40% of the land area of the contiguous United States. The floodplains of the Mississippi basin could merit an entire book focused on the system's biophysical processes, ecosystems, and loss and potential for recovery (for recent reviews, see Piazza [2014] and the case study in Wohl [2011]). For example, excess nutrients, particularly nitrogen, are a primary management challenge in the Mississippi basin. Much of the Mississippi basin is in row crop agriculture and the application of nitrogen fertilizer to these crops contributes to the elevated nitrogen levels in the Mississippi. When the Mississippi flows into the Gulf of Mexico, the elevated nitrogen promotes algal blooms that contribute to a "dead zone" that forms each year and ranges in size between 13,000 and 20,000 km² (comparable to the size of New Jersey). Restoration of wetlands and floodplains, systems that can process and remove nitrogen, has been recommended as partial solutions to this challenge (Zedler 2003). Within the wealth of interesting topics that emerge from the Mississippi basin, this case study will focus primarily on one part of the bigger picture—the integration of floodplain management into the flood-management system of the lower Mississippi.

The 1927 Mississippi flood exposed two primary limitations to river and floodplain man-

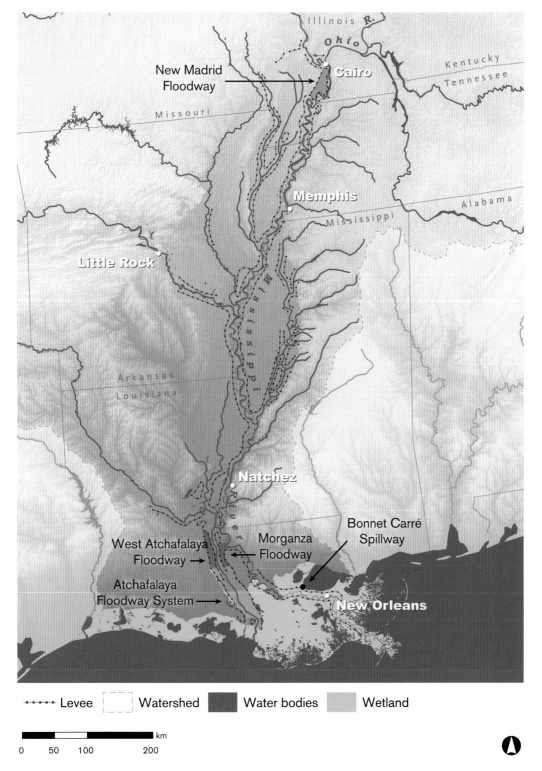

FIGURE 11.3 The Mississippi River and Tributaries Project (after US Army Corps of Engineers 2015; Gesch et al. 2002; Natural Earth 2016; US Census 2016; USGS 2016).

agement of that era. First, flood management relied excessively on levees. The so-called "levees-only approach" presumed that nearly the entire floodplain could be disconnected from floods (Barry 1998). Second, the management of rivers and floodplains was fragmented across states and other jurisdictions. The flood revealed the limitations of levees and the vulnerabilities of uncoordinated management, and in response, the US Army Corps of Engineers (USACE) developed the Mississippi River and Tributaries Project (MR&T). The MR&T coordinated levee placement and design, dam development and operations, floodplain management, and navigation for the lower Mississippi River basin, including several major tributaries (USACE 2008).

In addition to replacing a piecemeal approach to river management with a comprehensive system approach, the MR&T also moved floodplain management away from the levees-only approach and included floodplain storage and conveyance as critical components of flood-risk management. Five floodways were designated and developed. These were portions of the historic floodplain that would reconnect to the river and convey flood waters during the highest floods. The floodways include the New Madrid Floodway in southeast Missouri and the West Atchafalaya, Lower Atchafalaya, Morganza, and Bonnet Carré floodways in Louisiana (figure 11.3). During large floods, water can move from the river into the floodways through gates or "fuseplug" levees. Fuseplug levees have a section that can be removed (e.g., with dynamite, in the case of New Madrid) to allow flood waters to enter the floodway. During very large floods, floodways are intended to convey a significant portion of flow in order to lower flood stages elsewhere in the system and reduce pressure on levees. The MR&T is designed to handle a "project design flood," which is the maximum flood with a reasonable probability of occurrence. Thus, the New Madrid Floodway is intended to convey nearly a quarter of the project design flow below the confluence of the Ohio and Mississippi rivers.

In the lower reach of the river, near the Gulf of Mexico, 60% of the total flow is designated to move through floodways, primarily those in the Atchafalaya Basin. The main channel, which flows past New Orleans, is intended to carry the remaining 40% of flow (USACE 2008). Additionally, four "backwater" or "natural storage" areas exist at the gaps in the levees where major tributaries enter the Mississippi (St. Francis, Yazoo, White, and Red rivers). These are intended to store water and attenuate the peak of flows arriving from tributaries.

The effectiveness of connected floodplains as components in a flood-management system was demonstrated during major floods in the lower Mississippi Valley in 1973 and 2011. During these events, the floodways and backwater storage areas relieved pressure on levees protecting Cairo, Illinois, large areas of agricultural land, and major urban and industrial areas from Baton Rouge to the Gulf of Mexico. The 2011 flood was the largest that the MR&T has confronted. The system managed the flood without a single levee breach or fatality and three of the floodways were activated simultaneously for the first time since the system was built. Diversion of Mississippi waters into the floodways and allowing water to accumulate in the backwater areas shaved off the peak of the forecast flood stages (figure 11.4). The dramatically different outcomes of the floods of 1927 and 2011 emphasize the effectiveness of both system-scale approaches to river management and the value of hydrologically connected floodplains, as activation of floodways was essential for reducing flood risk for riverside cities such as Cairo, Baton Rouge, and New Orleans. The MR&T is also an example of integrating "green" infrastructure, such as floodways and backwater areas, with engineered infrastructure, such as dams and levees. In this case, some of the "green" infrastructure is comprised of heavily modified floodplain surfaces, within a system that is still heavily reliant on levees.

In addition to providing more effective flood management, the system-scale approach to

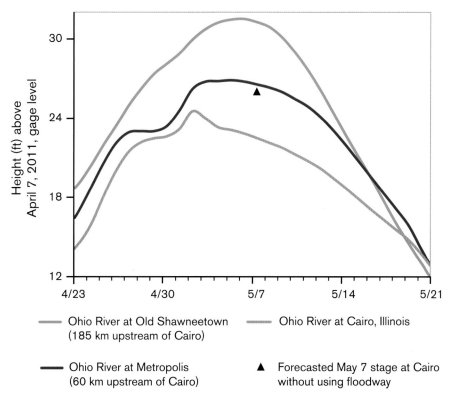

FIGURE 11.4 River stage on the Ohio River at Cairo, Illinois, adjacent to the floodway; Metropolis, Kentucky, approximately 60 km upstream of Cairo; and Old Shawneetown, Illinois, approximately 185 km upstream of Cairo. River stages at all three gages are plotted relative to their height on April 7, 2011. At Cairo, river stage began to decline on May 3 after opening of floodway and declined for the rest of the flood event. The gage at Old Shawneetown shows that the Ohio River flood did not crest upstream until May 6. Thus, the declining stage at Cairo reflects increased conveyance due to the opening of the floodway. The May 7 gage elevation at Cairo was 1.1 m below US Army Corps' forecast for Cairo without operation of the floodway (adapted from Opperman et al. 2013).

water-management infrastructure in the MR&T has the potential to contribute to greater environmental sustainability for the lower Mississippi River. During the 2011 flood, large volumes of water were stored and conveyed through backwater areas and floodways. Flooding these portions of the historic floodplain provided greater environmental benefits than would have been achieved through pre-1927 uncoordinated and levees-only flood management. In 2011, the floodways provided foraging habitat for fish and birds (D. Thomas, Illinois Natural History Survey, pers. comm., 2011) and an opportunistic study that took advantage of the inundation of the New Madrid Floodway found that several fish species had faster growth rates on the floodway than they did in the river (Phelps et al. 2015). Rantala et al. (2016) observed that the New Madrid Floodway did provide some ecosystem services during the period of inundation, but provision of services was relatively limited due to the floodway's lack of diverse vegetation and because it is designed to convey water and drain quickly, limiting residence time of water on the floodplain. Similarly, BryantMason et al. (2013) reported that the Atchafalaya River (Morganza and Atchafalaya Floodways) retained 7% of the nitrate that flowed into it during the 2011 flood, attributing the relatively low retention to the low residence time of water.

The original design of the MR&T constrains its ability to provide ecosystem services. The

MR&T was designed essentially only for flood control and navigation, reflecting the values and scientific knowledge of the time, and did not strive to promote floodplain biodiversity or ecosystem services. Additionally, the Mississippi Basin has experienced three historic floods in less than 20 years, and forecasts suggest that flood magnitudes will increase with climate change (Kundzewicz et al. 2008). River managers thus need to analyze whether the current floodways and backwater areas provide sufficient floodwater storage and conveyance to maintain the integrity of the MR&T. This combination of needs and opportunities suggests that reassessing the MR&T, and perhaps modifying its design and operation, could yield significant benefits in terms of reconciling flood risks with a broad range of river-floodplain benefits, reflecting current scientific understanding and societal values and expectations (Kozak et al. 2016; Rantala et al. 2016).

For example, while the floodways in the Atchafalaya are naturally vegetated, design and management of the system for flood control is damaging the floodplain forests because of altered flow and sediment dynamics. Improving flows to maintain the forests may require a combination of local-scale adjustments to floodway "plumbing" and system-wide changes in flow management (e.g., how flows are allocated between the Atchafalaya and the main stem Mississippi). In addition to improving forest conditions, these changes could promote greater retention of nitrogen within the Atchafalaya (Piazza 2014; Kozak et al. 2016).

CASE STUDY 5: NAPA RIVER—WATERSHED-SCALE APPROACH REDUCES FLOOD RISK

The Napa River runs through a California valley famous for its bucolic vineyards and high-quality wines. The alluvial soils responsible for its celebrated terroirs were deposited by the Napa River, which is 89 km long and drains 975 km^2. With a Mediterranean climate, the river's flow sometimes ceases upstream of tidal influence during the dry season and can swell to over 920 m^3/s in the wet season. Due to watershed characteristics and development patterns, residents have endured at least 22 damaging floods since record keeping began in 1865 (figure 11.5). The most destructive flood event was in 1986, with damage estimates ranging from $93 to $185 million (in constant US dollars) and three deaths (USACE 1998). The frequent and severe flooding in this relatively small community caused it to emerge as a priority flood protection project at the national level. Throughout the 1900s, flood management on the Napa employed traditional approaches such as dredging, levees, and floodwalls, but these interventions never reduced risk to an acceptable level. Following a paradigm shift driven largely by citizen activism, managers have pursued a "living river" strategy linking numerous projects and policies to address flood risk at the watershed scale.

Four distinct project areas along the Napa River are discussed here in an upstream-to-downstream progression that highlights system-wide restoration of geomorphic function (figure 11.6). Projects include two areas in the middle reach of the river, a keystone project in the city of Napa, and the restoration of tidal marshes in the estuarine zone. The variety of restoration efforts in the watershed shows that projects with disparate funding sources, lead agencies, stakeholders, and objectives can complement one another functionally.

Themes common to the projects include reach-appropriate scaling of engineered geomorphic features, transitioning land uses at key locations to be compatible with flooding, and local policies that promote retention of storm runoff on-site. Methods for addressing flood risk encompass many of the green infrastructure approaches discussed in chapter 10: (1) slowing in-stream water down with in-channel structures such as logs, root wads, and boulders; (2) increasing flood capacity by spreading water out in engineered bypass channels and connecting incised reaches of the river to constructed floodplains and backwaters; (3) replacing existing levees with

FIGURE 11.5 A Napa River flood 1940 (A) and low water conditions in 2016 (B). The 2016 photo shows flood bypass infrastructure crossing the neck of an oxbow at center right, outlined in red. In both photos, the main channel is outlined in blue. The direction of flow is from the top of the photo toward the bottom (US Army Corps of Engineers 1995; photo by Amber Manfree and Heather Davis 2016).

levees set back as far from the channel as possible (County of Napa 2008; USACE and Napa County Flood Control and Water Conservation District 1999; California Land Stewardship Institute et al. 2011); and (4) enacting policies for new development that eliminate or minimize increases in surface runoff and encourage groundwater infiltration (Napa County Department of Public Works 2008). In addition to flood-management benefits, in-channel structures, floodplains, backwaters, and setback levees are designed to meet habitat restoration objectives. As successive projects have been implemented, they are integrated with existing projects for habitat continuity and flood management.

Middle Reach

The upstream-most major restoration project on the Napa River is near the town of Rutherford, a small farming community. The landscape along this reach is primarily agricultural and, although property values are extremely high, the project included converting 12 ha of vineyard along 7.2 km of river to floodplain habitat. Conditions preceding restoration in this reach were similar to riparian conditions in many agricultural landscapes. Floodplains and braided channels were separated from the river with levees or filled in to create more space for crops, confining the river to a single narrow channel. Meanwhile, upstream dams caused sediment deficits, leading to incision that lowered the riverbed 2–3 m (Grossinger 2012).

The Rutherford Reach Restoration Project was initiated in 2002 by landowners who were interested in reducing bank instability and flood problems. The project also created a mechanism to bring landowners into compliance with government standards for sediment loads. Partnering with the County of Napa allowed landowners to receive otherwise unobtainable permits for in-channel modifications and opened doors to public funding sources. Landowners voluntarily contribute money to support long-term monitoring and adaptive management. This steady source of project-specific funding allows Napa County staff, in coordination with a landowner advisory committee, to respond quickly to complications such as alien plant species invasions or bank erosion events. Timely and organized responses from resource-management professionals prevent more serious problems from developing and also avoid piecemeal or improvised interventions that might disrupt watershed processes. Adaptive management and monitoring also allow staff to apply lessons learned to

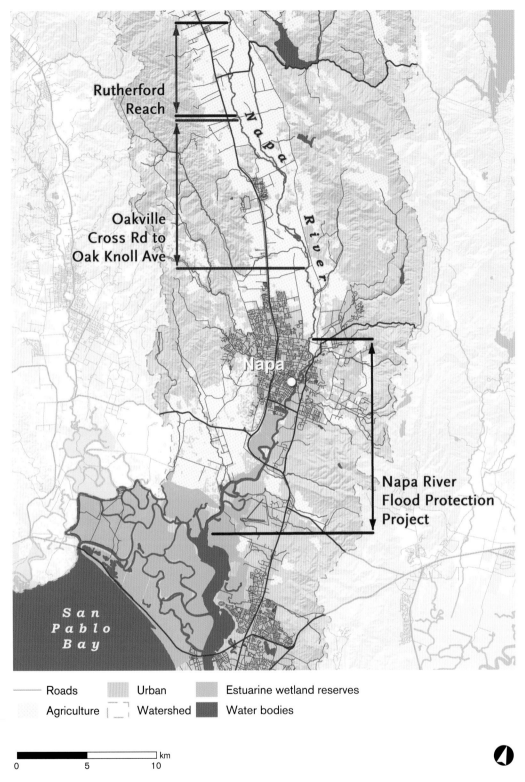

FIGURE 11.6 Numerous projects on the Napa River are restoring geomorphic function along 54% of the river's length (Gesch et al. 2002; CalAtlas 2012; GreenInfo Network 2016).

other sites throughout the county. For example, the design for the upcoming Oakville to Oak Knoll Restoration project, which runs along 14.5 km of the Napa River downstream of the Rutherford Reach, is very similar to, and draws heavily on, successes seen in the Rutherford Reach.

Floodplain design in both the Rutherford reach and the Oakville to Oak Knoll Reach includes floodplain terraces, secondary channels, in-stream habitat structures, and widened riparian zones in areas where levees have been set back. Most common are floodplain terraces that address the loss of floodplains that occurred when historical floodplains became disconnected from the river as its bed incised. These terraces are graded at the elevation of the 1–2 year flood interval to allow frequent flooding to support riparian vegetation, fine sediment deposition, and reduction in stream velocity and erosion. Floodplain terraces generally have a 3:1 gradient and are 3–10 m wide with length determined by site conditions. The middle reach of the river historically featured sections with braided channels. To mimic this feature, floodplain secondary channels were constructed by excavating abandoned channel beds at slightly higher elevations than the main stem. These secondary channels will be active during relatively frequent flood events (recurrence interval of 1.5–2 years). This allows frequent inundation, sometimes over long time periods, and creates high flow refuges as well as ephemeral wetlands as waters recede after flooding. Setback levees are constructed in areas where adjacent land use or facilities require flood protection during major inundation events. In some areas, setback levees may be part of a "managed retreat policy" that gradually expands the riparian corridor (California Land Stewardship Institute et al. 2011). Levees may be completely removed where local or upstream restoration actions have sufficiently lowered flood risk.

All three of the interventions described above allow water to slow down and spread out during flood events. Geomorphically, functional channel width has been increased, erosion potential reduced, and flood capacity increased. Ecologically, zones of lower velocity flow offer refuges for aquatic species, provide foraging opportunities for riparian animals, and encourage nutrient cycling and transport. These designs offer a compromise between meeting geomorphic objectives and protecting adjacent land uses. While present-day floodplains are not as extensive as they were historically, the habitat value of restored reaches has increased dramatically from conditions a decade ago.

Riverine-Tidal Transition Zone

The Napa River Flood Protection Project (the Project) focuses on the downtown area of the city of Napa, where the river is influenced by tides from the San Francisco Estuary. The Project attained national prominence with an innovative design that will increase flood capacity threefold, develop urban recreational amenities that bolster the economy, and restore fish and wildlife habitat. The USACE touts it as a pilot project for a new type of flood management, where "living river" principles guide design (Dickson 2015). The lengthy stakeholder process that fostered the Project brought the community together and set the tone for subsequent projects in Napa Valley.

Despite the clear need to address flood risk in Napa Valley, voters had previously rejected several proposals for traditional flood-management projects due to environmental and fiscal concerns. A 1975 USACE proposal calling for "channel widening and realignment, dredging, riprap on portions of the riverbanks, and construction of concrete step-walls through the central urban area of the city" (USACE 1975) was defeated twice. A similar 1995 proposal was also rejected at the polls (USACE 1995). Through years of citizen action and collaboration with agencies, an innovative proposal, focused on "living river" principles, emerged. This proposal was supported by two-thirds of voters in 1998. Planning documents define a "living river" as one that "conveys variable flows and stores water in the floodplain, balances sediment input with sediment transport, provides good quality fish and wildlife habitat, maintains good water quality and quantity, and lends itself to recreation and aesthetic values." Explicit Project objectives include attaining geomorphic equilibrium with sediment supply and transport and river-floodplain connectivity (Community Coalition for a Napa River Flood Management Plan 1996).

Within the city of Napa, the Project is designed to manage a 100 year flood event and expands the space available to the river, allowing it to build point bars, spill onto floodplains, and establish a vegetated riparian corridor. This required removal of dozens of buildings, extensive recontouring of banks, and reconstruction of five bridges and two railroad trestles (USACE and Napa County Flood Control and Water Conservation District 1998). Engineered floodplain terraces, floodplain secondary channels, and setback levees are incorporated into project design at a scale commensurate with the size of the river in this reach. For example, floodplain terraces on the east bank of the river are 20–300 m wide and 4 km long. An oxbow dry bypass channel, effectively a floodplain secondary channel, is 100 m wide and 400 m long (figure 11.5). Levees and dikes have been breached just downstream of the urban area, connecting both tidal inundation and flood flows to 490 ha of historical marsh floodplain. Reconnecting the river to its floodplain and tidal marsh, terracing of the marsh plain, and reestablishment of native vegetation in these areas have created a mosaic of productive habitat in an area that formerly featured warehouse yards and alfalfa fields (Napa County Resource Conservation District 2007; Stillwater Sciences 2013).

With the completion of the project, the Federal Emergency Management Agency will revise flood zone maps to represent reduced levels of risk, which will in turn lower flood insurance premiums for some residents. The enhanced security from floods and improved aesthetics have spurred an estimated US$1 billion in private business and real estate investment, resulting in dramatic redevelopment of the city center (Huffman 2014).

The Napa River Flood Protection Project is similar to the Dutch "Room for the River" program (earlier this chapter) in that it uses a mix of gray and green infrastructure to lower flood risk and features aesthetically pleasing urban design of flood-management infrastructure. While the green infrastructure features have gained most of the attention, the project also includes new levees and floodwalls, bypass culverts and channels, bank stabilization measures, pump stations, and water detention facilities.

In the downtown area, flood control, habitat, and open space for recreation are coincident at several sites. The oxbow bypass channel bed is landscaped with native plants and functions as a park, concert venue, and pedestrian-friendly thoroughfare. A city street crossing the upstream end can be closed off with massive flood gates during high-flow events. Floodwalls serve dual purposes as riverfront walkways. New infrastructure is both functional and attractive and has been built in a way that highlights the river and connects it with the city center (Dickson 2015; Urban Design and Aesthetics Workgroup 1997).

Estuarine Marshes

Downstream of the Napa River Flood Protection Project, the estuary meanders for 18 km to San Francisco Bay. In the 1900s, islands of tidal marsh in this area were diked, drained, and managed as industrial salt evaporation ponds or hayfields (Grossinger 2012). Today, the area is a reserve system encompassing over 14,000 ha of wetlands managed by a patchwork of agencies ranging from the US Fish and Wildlife Service to local nongovernmental agencies. Managers share the common goal of reactivating tidal processes throughout nearly the entire historical extent of estuarine marsh. Restored tidal marsh plains provide substantial additional space for flood waters to spread out. Critically, restoring tidal marsh will help counteract the impacts of sea level rise, which were not accounted for in the design of the Napa River Flood Protection Project.

Conclusion

In Napa Valley, projects and policies work together to reduce flood risk, restore habitat, and provide economic and civic benefits. Ongoing projects are returning 54% of the Napa River's length to a state of greater connectivity to floodplains with restored geomorphic processes. The increase in connectivity and dynamic processes are creating a diverse gradient of habitats that range from riparian forests to estuarine marshes. Floodplains have been created and restored in agricultural, urban, and industrial landscapes. One of the greatest successes has been how the engagement of citizens, supported by local agencies, has led to innovative project designs and brought the community together around shared interests. Projects on the Napa River are reconciling environmental goals with agricultural and urban land uses, and will have long-ranging benefits.

CASE STUDY 6: THE MURRAY-DARLING—ATTEMPTS TO RECONCILE RIVERS, FLOODPLAINS, AND WETLANDS IN AN OVER-ALLOCATED SYSTEM

While floodplain restoration typically involves a rising river flowing out into forests and wetlands, this case study describes efforts to reconcile a highly altered river system using a suite of approaches far removed from that "natural" image. Reconciling the Murray-Darling River in Australia involves water markets, impact investments, engineering refinements to system "plumbing," and targeted delivery of water to floodplain wetlands through canals and even pipes. Whether this approach will be sufficient to maintain healthy river-floodplain ecosystems is still vigorously debated (Pittock et al. 2015), but the Murray-Darling River example certainly illustrates the quest for innovative solutions amidst over-allocation, persistent drought, and social conflict.

Australia is the driest continent on earth with the most variable river flows (Wohl 2011). The Murray-Darling is the largest river basin in Australia (approximately a million square kilometers), has a relatively unpredictable flood regime compared to other temperate rivers (such as the Sacramento), and highly variable annual discharge, ranging from 7 billion cubic meters (bcm) to 118 bcm in the past century (Richter 2014). Within the basin, three primary rivers arise in the Great Dividing Range, near the coast of southeast Australia—the Murray, the Darling, and the Murrumbidgee—that converge as they flow westward across semiarid and arid flatlands. The Murray represents 5% of the basin area but half of total discharge (Wohl 2011). Due to the relatively unpredictable flooding regime, the Murray basin apparently lacked fish that were obligate floodplain spawners, but a number of native fishes opportunistically used inundated floodplains for spawning, rearing, feeding, and dispersal (Wohl 2010).

Paralleling the history of the Sacramento River of California (see chapter 12), the Murray basin experienced a gold rush in the 1850s that led to a dramatic increase in population of the basin along with the need to feed that growing population, setting the Murray on its path to becoming Australia's "bread basket" (Richter 2014). Agriculture in the basin required large-scale irrigation, so reservoirs and other infrastructure were constructed beginning in the 1920s (Wohl 2011). Today, reservoirs can store up to 3 years of mean annual flow (Richter 2014) and total diversions have averaged 40–50% annual outflow. Through this water management and conversion of floodplains to agriculture, the Murray has become extremely important to Australia's economy supporting a quarter of the country's cattle and dairy, half of the sheep, and three quarters of irrigated land producing 90% of irrigated crops, along with providing water to 3 million people (Richter 2014).

While promoting agricultural productivity, storage and diversion of river flows has had considerable impact on the basin's river-floodplain systems in terms of hydrology, geomorphic processes, and ecosystems. Low flows have been reduced and the flooding regime has been altered, producing lower flood peaks with faster recession rates. The frequency of spring flooding has declined, the frequency of summer flooding increased (Robertson et al. 2001), and the frequency and extent of overbank flooding has been dramatically reduced. Seasonally flooded wetlands have generally become consistently inundated or consistently dry (Wohl et al. 2011). By the 1980s, the total outflow from the Murray-Darling basin had been reduced by 40% and, again paralleling the trajectory of the Sacramento, native fish species had declined dramatically (Richter 2014). Nonnative fish species, such as common carp, were common (Wohl 2011) and floodplain forests were in decline (Mac Nally et al. 2011).

The flow of the Murray-Darling had become over-allocated. To address declining ecosystem function and other problems, such as increased soil salinity, diversions were capped in 1995 and a cap and trade market was introduced (Pittock 2013). Total water consumption was limited to 11 bcm, approximately the level of 1994 and, through an initiative called the Living Murray, the state and federal governments planned to buy back 0.5 bcm to dedicate to restoration. Restoration would focus on six iconic wetland sites that were important to Aboriginal people with managed delivery of water to wetlands through canals or pipes (Richter 2014).

Just as this attempt at ecological reconciliation got underway, Australia was hit by the onset of the Millennium Drought, so called because its timing flanked the turn of the millennium. For much of Australia, the drought persisted from 1997 to 2009. As the drought settled in, the Murray-Darling basin experienced depleted reservoirs, dramatically lower flows, and widespread ecosystem decline, with 80% of red gum floodplain forests showing stress (Richter 2014; Wohl 2011). In response to the drought, the Australian Water Act of 2007 created the Murray-Darling Basin Authority with a mandate to prepare a basin plan, along with US$9.7 billion allocated to be spent on infrastructure projects, water efficiency, and the "buyback" of water rights to provide water for restoration.

The basin authority ultimately called for 2.75 bcm of water buybacks after the original proposal of 3 bcm was met with considerable protest from agricultural communities. However, comprehensive studies had concluded that more than twice that much water would need to be reallocated to the river to maintain healthy river-floodplain ecosystems (Pittock 2013; Richter 2014). Conservation interests and scientists expressed further concern that too much emphasis had been placed on "Environmental Works and Measures," engineering projects facilitating targeted delivery of water to specific wetland sites, which were motivated, in part, by the goal of using less water to achieve restoration outcomes (Pittock 2013).

By 2014, 2.4 out of the 2.75 bcm had been bought back through the water market and reallocated to environmental purposes (Richter 2016). The water market also helped the agricultural economy weather the drought by incentivizing efficiency and promoting trading from lower- to higher-value water uses (Pittock 2013). Currently more than half of total allocations in the basin, on average, is actively traded (Richter 2016).

As described above, wetland restoration projects generally feature targeted delivery of water to a site through canals or pipes. This precise, managed flooding has allowed scientists to use those events as experiments to study floodplain ecosystem responses to interventions. For example, monitoring of one of the initial restoration projects, the Barmah Forest, found minimal benefits from initial application of 100 million m^3 and modeling suggested that, rather than consistent annual allocations, the environmental water should be saved for several years and then delivered as a larger flood event. This larger allocation, combined with a large natural flood, produced significant benefits for waterbirds in 2000–2001.

Robertson et al. (2001) studied 6 years of managed flooding and found that summer floods promoted growth of river red gums, while spring flooding promoted aquatic primary production, such as macrophytes and biofilms. Beesley et al. (2014) studied 26 discrete watering events, evaluating how fish abundance varied with wetland (e.g., habitat) and water delivery characteristics (e.g., source, timing, duration). They found that delivery mode of the source of water had a large influence, with fish abundance greatest when delivery was more "natural" (from the river, delivered from a channel, at a time appropriate for spawning) rather than artificial (sourced from an irrigation channel, delivered through pipes). They pointed out that watering events were often focused on a specific goal, such as restoring forests, without considering other resources such as fishes. For example, the precise delivery of water, through pipes and culverts, often prevented fishes from migrating to and from the floodplain.

Debate continues as to whether these efforts will be sufficient to maintain healthy river-floodplain ecosystems, particularly in the face of a changing climate (Pittock 2013). Clearly, the Murray-Darling represents a long-term attempt at reconciliation, with highly managed interventions attempting to increase ecological benefits within a system that is dramatically altered, including widespread nonnative species.

The active water markets within the basin allowed for the development of another tool for the reconciliation toolbox. In 2015, the Nature Conservancy, the Murray Darling Wetlands Working Group, and Kilter Rural (an asset management firm) launched a Water Sharing Investment Partnership (WSIP). The WSIP raised funds (US$21 million) to acquire, manage, and trade a pool of water rights. The funds were drawn primarily from impact investors (those who expect a return but invest funding to promote a socially beneficial outcome), although government or philanthropic money could be added. The majority of the WSIP's water will be sold or leased within the agricultural community, providing a return for the investors. The remainder of water will be donated to an Environmental Water Trust to implement specific "environmental watering" events. The WSIP is aiming for investments of US$77 million to manage 40 million m^3 of water and provide targeted flows to thousands of hectares of wetlands, particularly those not receiving water from government programs. This approach is self-funding and does not require continued government or philanthropic funding, and so can complement the larger effort to reconcile the rivers and floodplains of the Murray-Darling.

TWELVE

Central Valley Floodplains

INTRODUCTION AND HISTORY

CHAPTERS 12–14 present an extended case history of interactions between people and floodplains in California's Central Valley. One reason we feature this area is our own familiarity with this system. It is a large floodplain system about which we know a great deal, knowledge that can be applied to other systems. But the Central Valley floodplain system is of interest in its own right due to the economic, social, and environmental significance of its floodplains. The Central Valley has a flooding regime that is fairly predictable in annual timing but highly unpredictable in duration and volume, creating a shifting mosaic of floodplain wetlands, waterways, and forests. The predictability of flooding gave rise to impressive aggregations of fish and wildlife that relied on these seasonally inundated habitats: Central Valley floodplains are a major reason that the Pacific Flyway for waterfowl and other birds exists (Garone 2011) and huge runs of salmon and other migratory fishes (Moyle 2002) took advantage of the productivity in the vast wetlands. These abundant resources in turn helped to support dense populations of native peoples living in or near the Valley.

Focusing on the Central Valley also allows us to integrate all the major themes in the book into one case study: how biophysical processes create ecosystem services, how management has changed those processes over time, and how management and ecosystem services can be reconciled. By looking closely at one system—through various lenses, including biophysical, historical, and social—we can draw lessons that apply more broadly to temperate floodplains across the world.

In this chapter, we review Central Valley climate and hydrology, describe the immense, dynamic floodplain system that once existed there, and summarize the history of change since the Gold Rush (approximately 1850). We finish by estimating the current extent and condition of floodplain habitat.

CENTRAL VALLEY CLIMATE AND HYDROLOGY

California has a Mediterranean climate (Gasith and Resh 1999), with nearly all precipitation falling between October and April, and thus

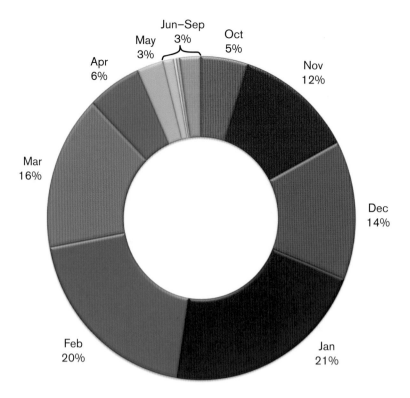

FIGURE 12.1 The average proportion of annual precipitation that falls each month in Sacramento, California, United States (CDWR CDEC 2016).

the state has a long dry season that extends over half the year (figure 12.1). Total precipitation can vary greatly between years (figure 12.2). Most years are either well above or well below the average, suggesting that "normal" is an elusive concept for California precipitation.

The Central Valley is the largest watershed in California, with the south-flowing Sacramento and north-flowing San Joaquin Rivers converging in an inland Delta and draining toward the Pacific Ocean through the San Francisco Estuary. The total area of the basin draining through the Delta encompasses approximately 160,000 km^2, or 42% of California's land area (The Bay Institute 1998; James and Singer 2008). Tributaries to the Central Valley drain numerous geomorphic provinces including the Sierra Nevada range along the eastern and southeastern margins, the Coast Ranges along the western and southwestern margins, and the Klamath and Cascade Mountains and Modoc Plateau to the north. The diverse range of physical environments and hydrological regimes within these contributing areas adds to the diversity of floodplain processes in the Central Valley.

Precipitation in the Central Valley watershed varies greatly spatially, within and between geomorphic provinces. In general, precipitation increases with elevation, exhibiting an orographic effect (figure 12.2). Precipitation also varies from west to east and from north to south. The Coast Ranges form a rain shadow such that the Central Valley receives less rain than the coastal areas to the west of the ranges. The Sacramento Valley in the north is relatively wet in contrast to the San Joaquin Valley in the south. However, the higher elevation southern Sierra Nevada receives more snowfall than the lower elevation northern Sierra, leading to prolonged spring snowmelt flows in rivers draining the southern mountains.

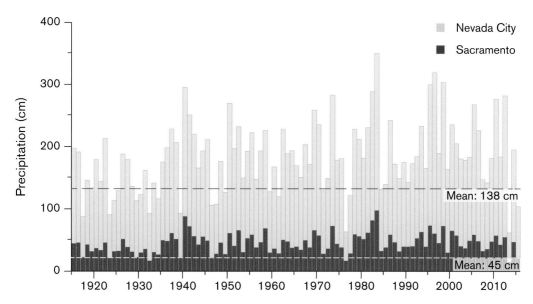

FIGURE 12.2 Annual precipitation between 1910 and 2010 for Sacramento and Nevada City (113 km north of Sacramento and 760 m higher in elevation) with horizontal lines indicating annual average precipitation (CDWR CDEC 2016).

Central Valley flood flows are generated by winter rainfall, or rain-on-snow events, and spring snowmelt. Winter storms generally produce distinctive high peak discharges with short durations, while snowmelt floods produce smaller, more gradual, and sustained floods (figure 12.3). Historically, floodplains could become inundated by winter rain beginning in November. These rain events could produce relatively rapid inundation, followed by somewhat longer draining periods. By contrast, the melting of the Sierra Nevada snowpack, which typically extended from April through June (Dettinger and Cayan 1995), historically led to sustained, low-magnitude flooding within the lowland flood basins and floodplains of the system.

The largest floods are produced by storms arriving from the southwest and this storm track is sometimes called the "pineapple express" because the storms begin in the Pacific near Hawaii. Because these storms originate in areas with warm water, they can have very high moisture content, leading to intense precipitation as the warm air masses are forced upward over the Coast Ranges and the Sierra Nevada. The large floods of 1964, 1986, and 1997 were all associated with this meteorological pattern (Roos 2006). In most winters, almost all precipitation comes in 2–4 large storms, delivered by "atmospheric rivers," which can create prolonged flooding if rain or snow falls over an extended period (Florsheim and Dettinger 2015). Particularly large flood events are generated when warm rain falls on previously accumulated snow—a "rain-on-snow" event (Carle 2004).

The Central Valley watershed can generate extremely high volumes of runoff and flood waters; the maximum recorded flood discharge for the Sacramento (17,500 m^3/s in 1986 [Roos 2006]) is approximately 20% of the magnitude of the Mississippi's maximum flood even though the Sacramento drainage area is only 2% of that of the Mississippi. Due to the considerably smaller drainage area, the Sacramento's flood peaks arrive much quicker than floods in the Mississippi (James and Singer 2008). The largest recorded flood on the San Joaquin was 2200 m^3/s in December 1950, just over 10% of the Sacramento's largest recorded flood (Roos 2006).

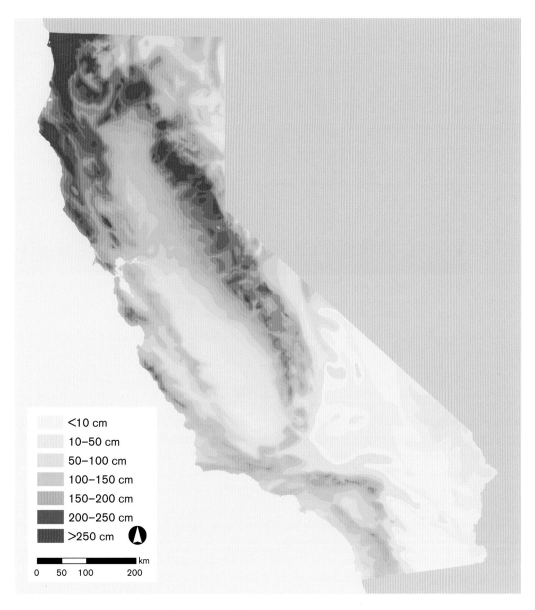

FIGURE 12.3 Average annual precipitation across California (CalAtlas 2012).

HISTORIC CONDITIONS OF CENTRAL VALLEY FLOODPLAINS

Central Valley floodplains displayed a changing continuum of surfaces, features, and habitats as rivers flowed out of adjacent mountain systems on either side of the valley, into the valleys and eventually to the Delta (see figure 1.3). Below we focus on two broad types of floodplain: the floodplain along the meandering northern Sacramento River ("meandering river floodplain") and the floodplain within the lowland flood basins along the Sacramento River downstream of Stony Creek ("flood basins"). In these sections, we describe major habitat features and synthesize the geomorphic and ecological processes that created them. We also describe which features were frequently inundated and capable of providing ecological benefits during "floodplain activation floods" (sensu Williams et al. 2009).

Meandering River Floodplains

The Sacramento River between Colusa and Red Bluff historically meandered across its floodplain, in a classic configuration of this floodplain type. The river here can be characterized as an alluvial river with noncohesive banks and moderate stream power that meandered across a floodplain; channel gradients are moderate (1–2%) but are higher than the channel gradients downstream. Other examples of this floodplain type include portions of the San Joaquin Valley and tributaries of the Sacramento and San Joaquin Rivers (e.g., Feather, Yuba, Tuolumne, and Merced Rivers). Meander migration (chapter 3) created a mosaic of successional stages supporting diverse, dynamic habitats (Larsen and Greco 2002). These included patches of forest with a range of age classes and species and various aquatic habitats, including open water, sloughs, and perennial marsh. Complex channels, with numerous side channels and links to off-channel water bodies, provided complex habitat for resident fishes such as Sacramento perch and thicktail chub.

The portion of the floodplain that was frequently inundated and connected to the river was composed of patches of riparian vegetation, secondary channels, and off-channel water bodies. Fish likely utilized this mosaic of habitats when floodplains became inundated and connected to the river. Sacramento splittail, Sacramento perch, and other native fishes spawn on submerged terrestrial vegetation (Moyle 2002) and thus would have utilized open areas of the mosaic for spawning. The complex channels, with numerous side channels and links to off-channel water bodies, likely provided diverse and complex habitat for rearing Chinook salmon and steelhead.

Flood Basins

Based on changes in slope and dominant substrate (moving from less cohesive gravel and sand to more cohesive silt and clay), these rivers transitioned from meandering channels to channels with fairly stable planforms (The Bay Institute 1998). South of Colusa, channel capacity of the Sacramento decreased dramatically from 7000 to 2000 m^3/s (Singer et al. 2008) such that the channel capacity was only approximately one-tenth of the largest flood volumes (James and Singer 2008). Due to the reduced channel capacity, during high flows a large volume of water left the main-stem channel and entered a series of flood basins that flanked the Sacramento (Kelley 1989). Gilbert (1917) and Bryan (1923) first described these flood basins, identifying six topographically distinct basins (Butte, Colusa, Sutter, Yolo, American, and Sacramento Basins; figure 1.3) that covered approximately 260,000 ha (The Bay Institute 1998). The basins occupied the low-lying areas between the natural levees along the river channels and the alluvial fans emanating from the Sierra and Coast Range mountains. For example, the Yolo Basin was 4 m lower than adjacent levee crests (Philip Williams and Associates 2003).

River flow into the basins historically began approximately 32 km north of Colusa into the Butte Basin. There were numerous breaks in the natural levees of the Sacramento River and major tributaries, providing connections between the main river channels and the channels and floodplains within the flood basins (Kelley 1989). The flood basins supported a diverse range of geomorphic features such as smaller alluvial levees, sand splays, channels, and wetlands (Florsheim and Mount 2002, 2003) and deltas formed by tributary flows (Florsheim et al. 2011).

The banks of the Sacramento River south of Colusa were characterized by broad, natural levees that formed during overbank flows because larger, heavier sediment particles were deposited first and closest to the channel (Katibah 1984; The Bay Institute 1998; Knighton 1998). Natural levees were 2–7 m in height (James and Singer 2008) and ranged between 3 and 5 km in width (Scott and Marquiss 1984).

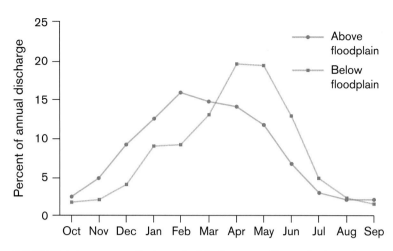

FIGURE 12.4 Average annual hydrograph of the Sacramento River at the rim stations (above the valley) compared to historic outflow from the Sacramento Valley, showing how the flood basins delayed the flood peak by 2–3 months (after The Bay Institute 1998).

Historically, these natural levees were not frequently overtopped and inundated, as evidenced by the fact that they were the locations of Native American and early European settlements (Stutler 1973; Kelley 1989). During winter and spring of most years, water left the main river through channels ("sloughs") and levee breaks that connected the main channel and the flood basins. Flow from the flood basins reentered the river (or other basins) through sloughs or flowed into the Delta; however, the basin's wetlands often held water for months.

The Feather River also built natural levees, which were breached during avulsions, creating sloughs that diverted flood waters into the basins. Early settlers along the Feather River described "a ridge of high ground running along the edge of the river all the way up with the exception of narrow passages or cuts, through which the water flows" (Kelley 1989, p.92). The principal "cut" the settlers described was Gilsizer Slough, which conveyed high flows from the Feather River into the Sutter Basin. The Sutter Basin also received flows from Butte Sink, flowing through a channel between the Sutter Buttes and the natural levees of the Sacramento and then Butte Slough, a channel that was 30 m wide and 10 m deep through which flood waters of the Sacramento flowed eastward into the basin (Kelley 1989, p.115).

The six basins acted something like a reservoir, temporarily storing and then releasing over 5 billion cubic meters (bcm; 4 million acre-feet) of water (The Bay Institute 1998; citing Grunsky 1929). This storage, which was roughly four times the capacity of modern-day Folsom Reservoir, reduced the stage height of floods in the main river channel and, through seasonal retention of large volumes of flood waters, dramatically increased residence time of water in wetlands. As a result of this storage in basins, peak outflow through the estuary was delayed by approximately 2 months (The Bay Institute 1998; figure 12.4). The basins had poorly drained soils with high clay content so surface water remained for long periods of time (The Bay Institute 1998). The early descriptions of the Sacramento River (in Davidson et al. 1896; Cook 1960) indicate that return flows from flood basins and wetland areas continued well into the late summer.

During periods when Sacramento River flows were too low to enter the basins, basins could still be inundated from tributaries that entered them through distributary channels. The natural levees of the Sacramento River helped to retain tributary flood waters within the basins. These tributaries included Putah and Cache Creeks in the Yolo Basin, the Cosumnes and Mokelumne Rivers in the Sacramento Basin, and Butte Creek in Butte Basin (Katibah 1984; The Bay Institute 1998). Inundation could also occur through direct precipitation and high water tables in the basins. Sacramento River floods that entirely filled the basins occurred in approximately 20% of years (Singer et al. 2008).

The lowest elevation portions of the flood basins were occupied by tule marshes. These were flooded very frequently and for long durations, often holding water through summer. Surrounding the tule marshes were fringes of willow forest (Philip Williams and Associates 2003). Riparian forests grew on the natural levees and other topographic high points on the floodplain (e.g., sand splays) in part because these surfaces were inundated less frequently.

These flood basins provided diverse and extensive habitat, including seasonal and permanent wetland, channels, and riparian forest of varying successional stages, for both terrestrial and aquatic species. Fish species, such as Sacramento splittail, would have entered the basins from the estuary or river and spawned on submerged vegetation (Moyle 2002). Juvenile Chinook salmon likely also entered from the river to feed and rear in the channels and shallow water of the marshes, similar to what is observed in today's Yolo Bypass (chapters 8 and 13).

Long-duration spring flooding can promote high production of algal carbon (chapter 7) and the historic system of flood basins provided an immense area with long residence times for algal productivity. It is likely therefore that the basins were an important source of carbon for the San Francisco Estuary (Sommer, Harrell, Kurth et al. 2004). For example, during wet years, the inundated Yolo Bypass (a remnant of the historic Yolo Basin) effectively doubles the wetted area of the Delta (Jassby and Cloern 2000) and, as it drains, the Bypass can potentially export water with high levels of biologically available carbon (Sommer, Harrell, Kurth et al. 2004). However, Jassby and Cloern (2000) suggested that, on a volumetric basis, the highly productive Yolo Bypass water today does not make a significant contribution to the carbon budget of the Delta. However, the historic flood basins had an area an order of magnitude greater than the current Yolo Bypass and these basins drained slowly compared to the Bypass. Thus, through this vast area and long residence time, the basins very likely boosted food webs both directly—fish and other organisms fed within the basins when flooded—and indirectly, with the downstream export of organic carbon, detritus, phytoplankton, invertebrates, and fish. This productivity helps explain the immense salmon runs, dense populations of resident fishes, and huge flocks of wintering waterfowl that once characterized the Central Valley (Cunningham 2010). These animal populations likely served as important sources of protein for California's indigenous people and the Central Valley once supported one of the densest populations of Native Americans in North America (Brown and Moyle 2005).

HISTORIC EXTENT OF CENTRAL VALLEY FLOODPLAINS

It is broadly acknowledged that the Central Valley once encompassed a considerable extent of floodplain, and that the current extent is a small fraction of the historic area (figure 1.3). However, deriving an estimate of the historic extent of floodplain compared to current extent is difficult because historical accounts and methods for characterizing floodplains differ and because floodplains are defined in numerous ways: hydrologically, geomorphically, and ecologically (chapter 1). Here we review estimates of historic floodplain extent, acknowledging limitations of comparing these estimates with current estimates.

Katibah (1984) estimated that in 1848 the Sacramento Valley had 324,000 ha (800,000 acres) of riparian forest, a key indicator of floodplain, and he conservatively estimated that the entire Central Valley contained approximately 373,000 ha (921,000 acres) of riparian forest. The Bay Institute (1998) estimated the Central

Valley (not including the Tulare Basin) encompassed 400,000 ha of riparian areas, based on soil types, with a smaller portion of that covered by riparian forest at any given time. The remainder would have been occupied by grassland, oak woodland, and oak savanna. Shelton (1987; citing Kuchler 1977) reported that historical riparian forest in the Sacramento Valley was 200,000 ha with the entire Central Valley (including Tulare Basin) encompassing 314,000 ha. The Chico Central Valley Historical Mapping Project (2003) estimated 408,000 ha of riparian forest in the entire Central Valley in 1900. Thus, valley-wide estimates for riparian forest range from 314,000 to 408,000 ha.

The flood basins of the Sacramento Valley were estimated to have occupied 260,000 ha (The Bay Institute 1998; citing Clapp and Henshaw [1911]). Flood basins were comprised of a mix of tule wetland, open water, and riparian forest. Large expanses of less frequently flooded floodplain wetlands existed outside the flood basins. Estimates of total extent of Central Valley wetlands, including the Tulare Basin, range from 520,000 ha (Mandeville 1857) to 1.6 million ha (Garone 2011). The latter estimate is about 28% of the total surface area of the Central Valley. Though these estimate vary widely, they consistently support the conclusion that floodplains and associated plant communities in the Central Valley once covered much of the Valley floor.

CENTRAL VALLEY FLOODPLAINS: A HISTORY OF CHANGE

Numerous studies have reported on the dramatic losses of various habitat types within the lowland Central Valley, with losses of several habitat types exceeding 90% (table 12.1). This section reviews the processes that converted or degraded Central Valley floodplain habitats.

Pre-Gold Rush Changes

The Central Valley once supported dense populations of native peoples who influenced floodplains by burning grasslands and wetlands during the dry season and hunting waterfowl and mammals such as beaver and elk. In fact the natural landscape of the Central Valley has been continuously influenced by people for at least 6000 years (e.g., Manfree 2014, for Suisun Marsh). The Spanish and Mexicans ran semiwild cattle throughout the Valley, which probably had grazing impacts on wetlands, and killed grizzly bears, which may have had important influences on floodplain ecosystems as a top predator and through their digging behavior. However, the systematic removal of beaver from the Valley by trappers may have had the most dramatic effects on wetlands. Tappe (1942) noted that the Hudson's Bay Company sent "fur brigades" of up to 200 trappers in the Central Valley in the first half of the nineteenth century and as a result the once abundant beavers were nearly extirpated from the state. Beaver dams create shallow, low-velocity habitat off the main channel in Central Valley rivers which provide favorable habitat for native fish, especially salmonids. Pollock et al. (2004) reported, for example, that the loss of beavers dramatically reduced floodplain rearing habitat for salmonids on the Stillaguamish River in Washington. The loss of beaver altered floodplains throughout North America, because their dams created wetlands that slowed runoff, retained water in ponds, deposited sediment, killed trees, and caused other habitat alterations (Naiman et al. 1988). The alteration by beaver of the landscape was so extensive—and their extirpation by trappers was complete—that it is difficult to fully appreciate their former importance to floodplain systems.

Late Nineteenth-Century Changes

Discovery of gold led to a dramatic increase in California's non-Indian population (Scott and Marquiss 1984). The Gold Rush, especially hydraulic mining, had a major influence on the Central Valley and its river-floodplains systems directly, through massive sedimentation, and indirectly, through population growth and subsequent land-use changes. Many miners recog-

TABLE 12.1
Estimates of floodplain and wetland habitat loss from the Central Valley

Habitat	Degree of loss	Reference
Central Valley riparian forests	89%	Barbour et al. (1991)
Central Valley riparian forests	99% destroyed within 100 years of settlement	Reiner and Griggs (1989)
Sacramento riparian and bottomland forests	90–98%	Garone (2011)
Tule marsh (includes Delta)	69%	Barbour et al. (1991)
Central Valley interior wetlands	94–96%	Reffalt (1985); Kreissman and Lekisch (1991)
All Central Valley wetlands	91%	Garone (2011)

nized that the Valley offered more reliable resources than the gold fields and agriculture greatly expanded, ultimately leading to demand for irrigation and flood control (Mount 1995).

As the Valley was settled, riparian areas were cleared for agriculture because of their high-quality soil and low frequency of inundation compared to the flood basins (Scott and Marquiss 1984). The steamships that began to serve Valley communities caused riparian forests to be felled for fuel (Katibah 1984). Snagging—the removal of logs and other large wood from channels to reduce obstructions to navigation—also reduced riverine habitats. Sedell et al. (1990) estimated that, to improve navigation, the US Army Corps of Engineers removed approximately 90 "snags" per kilometer from the Sacramento River. Based on the dynamics of wood in large rivers, each snag was likely not an individual log but a mass or jam of multiple logs (Swanson 2003).

Not only did removing wood reduce habitat complexity of river channels, but also loss of large volumes of wood may have reduced frequency of overbank flows. Williams (2006) quotes John Bidwell as testifying in 1883 that "I can speak generally that our streams there are less likely to overflow than formerly from the fact that we cut out the drifts from them," with "drifts" referring to logs and log jams. Bidwell was likely describing channels not affected by hydraulic mining sediment or to channels prior to advent of hydraulic mining. Removal of wood, particularly large wood jams, has reduced the frequency of overbank flooding in other rivers (Montgomery et al. 2003). As early as 1868, settlers commented on lack of available wood in the valley, indicating the extremely depleted status of riparian forests only 20 years after the Gold Rush began (Thompson 1961).

Initial Flood-Control Efforts and the Gold Rush

In his classic book "*Battling the Inland Sea: Floods, Public Policy, and the Sacramento Valley*," Robert Kelley (1989) chronicled the long struggle between the Central Valley's settlers and its floods. Kelley describes how the Valley population and governments were slow to realize the intensity and frequency with which their new home was subject to flooding. For example, Kelley wrote that after a massive flood inundated the new city of Sacramento in 1850, the experience did not leave city leaders "appalled and humbled by their experience but instead confident that the problem could be promptly mastered." The Sacramento newspaper, *Placer Times*, noted that "it will be seen that Sacramento City can be easily protected against inundations, and that, too, at comparatively small expense" (Kelley 1989, p.11). This confident statement would prove to be wildly unrealistic and, in fact, even today, Sacramento is considered to be one of the most flood-vulnerable major American cities (James and Singer 2008).

Immediately after California achieved statehood, the US Congress passed the Arkansas Act (1850), which gave California, and other states, title to all "swamp and overflowed" land within their borders, and required that the states invest funds from selling the land into actions to improve the land for agriculture, such as drainage. California did not pursue a centralized reclamation strategy but, instead, required purchasers of swampland—in the Central Valley, this referred to the flood basins—to sign an affidavit that they would undertake such activities (Kelley 1989).

Criticism of this policy emphasized that individual efforts at drainage and flood control had no chance at success against the large floods that could turn the Valley into an "inland sea." In response, the California legislature passed a bill in 1861, creating the Board of Swamp Land Commissioners. The board was tasked with promoting larger-scale drainage and flood-control projects that would be designed to correspond to natural features of topography and flow rather than corresponding to individuals' property boundaries. To do this, the board authorized the formation of districts, which were a new government entity that could raise taxes from within the district and apply the funds to drainage and flood control (Kelley 1989). The districts were to encompass a unit of land "susceptible to one mode or system of reclamation ... within natural boundaries," which included features such as an entire flood basin (Kelly 1989, p.48). Though able to work on much more systematic basis than individual landowners, the districts soon realized that they did not have sufficient funding, engineering expertise, or understanding of the Valley's hydrology to effectively prevent the basins from flooding (Kelley 1989).

The districts' failures combined with political changes led to a devolution of flood-control planning back to the local level, with counties given responsibility for flood-control within their borders. However, the legislature instructed counties to strive to align flood-control works with the natural flow and drainage features of the valley. The Green Act of 1868 further extended this local devolution, giving flood-control authority to local and self-defined districts. The Green Act had been written by William Green, a newly elected assemblyman, editor of the *Colusa Sun*, and former surveyor of Colusa County. The Green Act precipitated a rapid transfer of flood-basin land from the state to private landowners (Kelley 1989).

In a countervailing trend to the broader narrative of "*battling* the inland sea," Kelley describes that some landowners within the Valley actually perceived the floods as beneficial rather than harmful. Many landowners realized that the floods served as free irrigation, particularly for farms on former grassland or riparian lands that were somewhat higher than the flood basins. Others noted that periodic floods deposited nutrient-rich soil and killed rodents that, if left unchecked, became agricultural pests. Other landowners utilized the tule marshes of the flood basins as grazing land for cattle during the Valley's hot, dry summers (Kelley 1989). Thus, the Central Valley floodplains were being utilized for recessional agriculture and pasture, similar to how floodplains are currently used in Africa and Asia (Opperman et al. 2013).

Despite these beneficial uses of natural floodplains, the Green Act and subsequent dramatic expansion of agricultural landowners led to a period characterized by "levee wars"—complete with late-night dynamite raids and vigilante gangs—as various districts and individuals strived to prevent flooding of their lands, often to the direct detriment of adjacent landowners or those on the opposite bank.

Meanwhile, even as the settlers were failing to grasp the size of floods that the Sacramento and its tributaries could produce, flooding was being exacerbated by hydraulic mining. Hydraulic mining originated in the Sierra

Nevada foothills near Nevada City in 1853, approximately 100 km north of the city of Sacramento. This mining entailed directing high-pressure streams of water from cannon-like "monitors," or nozzles, onto hillsides, turning soil and rock into slurry from which gold could be extracted (figure 12.5). Hydraulic mining expanded throughout the Yuba, Bear, and Feather River basins, all tributaries to the Sacramento River, and immense volumes of sediment were washed into the river canyons of the Sierra Nevada mountains and foothills. The fine sediment moved quickly and continuously as suspended load and began to be deposited in San Francisco Bay and the low gradient portions of Valley rivers, impeding navigation as early as 1856. Coarser sediment moved during winter floods and was deposited in river canyons and on Central Valley floodplains (Kelley 1989).

It is estimated that hydraulic mining delivered over a bcm of sediment into the Central Valley. The Yuba, Bear, and Feather Rivers experienced rapid aggradation, as did the Sacramento below its confluence with the Feather (James and Singer 2008). Channel beds rose by as much as 6 m, greatly exacerbating flooding (figure 12.6), and the consequent deposition of sediment across farm fields removed nearly 20,000 ha of farmland from production (Scott and Marquiss 1984). Fragmented attempts to hold back the wave of sediment were no more successful than fragmented efforts at halting inundation of the flood basins (Kelley 1989).

Key legal decisions put a halt to both the levee wars and the hydraulic mining, setting the stage for the Valley's current flood-management system. In 1876, the Keyser Decision by a federal court held that landowners could not redirect flows in ways that harmed other landowners (Kelley 1989). In the Sawyer Decision of 1884, a federal court decided that hydraulic mining was causing undue harm to Valley landowners, and decreed that hydraulic miners could only continue if miners could contain their sediment. This decision effectively ended hydraulic mining.

Evolution of Current Floodplain Use

While the Keyser Decision facilitated transition to more systematic flood control for the Sacramento Valley, debate broke out over the best approach to accomplish this. Owners of large properties within the flood basins favored a system that would prevent overflow from the river into the basins, and this "levees-only" approach was consistent with standard engineering practice of the day. Flanking the main channels with constructed levees was also consistent with the goal of hastening the movement of hydraulic mining sediment downstream, and so levees were placed close to the channel to promote scouring and downstream transport of sediment (James and Singer 2008). Although most landowners and engineers favored this approach, Will Green asserted that the volume of water produced by the Sacramento and Feather Rivers during flood could never be contained between levees and, instead, portions of the flood basins should be designated for managing flood waters (Kelley 1989).

Several decades of continuing levee failures, exacerbated by inadequate design and poor construction materials (Scott and Marquiss 1984), made clear that the system was not working. In the mid-1890s, A. H. Rose was appointed State Commissioner of Public Works. He reviewed the Sacramento Valley's flood defenses and concluded, "the prime cause of failure has been the same throughout, namely: the endeavor to construct through individuals or small districts, and without unity of action, the integral parts of a vast drainage and protection system, itself without design or conception" (Kelley 1989, p.238). Rose commissioned a study, which found that Sacramento River floods could not be contained between levees but must be allowed to overflow in some locations. The report observed that this overflow was needed to accommodate relatively frequent flood events and outlined a system of weirs that allowed flood waters to enter bypasses, located within flood basins. The report solidified support for a multichannel approach that included bypasses

FIGURE 12.5 Hydraulic mining methods used in the Sierra Nevada range washed tremendous volumes of sediment into Central Valley streams (Watkins 1800s, courtesy of the Bancroft Library).

FIGURE 12.6 A street in Sacramento, California, during the flood of 1861–1862 (Lawrence and Houseworth ca. 1862, courtesy of the Bancroft Library).

and, in principle, outlined the coordinated system that eventually was constructed (James and Singer 2008).

Two large floods, in 1907 and 1909, galvanized state government to fund flood control. Then the US Congress passed the Flood Control Act in 1917, authorizing a direct federal role in flood control with a federal cost contribution of 50%. The act specifically authorized a Sacramento River Flood Control Project (SRFCP), which was mostly built between 1918 and 1944 (James and Singer 2008).

The current SRFCP consists of 1760 km of levees along river channels, including the Sacramento below Ord Ferry, and along portions of the Feather, Yuba, Bear, and American Rivers, and a few smaller channels. The system also includes five weirs that direct flood flows into a system of overflow areas that occupy portions of the historic flood basins, including the Butte Basin and the Tisdale, Sutter, Yolo, and Sacramento Bypasses (figure 12.7). The Tisdale and Sacramento Bypasses are relatively small features that convey water into the much larger Sutter and Yolo Bypasses, respectively.

From the 1940s to the 1970s, multipurpose reservoirs—providing flood control, hydropower, and water storage—were built on nearly every river entering the Central Valley, primarily through the Central Valley Project (federal government) and the State Water Project (California government; Hundley 2001, revised edition). These reservoirs have aggregate storage volumes of 12.9 bcm for the Sacramento Valley and 10.1 bcm for the San Joaquin, with 3.5 and 2.9 bcm of that storage dedicated to flood control in each basin, respectively. Flow regulation, storage, and diversion resulted in major alterations to the hydrology of Central Valley rivers, described in a subsequent section.

The cumulative effect of the constructed levees and flow regulation from the reservoirs has caused the majority of floodplains to become hydrologically disconnected from the rivers and converted to other uses—primarily agriculture but increasingly urban and suburban development (Burton and Cutter 2008; figure 12.8). In the next section, we summarize estimates of the degree of loss of Central Valley floodplain habitats and the current extent, and the effects that this loss and flow alteration have had on floodplain habitats and processes.

Managing flood risk in the Central Valley is challenging. Flood recurrence intervals have been generally underestimated (James and Singer 2008, citing NRC and James 1999) even as flood magnitudes are forecast to increase due to climate change (Hayhoe 2004; Das et al. 2011). Sedimentation in the bypasses (Singer et al. 2008), deterioration of constructed levees, erosion, and a maintenance backlog for levees and other structures contribute to rising flood risk. Finally, population growth in the Central Valley places more people at risk from flooding. For example, the population living behind levees in three Central Valley counties near Sacramento in 1990 was approximately 800,000. By 2005, this population had increased to over a million and is forecast to reach 1.3 million people by 2020 (Galloway et al. 2007; Burton and Cutter 2008).

CHANGES TO CENTRAL VALLEY FLOODPLAINS

Changes to Hydrology

The system of dams and reservoirs built for water supply, flood control, and hydropower have dramatically changed the flow regime of Central Valley rivers, including the flood regime. Reservoirs in the Sacramento and San Joaquin Valleys collectively maintain 6.4 bcm of flood storage. Nearly all locations in the Sacramento Valley show a significant reduction in the peaks of the small, frequent flood events, because these can be easily stored within reservoirs. Many of the dams, such as Shasta on the Sacramento River and Oroville on the Feather River, are capable of storing moderate to large floods and thus reduce the peaks of larger, less frequent floods (see figure 13.1). For example, the magnitudes of the 10 and 100 year floods on the Sacramento River at Bend Bridge (83 km

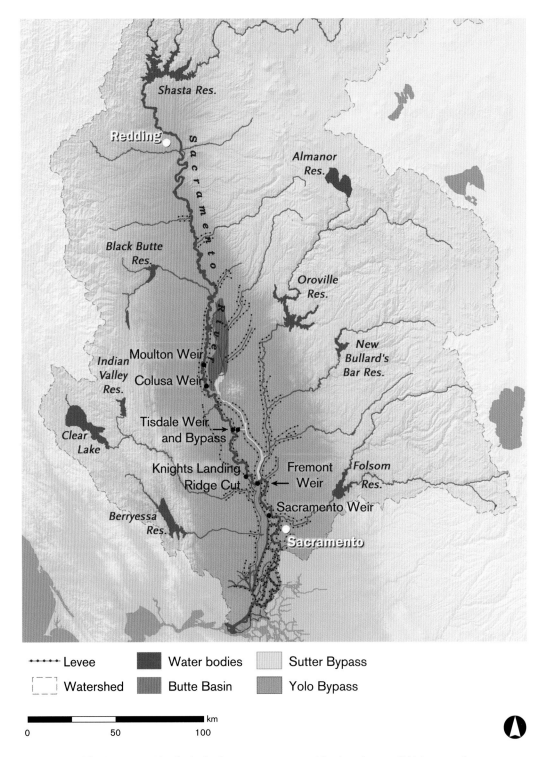

FIGURE 12.7 The Sacramento River basin flood-management system (Gesch et al. 2002; CalAtlas 2012; Sacramento Area Council of Governments 2016).

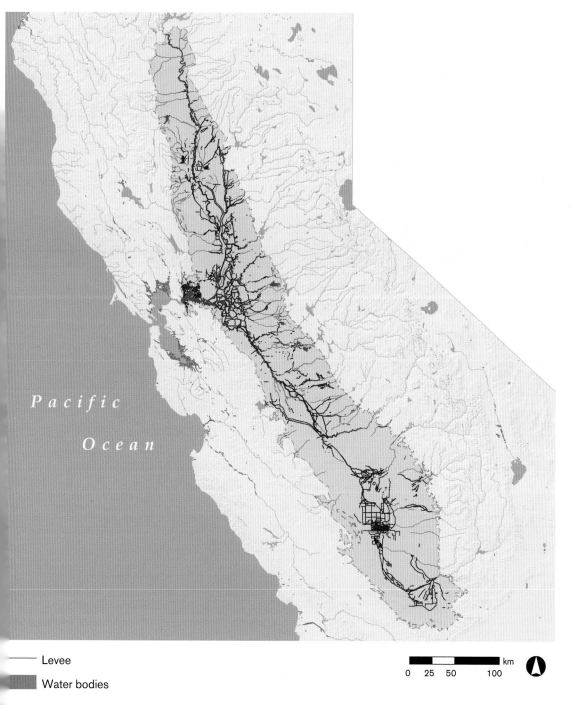

FIGURE 12.8 Constructed levees in the Central Valley (Gesch et al. 2002; CDWR 2010; CalAtlas 2012; Sacramento River Forum 2016).

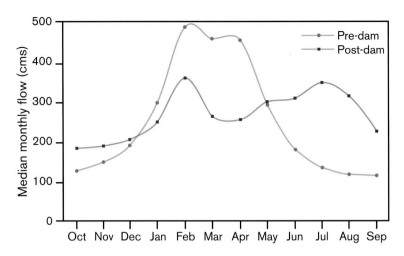

FIGURE 12.9 Monthly hydrographs for the Sacramento River, before and after construction of Shasta Dam (after Williams 2009).

downstream of Shasta Dam) have been cut approximately in half by flood regulation at Shasta Dam (Singer 2007). The attenuation of flood peaks by reservoirs becomes much less pronounced lower in the valleys. In contrast to Bend Bridge, the more downstream gages at Colusa and Verona (330 and 440 km downstream from Shasta, respectively) show little influence of regulation on the peaks of moderate to large floods. Thus in the lower portion of the Sacramento Valley, flood management depends largely on the system of weirs and bypasses (Singer 2007).

In addition to reducing flood peaks, the large, multipurpose reservoirs are also operated to store water for irrigation and municipal water supply. The reservoirs generally begin to fill in spring as risk of floods declines (Roos 2006), resulting in lower spring flows compared to the natural hydrograph (Williams et al. 2009). In summer and fall, flows in rivers of the Sacramento Valley are elevated due to in-channel deliveries for irrigation, resulting in an overall "flattening" of the hydrograph of the Sacramento River (figure 12.9). In addition to lower winter and spring flows, rivers in the San Joaquin Valley also have reduced summer and fall flows because irrigation deliveries are made through a canal system (Brown and Bauer 2010).

While flood management reduces the magnitude of geomorphically important peak flows in much of the Central Valley, the pattern of refilling reservoirs in spring greatly reduces the magnitude and duration of the preregulation snowmelt flood pulse (Yarnell et al. 2010). This spring flood pulse, described as the "floodplain activation flow" by Williams et al. (2009), can be particularly important for supporting environmental benefits from the Central Valley river-floodplain system, such as spawning and rearing habitat for native fish and food-web productivity (Opperman et al. 2010).

In short, flow regulation from reservoirs has reduced the magnitude of small and moderate floods throughout the Central Valley. In some locations, such as below Shasta and Oroville Dams, the magnitude of large floods has been reduced, although the lower areas of the Valley show much less change from historical levels. Due to operation for water deliveries, flows in the Sacramento Valley are elevated in summer and fall, but are greatly reduced in the San Joaquin Valley because of diversions. Throughout the Valley, the ecologically important snowmelt pulse, which provided a frequent, long-duration inundation of floodplain habitats, has been greatly reduced because reservoirs typically refill during spring (Yarnell et al. 2010).

Changes to Meandering River Floodplains

Meandering river floodplains have been primarily affected by two processes: upstream dams, which regulate flow and levees, and riprap, which reduces meander migration and connection to floodplains. Regulation from dams has generally decreased peak flows—dampening most winter floods and reducing or eliminating the snowmelt pulse, in some rivers dramatically so, particularly in the San Joaquin Basin (The Bay Institute 1998). As a result, remaining floodplain surfaces are inundated much less frequently; they still receive flood waters in very large, infrequent events, but the frequent events are much smaller. For example, on the Stanislaus River, the current Q_{25} (flow with a 25 year recurrence interval or a flow with a 4% exceedance probability) is lower than the pre-dam Q_2 (Schneider et al. 2003). On the Sacramento River at Bend Bridge, Shasta Dam has decreased the Q_2 from 117,000 to 76,900 ft³/s. The pre-dam Q_5 (157,000 ft³/s) is now equivalent to the present-day Q_{30} (LASRL 2003).

Capture of sediment by reservoirs disrupts the balance between deposition and erosion, which leads to channel incision and increased isolation of the channel from the floodplain, even at flows well above base levels (Kondolf 1997; Schneider et al. 2003). Thus, the combination of flow regulation, bank armoring, and channel incision results in a much smaller area of floodplain being inundated on a frequent basis. Stella et al. (2003) estimated that the frequently inundated surface of a reach of the Merced River is one-fifth that of the pre-dam river. Due to constriction between levees, when flows become high enough to exceed channel capacity, they move across the confined floodplain with greater depth and velocity. Lower flood magnitude, along with bank armoring, has also diminished the rate of channel migration, which reduces the amount of floodplain that is reworked and deposited as new surfaces. For example, due to flow regulation, the amount of exposed gravel bar declined by more than half between 1938 and 1997 in a 31 km study reach of the Sacramento River (Greco and Plant 2003). Reduced flows have also decreased the frequency of channel cut-off and oxbow lake creation, both of which provide valuable habitats. The loss of these dynamics reduces heterogeneity of vegetation types and impacts species dependent on specific vegetation structures. Changes in flow regime and reduction of geomorphic processes that create specific habitat features—such as cut banks and channel bars—have been implicated in decline of bank swallows (*Riparia riparia*; Moffatt et al. 2005; Girvetz 2010) and extinction of Sacramento Valley tiger beetle (*Cicindela hirticollis abrupta*; Fenster and Knisley 2006).

While much of the Sacramento River's banks between Colusa and Red Bluff are armored, reducing channel migration, the remaining unarmored portions may actually have a greater than historic rate of bank erosion and meander migration due to the conversion of riparian forests to agriculture. Forested banks have greater resistance to erosion due to increased bank strength or increased channel roughness. Conversion of floodplain from riparian forest to agriculture has increased meander migration rates and erosion of unforested banks by 80–150% (Micheli et al. 2004).

Today, the portion of the floodplain that is frequently inundated is greatly reduced. What remains is composed primarily of remnant riparian forest and agricultural land not protected by levees or behind very low levees. In general, because of loss of channel migration and constriction by levees, frequently inundated floodplain is generally no longer a mosaic of patches but is composed largely of mature riparian forest (Stella et al. 2003). Loss of channel migration and creation of new surfaces has also resulted in a reduced amount of vegetation in early successional stages, which has implications for animal species dependent on these successional stages. Fishes that utilize the floodplain can no longer do so with the same frequency. Chinook salmon have fewer options for slow-water habitat most years and so may move downstream faster than previously, perhaps to rear in the San Francisco Estuary. Because there is evidence that native fishes benefit from a mosaic of patch types on the floodplain (Crain et al. 2004), the current floodplain provides less habitat diversity and less spawning and rearing habitat for native fishes. Reduced frequency and area of floodplain inundation has undoubtedly reduced exchange of nutrients, energy, material, and organisms between floodplain and river and downstream habitats (e.g., large wood, organic carbon, invertebrates, juvenile fish).

Changes to Flood Basins

Two of the basins (Sutter and Yolo) have been partially converted to engineered bypasses that now frequently convey high flows. The other flood basins have been primarily converted to agriculture although some, such as Butte Basin, still contain large areas of wetland that are managed as wildlife refuges or duck clubs. Rice fields within the former basins are often intentionally flooded in winter and provide extensive areas utilized by waterfowl and wading birds (O'Malley 1999; Bird et al. 2000).

The two bypasses retain some of the characteristics of flood basins and lack others. Flood waters enter the bypasses when river stage reaches a certain height and water is conveyed across weirs rather than through sloughs. The Yolo Bypass begins to flood when combined flows of the Sacramento and Feather Rivers and the Sutter Bypass exceed 1600 m³/s (Sommer, Harrell et al. 2001). The natural levees of the main channels have been augmented with additional constructed levees and armored with riprap and, thus, overbank flow from the main channel across or through the levees is now an extremely rare event. The bypasses receive flow in about 60% of years and, similar to historical patterns, the bypasses can achieve partial inundation from west-side tributaries in years that they don't receive flow from the Sacramento. The topography and surface cover of the bypasses are maintained to promote rapid drainage (Sommer et al. 2001). Thus, rather than representing a "reservoir-like" system that stored and delayed the large flows, the stage in the bypass very closely follows that of the river, and residence time on the bypass floodplain is much reduced relative to the historic basins. However, the bypasses still have slower water and longer residence times than the main-stem channel (Sommer, Nobriga et al. 2001). Due to upstream flow regulation and the refilling of reservoirs in spring, bypasses are inundated less frequently and for shorter durations in spring than prior to construction of the large reservoirs of the Sacramento River watershed (Williams et al. 2009).

To promote rapid conveyance, the bypasses are managed to minimize hydraulic roughness. Thus, land cover such as riparian forest or even extensive patches of tules are discouraged. Most of what is flooded is annual vegetation, bare ground, or wetland (Sommer, Nobriga et al. 2001). Fringes of riparian forest exist along a primary drainage channel and in other isolated pockets. Riparian forest along the natural levees of the Sacramento River and its tributaries, which are now behind engineered and maintained levees, is minimal and linear. Whereas the historic basins were estimated to occupy 260,000 ha (The Bay Institute 1998), the bypass system occupies a total of 31,300 ha (Sommer, Harrell et al. 2001), along with more acreage in the Butte Sinks/Butte Basin (see below).

CURRENT EXTENT OF FLOODPLAIN HABITAT

As described above, Central Valley habitats were rapidly and extensively converted to agriculture and other land uses (table 12.1) and current estimates of floodplain extent are a fraction of historic estimates. Katibah (1984) estimated that, in the mid-1980s, there remained 41,300 ha of riparian forest in the Central Valley, with approximately half in disturbed or degraded condition. This represents 11% of his estimate of historic riparian forests in the Central Valley. Further, much of the remaining riparian forest is on the "dry side" of levees or on terrace surfaces that are no longer hydrologically connected to the river (Philip Williams and Associates 2003) and thus don't represent riparian forest ecosystems that are likely to regenerate on their own in response to flooding.

The majority of the flood basins have been largely disconnected from river flows. Portions of the historic basins still receive frequent inundation, including the Butte Basin and the Yolo and Sutter Basins because of Yolo (24,000 ha) and Sutter (7300 ha) bypasses. The Butte Basin still encompasses significant wetland habitats with 7200 ha in the Butte Sink Wildlife Management Area (US Fish and Wildlife Service 2012), 3700 ha in the Upper Butte Basin Wildlife Area, and 50 ha in the Colusa Bypass (California Department of Fish and Wildlife 2012a, 2012b).

Estimating the remaining area of Central Valley floodplain habitat is thus not a simple question. For example, waterfowl habitat, which can be produced on flooded agricultural fields, is produced on a much larger area of historic floodplain than is native fish habitat, which requires long-duration spring flooding of areas directly connected to rivers. For a ballpark estimate of historic versus current floodplain extent, we can draw on two mapping efforts (Shelton 1987; Chico State University 2003). They report that current total floodplain habitats (riparian and wetland) occupy 100,000–200,000 ha, compared to 840,000–1,600,000 ha historically although much of what remains has altered flow regimes or lacks connectivity to the river.

CONCLUSIONS

Floodplains of the Central Valley and their associated wetlands—like temperate floodplains and wetlands everywhere—have a long history of being managed for everything but naturally functioning floodplain. They have changed dramatically as the result of many interacting factors, from beaver trapping to mining to dam and levee construction. Even much of the remaining floodplain or wetland habitat is degraded or highly managed, and is largely disconnected from the floodplain or river. From a habitat perspective, the Central Valley is fortunate that the flood-control system features bypasses, whose ecological values have recently become apparent. As discussed in the next chapter, managers and stakeholders are exploring how to improve the bypass for both their flood-risk-management functions and their ecological functions. This pursuit of multiple benefits within flood-management systems has great potential to restore or maintain the biodiversity and ecosystem services of floodplains around the world. Paralleling the history of the Central Valley, other river-floodplain systems have long histories of both failed and successful management. These lessons can be applied to the creation and management of innovative reconciled floodplains (chapters 14, 15).

THIRTEEN

Central Valley Floodplains Today

> Thousands of antelope, tule elk, and deer grazed the Valley floor in drifting bands ... and the Valley's many small and larger watercourses were full of fish ... birds of all descriptions swept overhead in flocks that could darken the sky ... The river's channel could never contain within its natural banks the huge flows of water that almost annually poured out of the canyons of the northern Sierra Nevada. Signs of yearly flooding were everywhere apparent ... Together, the ponds in the basins annually created a vast inland sea a hundred miles long occupying the centerline of the Sacramento Valley which slowly drained back into the river channels and down through the delta during the spring months.
>
> KELLEY (1989, P.3)

THE FLOODPLAINS OF CALIFORNIA'S Central Valley today are very different from the floodplains that Robert Kelley described. The "vast inland sea" has been leveed, drained, and converted to farmland, with only relatively small remnants allowed to flood. This follows the trajectory that most temperate floodplains have followed, particularly those in Mediterranean regions such as California. Fortunately, much can be learned from studying the few remaining functioning floodplains to understand natural processes (Bayley 1991; Ward et al. 2001). Some of the best remaining opportunities to study how temperate floodplains function are in fact on reconciled floodplains, where humans and alien species are active players in the ecosystem, but flood regimes follow natural or seminatural patterns. While studying "natural" floodplains will remain important for improving our understanding of fundamental biophysical processes, studying reconciled floodplains provides insights into how managers can achieve a broader range of benefits out of the temperate floodplains that are highly managed. For example, Bond et al. (2014) show that even heavily engineered floodplains can be used to experimentally test and demonstrate floodplain restoration actions.

Floodplain-management actions, such as reconnection with river flows, are often labeled as restoration, but these projects are rarely well documented, monitored, and studied (Sparks et al. 1990). Lack of such monitoring and research has undoubtedly contributed to the high rate of failure, or suboptimal performance, of river and floodplain restoration projects (Kondolf 1995; Ward et al. 2001; Downs and Kondolf 2002). There are, however, success stories in California of well-documented examples of reconnected floodplains that provide a broad range of ecosystem services. In this chapter, we present three case histories of Central Valley floodplains: the Cosumnes River Preserve (CRP), the Yolo Bypass, and the Sacramento River. The first two are remnant floodplains that have been studied extensively in the past decade, while the Sacramento River is an example of a planning effort for reconciliation on a fairly large scale. The CRP and Yolo Bypass are quite different from each other. The CRP is relatively small and is arguably the most "natural" floodplain in the Central Valley. The Yolo Bypass is considerably larger and bears only superficial resemblance to a natural floodplain. Yet both support floodplain ecosystems that contain many native species and provide a range of ecosystem services. Both are novel ecosystems that rely on reconciliation, rather than restoration, to function. The Sacramento River case focuses on a chain of restoration projects in the valley that also includes the Yolo Bypass.

COSUMNES RIVER PRESERVE

The Cosumnes River is the only river draining the western slope of the Sierra Nevada that does not have large dams on its main stem; therefore, high flows on the Cosumnes are mostly unchanged from their natural condition (Moyle et al. 2003; see figure 9.1). Because of this, Moyle and Yoshiyama (1994) recommended that the Cosumnes River be managed as a "key watershed" to protect the landscape-scale processes that still operate there but are rare elsewhere in the Central Valley. The Nature Conservancy (TNC), along with numerous partners, established the CRP along the lower river in 1987. Within the preserve, the Cosumnes River can connect to its floodplain and the CRP thus provides one of the few opportunities in the Central Valley to study a near-natural floodplain with relatively unimpaired hydrology and a mosaic of habitats (Mount et al. 2003; Swenson et al. 2003).

The watershed encompasses 3000 km² with elevations ranging from 2400 m to near sea level at its confluence with the Mokelumne River in the Sacramento-San Joaquin Delta (Swenson et al. 2003). The Cosumnes watershed is relatively low elevation, with only 16% of its area above 1500 m, so the river's hydrograph is primarily driven by winter rainstorms, resulting in higher winter flood pulses and lower spring snowmelt flood flows compared to other rivers draining the Sierra Nevada (Moyle et al. 2003). The Cosumnes River watershed contains just one reservoir, Sly Park Reservoir, located on a tributary that provides approximately 10% of the average annual yield of the watershed. Other than this modest alteration, winter and spring flows of the Cosumnes River are unimpaired. Summer flows are reduced, however, due to diversions and groundwater pumping near the river.

Historical Conditions

Historically, the Cosumnes River emerged from the foothills and meandered across a broad floodplain (Swenson et al. 2003). The river was characterized by a complex of multiple channels that shifted over time as a function of sediment deposition and other processes. The floodplain featured abandoned channels, marshes, and seasonal lakes or "lagunitas." Riparian forest was established on higher elevations, including natural levees, with tule marshes occurring in lower areas (Florsheim and Mount 2002). The river was tidal at its downstream end (Swenson et al. 2003). The Cosumnes floodplain historically was part of the Sacramento flood basin, the southernmost of the six basins described by Gilbert (1917; chapter 12). During wet years, overbank flow from the Cosumnes and the adjacent Mokelumne River inundated the basin along with overflow from the Sacramento and American rivers. These flood waters filled the marshes and lakes and slowly drained to the Delta through multiple channels (Florsheim and Mount 2003).

Flow in the Cosumnes River varies dramatically between seasons, and the floodplain is generally inundated between 1 and 5 months a year in winter and spring. The floodplain surface and the lower Cosumnes River itself generally become dry during the summer. However, permanent water persists in the many sloughs that drain the floodplain; these sloughs are weakly tidal, are fringed by riparian forest, and support large populations of resident warm water fishes (mostly alien species).

Diverse land uses have altered the river and its floodplain, especially by increasing the river's sediment supply through mining, grazing, timber harvest, and conversion of natural vegetation to vineyards and other agriculture. This increase in sediment supply is reflected in an anthropogenic layer of sediment in soil cores of the floodplain. Although mining was not as intense in the Cosumnes River as in other Sierra Nevada watersheds, mining debris raised the bed of the Sacramento River by 3 m, which likely had a backwater effect on the lower Cosumnes River and its floodplain (chapter 12). In the nineteenth and twentieth centuries, channel migration, avulsion, and overbank flows were greatly reduced by extensive levees, the floodplain was leveled for agriculture, and many of the standing water bodies were drained (Florsheim and Mount 2003).

TNC began purchasing land on the Cosumnes River floodplain in 1984, initially to protect large groves of valley oak (*Quercus lobata*) that grew on a small remnant of natural floodplain. The CRP was formally established in 1987 and has grown to encompass 16,000 ha, most of which is managed as wildlife-friendly agriculture, such as grazing and organic rice production, compatible with periodic flooding (figure 13.1). Habitat types on the preserve include riparian forest, lakes, wetlands, and annual grassland with vernal pools. Initial management of the preserve focused on active planting of riparian trees and 500 acres were hand planted, with mixed success. However, it soon became obvious that riparian trees, such as willow (*Salix* spp.), Fremont cottonwood (*Populus fremontii*), and valley oak, were establishing successfully on their own, particularly in areas with deposition from floods. TNC now emphasizes restoring riparian vegetation through natural processes of flooding (Reiner 1996).

Floodplain Renewal

In 1985, a levee breached naturally during a flood, allowing the river to flow onto a farm field. The farmer fixed the levee but did not remove the splay of sediment that had been deposited on his field by the flood nor did he try to plant a crop in the freshly deposited sediment. Cottonwoods and willows rapidly established in the new substrate. This area, now referred to as the "accidental forest," became part of the CRP in 1987 and currently has a tall canopy of cottonwoods with an understory of valley oaks (Tu 2000; Swenson et al. 2003; figure 13.2). Due to the successful riparian regeneration following the 1985 breach, TNC sought to increase connectivity between the river and floodplain through additional breaches. After hydrological modeling indicated that a new breach would not increase flooding for neighboring landowners, in 1995 TNC opened a 16.5 m breach in the levee downstream of the accidental forest, allowing the river to access an additional 80 ha of floodplain. Following the floods of 1997, TNC implemented another breach in the main levee upstream of the accidental forest (the "Corps Breach"), with additional breaches in an internal levee to allow flood waters to drain into the accidental forest and the new floodplain created in 1995 (Swenson et al. 2003). At the Corps Breach, TNC built a low setback berm to protect nearby rice fields and also dug a 2 m deep pond on the newly connected floodplain for waterfowl.

Since levee breaches reconnected the floodplain, the CRP has supported diverse studies on floodplain physical processes, primary and secondary productivity, fishes, and birds, discussed in the next sections.

Floodplain Processes

Prior to the breaches, the levee contained floods up to about the Q_5 (0.20 exceedance probability;

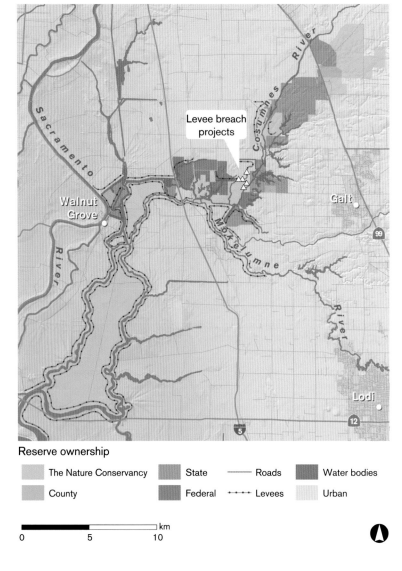

FIGURE 13.1 The Cosumnes River Preserve (Gesch et al. 2002; CalAtlas 2012; GreenInfo Network 2016).

see chapter 2). Mount et al. (2003) calculated that hydraulic connectivity through the present breaches is established at flows only half the size of the mean annual flood, with an annual exceedance probability of 0.95. Water begins moving onto the floodplain when discharge exceeds approximately 21 m³/s at a gage upstream at Michigan Bar. The floodplains at both breach sites have been inundated repeatedly, allowing sediment to deposit on the floodplain. The length of time the floodplain is connected to the river can vary widely between years. For example, the river and floodplain were connected for 128 days in 2005 but only 22 days in 2002 (Ahearn et al. 2006). Booth et al. (2006) provide a thorough review of hydrological variability and flood types at the CRP.

Florsheim and Mount (2002) and Mount et al. (2003) documented how high-magnitude flow events created complex floodplain

FIGURE 13.2 Cosumnes River forest regeneration following levee breaches. The "accidental forest" (AF) is a stand of riparian trees that generated on a sand splay deposited during an unintentional levee breach in 1985. Solid arrows indicate both the river's flow direction and the location of the 1995 intentional levee breach. (A) During a 1996 flood, a crevasse sand splay formed due to sediment transport and deposition (dashed arrows in both photos). (B) In 2006, a forest is colonizing the splay (photos by Mike Eaton).

topography, including sand splays formed at breach openings. Sand splay complexes influence floodplain-river connectivity, increase local high hydrologic residence time, and provide bare ground for establishment of riparian plant communities (Trowbridge 2002). The higher elevation surfaces of the sand splays, which are higher than most of the surrounding floodplain surfaces, support growth of riparian trees (figure 13.2; Tu 2000; Trowbridge 2002; Mount et al. 2003).

Primary and Secondary Productivity

The topographic complexity of the Cosumnes floodplain and the river's variable hydrograph has led to high spatial and temporal heterogeneity of the floodplain's hydrology, water quality, and productivity. Ahearn et al. (2006) found that, upon initial inundation, water quality on the floodplain was similar to that of the river, but diverged upon disconnection. Primary productivity (largely from algae) peaked 2–5 days following cessation of river inflow. Concentrations of phytoplankton peaked during the draining phase (see discussion of phases in chapter 4) and were measured at 19 and 18 μg/L during two draining periods in 2005, approximately four times the level found in the river. Areas more distant from primary flow paths through the floodplain had higher residence time and greater productivity (Ahearn et al. 2006); this is similar to the "inshore retention concept" (Reckendorfer et al. 1999), which emphasizes the importance of littoral backwater areas to the productivity of large rivers. Export of algal biomass was high during the draining phase both because longer residence time led to greater overall productivity on the floodplain and because more distal portions of the floodplain, which had longer residence times and productivity throughout the flood event, drained and mixed with floodplain water being exported (Ahearn et al. 2006).

Ahearn et al. (2006) also found that the ability of the Cosumnes floodplain to produce and export algal biomass varied with the sequence and characteristics of flood events. For example, a long interval between flood events allowed water on the floodplain to develop high algal biomass with the second flood exporting much of this productivity to the river. In contrast, a rapid sequence of floods did not allow for high algal productivity and thus exported water with lower algal biomass. The Ahearn et al. (2006) study focused on phytoplankton; however, periphyton (see figure 7.1)

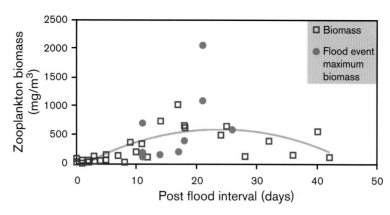

FIGURE 13.3 Zooplankton abundance in relation to number of days since the last flood event on the Cosumnes River Preserve (after Grosholz and Gallo 2006).

and detritus from terrestrial plants likely also contribute major sources of carbon and energy for floodplain food webs and can also be exported back to the river.

Due to a variety of factors, including greater algal productivity, zooplankton biomass was 10–100 times greater on the Cosumnes floodplain than in the adjacent river. Flood events initially reduced zooplankton biomass by flushing organisms from the floodplain; biomass then increased following the flood event and generally peaked between 10 and 25 days after the flood (figure 13.3). Water on the floodplain also tended to be warmer than the adjacent river, which may also have increased zooplankton growth rates (Grosholz and Gallo 2006).

In short, primary and secondary productivities (chapter 7) are closely tied to residence time of the water on the Cosumnes floodplain, which in turn depends on the frequency and magnitude of flooding. A series of moderate flood events spaced over a 2–3 month period appear to be optimal for maximizing productivity, which in turn can result in high production of fishes and birds.

Fishes

The high zooplankton and invertebrate productivity of the floodplain provides abundant food for larval and juvenile fish. Moyle et al. (2007) sampled fishes on the floodplain for 5 years (1998–2002) during the winter-spring flooding season. They sampled 32 species of fish in the seasonal floodplain, the river, and a neighboring slough; 18 were present in all years in all three habitats. Eight of the 18 abundant species were natives, while the rest were aliens (see table 8.1).

Moyle et al. (2007) classified the fishes into six guilds based on how they used the seasonal floodplain: (1) floodplain spawners (four species), (2) river spawners (six species), (3) floodplain foragers (five species), (4) floodplain pond fishes (two species), (5) inadvertent users (eight species), and (6) nonusers/avoiders (seven species). See chapter 8 for definitions of the guilds. There was a consistent pattern of seasonal floodplain use over the 5 year period, with some annual variability due to the timing and extent of flooding. The first fishes to appear on the floodplain were floodplain foragers, inadvertent users, and juvenile Chinook salmon (river spawners). The next fishes to appear were adult floodplain spawners, principally Sacramento splittail and common carp, although small numbers of foragers and inadvertent users from neighboring sloughs and ditches were also present. Juvenile splittail and carp quickly grew large enough to dominate floodplain fish samples, along with juvenile Sacramento sucker and pikeminnow (river spawners). The adult floodplain spawners left the floodplain when inflow decreased and their

FIGURE 13.4 Rapid growth of Chinook salmon juveniles on Cosumnes River floodplain. Fish on the left were reared in the river, while those on the right were reared in the floodplain (photo by Jeff Opperman).

juveniles persisted only as long as occasional new flood pulses kept water levels high and temperatures low. Most juveniles left the floodplain either with the pulses or with declining inflows. Usually, the floodplain disconnected from the river by mid-May. In two large shallow ponds of residual water, alien species, mainly juvenile centrarchids, western mosquitofish, and Mississippi silverside, dominated catches by June.

Native fishes that were most abundant each year were those with juveniles that used the floodplain for rearing. The Cosumnes floodplain often provides improved rearing conditions for native fishes, compared to the river, likely due to low water velocity and greater productivity. For example, Ribeiro et al. (2004) reported that juvenile Sacramento splittail rearing in the Cosumnes floodplain generally was in better condition than fish rearing in the river or in slough habitats. Jeffres et al. (2008) reported that juvenile Chinook salmon grew faster in enclosures within floodplain habitats than within enclosures in adjacent river habitats; highest growth rates occurred in floodplain areas where the water had the highest residence time (figure 13.4).

Driven by the natural flood regime of the Cosumnes, the timing of floodplain inundation favors native fish and common carp, species which tend to spawn earlier in the year. Conversely, the seasonal floodplain is generally dry when most nonnative fish spawn, in late spring or summer (Crain et al. 2004). Native species' synchronization of spawning with the floodplain inundation—and nonnative species' lack of synchronization—is supported by the fact that the floodplain is dominated by native fishes through the end of April, after which time connection to the river is usually lost (Moyle et al. 2007). Most native fishes leave the floodplain as it drains and remnant water (in pools) becomes dominated by short-lived alien fishes such as Mississippi silversides, western mosquitofish, and golden shiner, or by juvenile carp. Most fish that become stranded are nonnative species, although some native fish do become stranded in artificial ponds, ditches, or behind roads and levees—structures that interfere with drainage. Colvin et al. (2009) reported a similar pattern of fish use of floodplains of the Willamette River, with native fishes dominating seasonal floodplains and nonnatives dominating permanent water bodies. Nonnative fishes were not found in floodplain habitats until water temperature exceeded 20°C in the spring.

Permanent floodplain water bodies such as ponds, oxbow lakes, and sloughs are also dominated by alien fish species, mainly from the eastern and Midwestern United States, including centrarchids (bluegill, largemouth bass, etc.) and ictalurid catfishes (*Ameiurus* spp.). Also important are cyprinids such as common carp, golden shiner, and goldfish (Moyle et al. 2007). This pattern is consistent throughout the Central Valley (Moyle 2002).

Birds

When flooded, the CRP attracts large flocks of ducks, geese, swans, and shorebirds, as well as several thousand sandhill cranes (*Grus canadensis*). The CRP is also a "hot spot" for migratory and resident passerine songbirds because of its diverse habitats, from open floodplain to old growth riparian woodland (Dybala et al. 2015). Wood et al. (2006) monitored the use of the floodplain by 22 species of passerine birds for 11 years. Most of these birds responded positively to development of floodplain forest, reflecting the increase in habitat complexity. Some birds appear to benefit from an increase in food supply of invertebrates from floodplains; abundance of tree swallows (*Tachycineta bicolor*) had a positive correlation with the extent of flooding. Song sparrows (*Melospiza melodia*) displayed a complex response to flooding. The abundance of adults in the spring was negatively correlated with winter flooding, but reproductive success in the spring was positively correlated with winter flooding. Ultimately, the winter flooding may be beneficial to song sparrow populations, because populations in the year following winter flooding were larger (Wood et al. 2006). Surprisingly, reproduction of birds specialized for living in mature riparian forest, rather than wetlands, declined over the study period, likely due to an invasion of alien black rats (*Rattus rattus*). Because the rats are arboreal, they can escape flooding and easily access bird nests (D. Whisson, UCD, unpublished data). The patch of riparian forest on the preserve is apparently too small to sustain songbird populations under high predation rates, without predator control.

Conclusions

The Cosumnes River floodplain, although small in size compared to historic floodplains of the Central Valley, demonstrates the importance of river-floodplain connectivity for floodplain ecosystems. Studies on algae, zooplankton, aquatic invertebrates, and juvenile fish all emphasize the importance of residence time for the productivity of floodplain food webs. Flood events at the Cosumnes vary dramatically from year to year in timing, magnitude, and duration, contributing to diverse and dynamic floodplain habitats. The Cosumnes River floodplain also clearly supports a novel ecosystem, with native species strongly dependent on a natural flooding regime but alien species remaining a large part of the fish biota. Because the Cosumnes floodplain is relatively small, active management is required to maintain the mosaic of habitats that benefit native fishes. For example, mature forest does not appear to be particularly valuable habitat for fishes. The present mix of floodplain forests in various successional stages and open floodable farmland is probably a beneficial pattern of habitats for native fishes and birds, although at some point it is likely that active clearing of some forest may be needed to maintain sufficient area of the early successional stages of open water and annual plants most favored by native fishes and waterfowl. This need for active management, along with the recognition that floodable farmland is part of the mosaic for maintaining native species, illustrates that reconciliation ecology is an appropriate management paradigm for the Cosumnes River floodplain.

YOLO BYPASS

The Yolo Bypass has been a critical component of the Sacramento River Flood Control Project since the early 1930s (Sommer, Harrell et al. 2001; chapter 12). The Bypass is a 24,000 ha portion of the historic Yolo Basin that receives and conveys flood waters from the Sacramento River and Sutter Bypass through a system of weirs (see figures 12.7 and 13.5).

The Bypass was adopted as an essential feature for flood management because levees and upstream dams simply could not contain the volumes of flood waters that were causing damage to Sacramento and other communities (chapter 12). While not described as such, the creation of the Yolo Bypass was basically a major floodplain reconnection project. Creation of the Bypass resolved more than 50 years of debate over whether Sacramento River flows could be contained within levees or whether flows would require access to historic floodplain (Stutler 1973; Kelley 1989). During major flood events, the Bypass now carries 80% of the flow of the Sacramento River, with the rest remaining in the incised and leveed channel that flows past Sacramento (Sommer, Harrell et al. 2001).

The Yolo Basin is one of the six basins that flanked the Sacramento River (figure 1.3). Historically, the Basin was bounded on the north and east by the natural levees of the Sacramento, on the west by the coalesced alluvial fans of Cache and Putah Creeks and on the south by tidal tule marshes. The basin became inundated when Sacramento River flows reached a sufficient stage to flow over natural levees, similar to flows cresting the Fremont Weir today. Other sources of inundation included tributary flows from Cache and Putah Creeks, groundwater inflow from the Sacramento River, and direct precipitation (Philip Williams and Associates 2003).

Set within that historical flood basin, the Yolo Bypass runs parallel to the western side of the Sacramento River, connecting the river at Knights Landing to the Sacramento-San Joaquin Delta near Rio Vista (figure 13.5). It is 68 km long, 2–4 km wide, and has a flood capacity of 9700 m^3/s at its upstream end, which increases to 13,900 m^3/s at the downstream end where flood waters flow into Cache Slough (Stutler 1973). Flow from the Sacramento and Feather Rivers and Sutter Bypass enters the Yolo Bypass by way of the Fremont Weir, a low cement wall that forms the northern boundary of the Bypass. Inundation of the Bypass generally begins when flow at the Fremont Weir exceeds 2000 m^3/s, which occurs in approximately 60% of all years (Sommer, Harrell, Solger et al. 2004); in about half of those years, the Bypass is inundated for a month or longer (Schemel et al. 2004). During larger flood events, additional flow from the Sacramento and American Rivers can enter the Bypass through the Sacramento Weir, a few kilometers downstream from the Fremont Weir. Other inflows to the Bypass include Putah and Cache Creeks, Willow Slough, and the Knight's Landing Ridge Cut (Schemel et al. 2004). The Bypass is graded to drain toward a perennial channel, the Toe Drain, so-called because it follows the toe of the major levee at its eastern edge.

Much of the Bypass is in seasonal agriculture with crops including rice, sugar beets, safflower, tomatoes, and corn. The State of California has flow easements allowing flooding in the bypass in any month but, historically, the Bypass has rarely had extensive inundation beyond May. The Bypass floods in more than half of all years, although larger flows have been reduced since closure of Shasta Dam in 1945.

While the bypass was originally constructed to protect cities from flooding while also supporting agriculture, more recently it has been recognized as valuable floodplain habitat that

FIGURE 13.5 The Yolo Bypass within the lower Sacramento Valley, California. A mosaic of land uses is shown including agriculture, wetlands, and other natural habitats (USGS 2012; CDWR 2016; Sacramento Area Council of Governments 2016).

supports significant populations of fish and wildlife (Sommer, Harrell et al. 2001). Approximately one-third of the bypass is now in nonagricultural habitats, including ponds, wetlands, grassland, and riparian forest. Recognizing these fish and wildlife benefits, the 65 km² Yolo Bypass Wildlife Area (YBWA) was established within the Bypass in 1991. However, the Bypass cannot be neatly divided into "natural" and "agricultural" land cover. Illustrating the concepts of reconciliation and novel ecosystems, portions of the YBWA are grazed to help manage vernal pool habitat and other portions are planted in rice—both because rice is highly valuable to migratory waterfowl and because the crop helps fund management costs. When the Bypass is flooded, native fish and birds use its full extent, benefitting from both agricultural and natural habitats.

Interstate 80 (I-80), a major highway connecting San Francisco to the eastern United States, crosses the upper end of the YBWA on the 5 km long Yolo Causeway. Interstate 5, the major north-south freeway of the US west coast, crosses the Bypass on another causeway 11 km to the north. A major railroad line, carrying Amtrak trains, also crosses the Bypass parallel to I-80. With tens of thousands of people using these arteries every day, the Yolo Bypass is one of the most visible flood control features—and one of the most visible wildlife areas—in the United States, with massive flocks of birds quite obvious from the freeway. During summer, when birds are much less numerous, bats from a colony of Mexican free-tail bats emerge each evening from their roost beneath the causeway, providing drivers with the unusual sight of hundreds of thousands of bats pouring into the sky each dusk.

The Sutter Bypass, upstream of the Yolo Bypass (figure 13.5), shares many features with the Yolo Bypass (Feyrer et al. 2006) but has been much less studied. However, because of their similarities, many of the broad trends described below likely also apply to the Sutter Bypass. During major flood events, water from the Sutter Bypass flows directly into the Yolo Bypass.

Due to its prominent role in Sacramento Valley flood management, the Bypass has always been of great interest to the hydrologists and engineers of the California Department of Water Resources (CDWR), the agency with primary management responsibilities for the Bypass. As the agency's requirements to protect the environment expanded, its attention to the Bypass has grown to include research from fish biologists and other aquatic scientists. Along with other agencies, academic institutions and consultants, the Bypass has become a major area for research on Central Valley floodplain ecosystems. Below we review this research and summarize insights gleaned from the Bypass about basic floodplain processes and the management of reconciled systems.

Hydrology

Recent research has demonstrated that the Yolo Bypass provides many functions similar to those of a natural Central Valley floodplain and, at 24,000 ha, the Bypass represents the largest floodplain in the Central Valley. Because of limited surface connections to the main river (the Fremont Weir and, occasionally, the Sacramento Weir), the Bypass has relatively simple hydrology compared to a natural floodplain such as the Cosumnes and processes within the Bypass can be measured without accounting for multiple connections from the river (Schemel et al. 2004; Sommer, Harrell, Solger et al. 2004).

When the Bypass is inundated, it contains an immense amount of shallow water habitat. Even relatively small flood events produce extensive areas of inundation, including more than 10,000 ha of shallow water habitat (<2 m deep), which is an order of magnitude higher than the amount of shallow water habitat found along the adjacent, channelized Sacramento River (Sommer, Harrell, Solger et al. 2004). Water in the bypass is shallower, slower, less turbid, and generally warmer than water in the river. These differences are magnified as residence time increases (Sommer, Harrell, Solger et al. 2004).

The residence time of water entering the Yolo Bypass at the Fremont Weir during a typical winter flood is between 5 days and 4 weeks until it returns to the Sacramento River near Rio Vista. Flood waters may persist longer in the lower half of the Bypass because tidal action can impede drainage. Residence time can be quantified by dividing the total volume of water within the Bypass by the inflow rate. When flows are small relative to total volume within the Bypass, the residence time is high. Conversely, when flows are large relative to total volume, residence time is low.

Although the Sacramento River provides most of the water to the Bypass, west-side tributaries (Ridge Cut, Willow Slough, Cache, and Putah Creeks) provide significant amounts of water during drier years when no or limited flow enters via the weirs. These tributaries can also provide important sources of water prior to and after periods of inundation from the Sacramento River, and be sources of nutrients during the draining phase, which can be important to productivity of phytoplankton (described below). Flow from these tributaries also aids in transporting highly productive water across the Bypass and downstream. This is an example of the ecological importance of diverse sources of water on a floodplain (chapter 2; Tockner et al. 2000; Ward et al. 2001).

Primary and Secondary Productivity

As is true of the Cosumnes River floodplain, aquatic productivity in the Bypass is strongly influenced by hydrologic patterns such as season of inundation and hydrologic residence time. Schemel et al. (2004) studied nutrient levels and productivity during floods in the Bypass, using chlorophyll *a* (Chl *a*) as an indicator of phytoplankton biomass. They found that Sacramento River water had relatively low concentrations of nutrients and thus, during initial inundation, nutrient levels in flood waters of the Bypass were low. Consequently, productivity was low due to low nutrient levels along with short residence time. Later in the flood, as inflows from the Sacramento were dropping, nutrient levels rose due to inputs from west-side tributaries and breakdown of organic matter on the Bypass. phytoplankton biomass increased during the draining phase because of these increased nutrient levels and because residence time increased, while water velocity, depth, and turbidity declined.

The Bypass can be a source of water with high algal biomass that can promote food webs both within the Bypass and, when the biomass is exported, to downstream ecosystems. Phytoplankton density can be high enough to produce a Chl *a* concentration of up to 23 μg/L, considerably higher than the value reported as producing growth in the zooplankton cladocerans (10 μg/L; Muller-Solger et al. 2002). Because Chl *a* concentrations in the Sacramento-San Joaquin Delta rarely exceed 10 μg/L (Sobczak et al. 2002; Schemel et al. 2004), the water draining from the Bypass can, under some circumstances, provide high levels of biologically available, nutritious phytoplankton for downstream consumers, at least in a limited area (see next section).

Within the Bypass food web, primary consumers of phytoplankton included cladocerans and copepods. Initial studies did not find differences in zooplankton density between the Bypass and the Sacramento River, because the low residence time of water in both habitats reduced reproduction (Sommer, Harrell, Solger et al. 2004). However, the density of zooplankton in the Bypass did rise substantially, and zooplankton became a substantial part of the diet of juvenile Chinook salmon, during the draining phase of a long-duration flood event (February to March 1998; Sommer, Nobriga et al. 2001). This suggests that zooplankton densities in the Bypass can become ecologically significant during long residence time draining phases.

Density of dipteran larvae, primarily chironomid midges, was significantly higher in the Bypass than in the Sacramento River (Sommer, Nobriga et al. 2001). During initial inundation, aquatic invertebrates in the Yolo Bypass can be dominated by the endemic chironomid midge (*Hydrobaenus saetheri*), which emerges from diapause from floodplain soils (Benigno and Sommer 2008). Chironomid midges in some years form a major part of the diet of juvenile Chinook salmon and splittail on flooded areas. Alternately, when water originates from other sources such as the Ridge Cut drain, a large planktonic crustacean, *Daphnia magna*, can dominate the invertebrates available as food to fish (Katz 2015). Other invertebrates captured in the drift of the Yolo Bypass included oligochaete worms (Naididae and Enchytraeidae), snails (Physidae), and cnidarians (Hydridae). In general, drift invertebrates, including terrestrial invertebrates, were more abundant on the floodplain than in the river and their abundance was positively correlated with flow. Significantly, the development of large populations of midges and other food-web organisms can be very swift—on the order of days—and, thus, even relatively short inundation periods may have ecosystem or population-level benefits for fish that can access the Bypass (Sommer, Harrell, Solger et al. 2004).

Export of Productivity

Although during the draining phase water from the Bypass has a significantly higher concentration of Chl *a* than does the Sacramento River, the ability of the Bypass to produce significant levels of biologically available carbon for downstream ecosystems, especially the Sacramento-San Joaquin Delta, is uncertain. Concentrations of phytoplankton were higher in Bypass outflow than in water in the Delta, but the period of this input is relatively brief (Schemel et al. 2004). Jassby and Cloern (2000) concluded that the Bypass is unlikely to be a primary source of carbon for the downstream Delta and estuary due to the relatively small proportion of water passing through the Bypass on an annual basis. The median proportion of water moving through the bypass during winter was only 10% of total Sacramento River flow and 1% during spring. Flows from the Bypass during other seasons are generally negligible (1% or less of total flow in spring through fall). Thus, concentrations of dissolved organic carbon (DOC) would have to be an order of magnitude higher in Bypass water than in river water for it to equal inputs from the river during winter. Further, when the proportion of water flowing through the Bypass is at its greatest, the proportional increase in Bypass contribution is offset by the lower residence time (and thus lower productivity) of water and the lower residence time of water flushing through the Delta during high flows. The findings of Mount et al. (2014), who reviewed the vast amount of information available on this issue in the draft Bay Delta Conservation Plan, support this assessment. The greatest benefit that the Bypass provides to the Delta ecosystem is therefore that, when flooded, it essentially doubles the area of the Delta (Jassby and Cloern 2000). Organisms within the Delta can then access resources available in the Bypass during periods of inundation.

Although under current conditions, the annual productivity contribution from the Bypass to the Delta seems to be negligible from a mass balance perspective, it may provide contributions at critical time periods at specific locations (Schemel et al. 2004). Further, active management of Bypass flooding—controlling timing, duration, and frequency of inundation—could greatly increase its contribution to downstream productivity. For example, managed flooding of the Bypass could promote a series of relatively short pulses with long draining times that would produce pulses of productivity to the Delta. Such managed flooding could also be timed to maximize the value of the Bypass to native fishes, with the optimal period of managed inundation being March and April, because this would coincide with peak Chinook salmon rearing on the floodplain and splittail spawning and larval rearing. As discussed above, while the direct export of algal carbon to the Delta may not be significant on an annual basis, productivity within the Bypass during inundation may be important, for example, in the form of juvenile fish that were spawned or reared on the Bypass and are consumed by downstream predators (see chapter 8).

The extent of the historic flood basins was an order of magnitude greater than the extent of current bypasses, and the basins experienced frequent and prolonged flooding. Thus, the system-scale benefits of the basins in terms of productivity and export to the Delta were likely significant.

Fishes

Fifteen native fish species and 27 alien species have been recorded within the Bypass (Sommer, Harrell et al. 2001). As in the Cosumnes floodplain, perennial aquatic habitat is dominated by alien species; surveys within permanent water bodies, such as ponds, collected 20 species of nonnative fish and only 3 native species (Sacramento blackfish, Sacramento sucker, and prickly sculpin). The native species represented <1% of the total number of individuals collected and <3% of total biomass (Feyrer et al. 2004). Both native and alien fish species use the Bypass while inundated, taking advantage of food resources and physical habitat provided by inundated annual vegetation and crop stubble. Native fishes tend to use the bypass earlier (January–April) than do alien fishes (April–June) (Sommer, Harrell, Kurth et al. 2004), similar to the way fishes use the Cosumnes River floodplain (Crain et al. 2004; Moyle et al. 2007). Sommer, Harrell et al. (2001) hypothesized that the inundated Bypass may have particular importance to native species because the spawning period of native fishes coincides with spring flooding, while the spawning period for many nonnative species occurs when the Bypass is generally dry.

Common carp are the primary alien fish that spawns in the Bypass, while Sacramento splittail appear to be the only native fish that relies on floodplain for spawning and is the primary native species that spawns in the Bypass. When flooded, the Bypass provides preferred spawning substrate of splittail—inundated annual vegetation—as well as feeding opportunities for adults and for larval fish (Moyle 2002). Strength of splittail year classes (age-0 abundance) is highly correlated to the length of inundation of the Bypass (Sommer, Nobriga et al. 1997), although wet years also provide more spawning habitat for splittail outside the Bypass (Moyle et al. 2004). Sommer et al. (1997) reported that splittail sampling programs had the greatest catch within the Bypass compared to other locations in the Delta and lower Sacramento River system. Splittail spawning in the Bypass occurs whenever flooding occurs, from February through April.

The flooded Bypass also provides important rearing habitat for juvenile Chinook salmon, with a greater prey base and water with lower velocities than the main-stem Sacramento River. In one study, juvenile Chinook within the Bypass fed primarily on chironomid larvae emerging from the soil (Sommer et al. 2001) although a study using artificially flooded fields on the Bypass found they fed mainly on zooplankton, especially *Daphnia magna* (Katz 2015), suggesting that juvenile salmon can rely on a range of prey sources depending on availability. Due to the combination of highly abundant food and near-optimal physical conditions, juvenile Chinook in the bypass grow faster with likely higher survival than those rearing in the river (Sommer, Nobriga et al. 2001; Henery et al. 2010; Katz 2015), as noted for the Cosumnes River (figure 13.4). Higher survival rates likely continue as the juvenile salmon transition from fresh water to salt water (smolting). Smolt survival into adulthood has been found to be positively correlated with smolt size upon entering the marine environment (Ward and Slaney 1988) and the higher survival rates of smolts can have significant positive influences on population viability of salmonids (Kareiva et al. 2000). By providing habitat that produces larger smolts, the Bypass and other Central Valley floodplains are thus likely contributing to greater survival of smolts that enter the San Francisco Estuary and ocean.

Within the Bypass, juvenile salmon appear to prefer slower water, including shoals and areas downstream of obstructions such as causeway, levees, and trees (Sommer, Harrell et al. 2001). Juvenile salmon also appear to use a range of inundated land cover types. For example, Katz et al. (2015) found that juvenile salmon would rear, with fast growth rates and low mortality, on artificially flooded fields in the Bypass, with flooded rice stubble creating particularly favorable habitat.

Juvenile Chinook salmon begin to enter the Bypass in January and they generally leave by April. The Chinook salmon that use the bypass are likely mostly fall run. The Central Valley hosts three other runs of Chinook, and the extent to which juveniles of those other runs—spring, late fall, and winter—use the Bypass is not well understood. Analyses of lengths of juvenile salmon caught in screw traps at Knight's Landing, just above the Fremont Weir, suggest that juveniles of all four runs may be using the Bypass to some extent during periods of flooding. The bypasses are much smaller in area than the historic flood basins, and these massive areas of floodplain may have been extremely important as rearing habitat for all four runs of Chinook salmon, all of which have outmigration of juveniles that would have allowed floodplain rearing (Moyle 2002).

Stranding of both juvenile splittail and salmon appears to be a minor issue primarily because the Bypass is graded to drain quickly toward the perennial Toe Drain. Native fish are rarely observed stranded in the few permanent water bodies on the Bypass (T. Sommer, pers. comm.). Sommer et al. (2005) reported that the majority of juvenile salmon within the Bypass successfully emigrated to the Delta.

The Bypass is a reservoir of mercury, a legacy of nineteenth-century mercury mining, primarily in the upstream Cache Creek watershed (Springborn et al. 2011). As a result, juvenile Chinook salmon rearing in the Bypass accumulate significantly more methylmercury than do juveniles rearing in the Sacramento River. However, because the salmon will grow three orders of magnitude larger as adults, the increased accumulation of methylmercury for those rearing on the floodplain represents an insignificant difference in the total mercury found in adults (Henery et al. 2010).

Birds

The Yolo Bypass supports a wide array of birds and is especially important habitat for migratory waterbirds. Because of its bird diversity and abundance, the YBWA is classified by the National Audubon Society as a Globally Important Bird Area (Suddeth 2014; Grimm and Lund 2016). So far, 112 species have been recorded from YBWA.

Most of the time, the Bypass is either not flooded or only partially flooded, so there is a strong resident component to the bird fauna, including (a) species that can forage in wetlands, ditches, and rice fields, such as various herons and egrets (Ardeidae), white-faced ibis (*Plegadis chihi*), and double-crested cormorants (*Phalacrocorax auritus*); (b) resident wetland songbirds, such as song sparrows and tricolored blackbirds (*Agelaius tricolor*); (c) raptors that forage in wetlands or agricultural fields, such as the rare Swainson's hawk (*Buteo swainsoni*) and northern harrier (*Circus cyaneus*); and (d) diverse upland species that use grasslands and other habitats when not flooded (Dybala et al. 2015). Tropical migrant songbirds are found in the limited habitats with trees and bushes along ditches and sloughs, in spring and fall.

Despite the high diversity of birds on the Bypass, most attention is paid to the migratory waterfowl (ducks, geese, and swans) and shorebirds, due to their high visibility, their impressively large flocks, and popularity with hunters and birdwatchers. Large numbers of shorebirds start arriving in the Bypass in August and continue through the winter. Migration patterns are complicated by the fact that some species overwinter in the Bypass and Central Valley, while others are moving through to more distant wetlands in Mexico or South America. Species such as cinnamon teal (*Anas cyanoptera*) and gadwall

(*A. strepera*) will forage in the Bypass for a few weeks to gain "fuel" for their extended migration further south. Others, like northern pintail (*A. acuta*), white-fronted geese (*Anser albifrons*), and snowy plover (*Charadrius nivosus*), spend the winter in Central Valley, often with complicated local movements related to foraging and avoiding hunters. Each species has its own distinct seasonal and local patterns of movement and so areas within the Bypass that provide prime habitat, such as the YBWA, offer a kaleidoscope of species during the peak migration period, November through March.

The migratory birds all share a common need for flooded land, and readily use habitats inundated by both natural and managed flooding. Within flooded habitats, the primary food for the migratory birds—seeds and invertebrates—is generally found on the bottom or close to it, and the birds have a range of behaviors for accessing the food. For example, different species have different optimal foraging depths (Suddeth 2014; Grimm and Lund 2016). Shorebirds (aka wading birds) require shallow (5–8 cm) areas such as mudflats for foraging, with preferred depth varying with each species' length of bills and legs. Dabbling ducks such as northern pintail, various teal, and mallards (*A. platyrhynchos*) are largely unable to forage in water greater than 45 cm deep and prefer water less than 25 cm deep (Taft et al. 2002; Petrik et al. 2012). The complex topography of natural floodplains historically provided a diversity of depths that supported a diversity of species' preferred foraging habitat. Managed floodplains, such as the Bypass, generally have less complex topography and so may require manipulation of the water supply or creation of flooded areas of appropriate depths to maintain a diversity of birds (Suddeth 2014; Grimm and Lund 2016).

The timing and duration of flooding also affect the growth and availability of plants preferred by waterbirds for food, such as swamp timothy and water grass. These species are often planted in fall or winter by duck hunting clubs to encourage waterfowl to forage in accessible places. Flooded rice fields also can serve this purpose. Within the Bypass, agricultural fields in general are important for ducks, providing nearly 80% of their food energy. In addition to seeds and vegetation, agricultural fields support high densities of invertebrates that are available to ducks after the plant material has been depleted (Suddeth 2014; Grimm and Lund 2016).

Mammals

The mammals of the Yolo Bypass have not been well studied. Truan et al. (2010) found 34 species of mammals along lower Putah Creek near its entry to the Bypass, and presumably most of these species also occur in the Bypass, especially within the narrow corridors of trees along the Toe Drain or creek entry points. Similar to other floodplains, populations of small mammals likely build up during long periods of low or no flooding but then crash when flooding drives them from their burrows. Common aquatic mammals include river otter (*Lontra canadensis*), beaver (*Castor canadensis*), and nonnative muskrat (*Ondatra zibethicus*). River otter in the Bypass and nearby areas feed on crayfish (mainly the alien *Procambarus clarkii*), fish, and, occasionally, ducks (P. Moyle, unpublished data). In summer, the most conspicuous mammals, at least at dusk, is the Mexican free-tailed bat (*Tadarida brasiliensis*), which has a nursery colony of over 250,000 roosting under the Yolo Causeway (Taylor 2013). These bats emerge at dusk to feed on the abundant insects produced in the Yolo Bypass.

Endangered Species

The US Endangered Species Act of 1973 is perhaps the most powerful piece of environmental legislation in the world, because it is unequivocal about the obligation to keep species from going extinct and the need to protect habitat for species listed as Threatened or Endangered. California's Endangered Species Act is less stringent in its requirements but still quite powerful in its effects. Therefore, the welfare of species

TABLE 13.1
Species that use Central Valley floodplain habitats with state or federal status
E = ENDANGERED; T = THREATENED; C = CANDIDATE FOR LISTING; SSC = CALIFORNIA SPECIES OF SPECIAL CONCERN

Species	Latin name	Federal status	State status
Greater sandhill crane	*Grus canadensis tabida*	–	T
Least Bell's vireo	*Vireo bellii pusillus*	E	E
Swainson's hawk	*Buteo swainsoni*	–	T
Tricolored blackbird	*Agelaius tricolor*	–	SSC
Western yellow-billed cuckoo	*Coccyzus americanus occidentalis*	C	E
Willow flycatcher	*Empidonax traillii*		
Yellow-breasted chat	*Icteria virens*	–	SSC
Chinook salmon, fall run	*Oncorhynchus tshawytscha*	–	SSC
Chinook salmon, spring run	*O. tshawytscha*	T	T
Chinook salmon, winter run	*O. tshawytscha*	E	E
Central Valley steelhead	*O. mykiss*	T	
Delta smelt	*Hypomesus transpacificus*	T	E
Southern green sturgeon	*Acipenser medirostris*	T	T
White sturgeon	*A. transmontanus*		SSC
Sacramento splittail	*Pogonichthys macrolepidotus*		SSC
California red-legged frog	*Rana aurora draytonii*	T	SSC
California tiger salamander	*Ambystoma californiense*		
Giant garter snake	*Thamnophis gigas*	T	T
Western pond turtle	*Actinemys marmorata*	–	SSC
Valley elderberry longhorn beetle	*Desmocerus californicus dimorphus*	T	–
Riparian brush rabbit	*Sylvilagus bachmani riparius*	E	E
Riparian woodrat	*Neotoma fuscipes riparia*	E	SSC

that are formally listed under the state and federal acts has to be taken into account by managers of the Yolo Bypass (table 13.1). In recent years, the most attention has been paid to the listed fishes, especially winter-run and spring-run Chinook salmon, both listed under state and federal Endangered Species Acts. Juveniles of the two runs use the flooded Bypass at least in some years, while adults occasionally get trapped when they move up the dead-end Toe Drain rather than the Sacramento River. Both adults and juveniles of Central Valley steelhead and southern green sturgeon, both listed as Threatened, are found in the Bypass on occasion, but it does not appear to be important habitat for either of them, although if even a few become stranded and die, there is great concern. The Bypass is probably the most important floodplain area for spawning Sacramento splittail, an endemic species of special concern.

Other species (table 13.1) are part of a habitat conservation planning effort by the Yolo Habitat Conservancy (Yolo Habitat Conservancy 2016). Of those species, the giant garter snake is probably of most concern in the Bypass because it lives along ditches, foraging on alien frogs and fish. Flooding may force individuals to evacuate underground dens where they overwinter, but how different flood regimes might affect this aquatic snake is not known.

Balancing Conflicting Uses of the Bypass

The overriding purpose of the Yolo Bypass is flood management, which means it lacks permanent buildings and its land cover is managed to maintain low hydraulic roughness and rapid draining. The farming of annual crops—such as rice, wild rice, tomatoes, and corn—maintains low roughness and, thus, annual crops are very compatible with flood management. Without farming (or some other form of active vegetation management), the Bypass would likely become a mosaic of floodplain forest and wetlands, as shown by the history of the Cosumnes floodplain. Although agriculture is often equated with loss of natural ecosystems, farming the Bypass actually results in high-quality habitat for migratory waterfowl, which use harvested fields that are inundated during winter. Due to high densities of waterfowl, hunters created duck hunting clubs in the Bypass, mostly on agricultural lands, and the precedent of managing the Bypass for both agriculture and waterfowl ultimately lead to creation of the Yolo Bypass Wildlife Area, which allows some hunting (Suddeth 2014).

More recently, the potential importance of the flooded Bypass to fishes has been recognized, as has its role in providing habitat for a range of other endangered species. While these diverse uses can coexist, management actions that are beneficial for some resources or values can be detrimental to others—such as the managed increase in the extent or duration of flooding—setting up the potential for conflict among various managers, users, and stakeholders (Salcido 2012). To illustrate, various parties have specific preferences for the timing of flooding and draining:

a. Farmers prefer to have their fields drain as quickly and as early as possible (preferably by early March), so there is plenty of time for fields to dry before tilling and planting. Delayed planting can postpone harvest until autumn when early rains can interfere with harvest. In general, erosion and deposition associated with flooding increases maintenance time and costs for farmers. Thus, from an agricultural standpoint, farmers presumably prefer minimizing flooding.

b. Native fishes, especially salmon, benefit from having the floodplain inundated as long as possible, ideally from January through April, after which temperatures become too warm. Long inundation creates high residence time of the water and high food production.

c. Dabbling ducks generally prefer less water than is ideal for fish, in order to maximize the extent of relatively shallow flooded areas that allow easy feeding, and the period that ducks benefit from a flooded bypass extends into the period that farmers prefer the Bypass to be dry. As dabbling ducks are highly valued for hunting, hunters share the ducks' preferences.

Adding complexity, the flood-management agencies must maintain vegetation with very low roughness to maintain conveyance and thus discourage land cover that impedes flood flows. While this need is compatible with farming, it can conflict with management objectives to maintain diverse vegetation for fish and wildlife.

Two additional issues that may affect management in the future are mosquito control and terrestrial species that are listed as Threatened, Endangered, or of special concern. Areas of standing water in the Bypass allow mosquitoes to breed in large numbers. Recently, managers have become increasingly focused on mosquitoes as vectors of West Nile Virus, which is fatal to many birds and some humans. The Bypass is part of Yolo County, which is currently developing a Habitat Conservation Plan and a Natural Communities Conservation Plan to ensure

TABLE 13.2
Summary of potential operations for a gated notch in the Fremont Weir

	Dec 1–Feb 15	Feb 16–28	March 1–23	Mar 24–April 10	April 11–May 15
Current % of years with Fremont Weir overflow	61	50	47	22	17
Potential % years with modified weir	69–89	67–75	72–81	61–67	19
Proposed volume (ft³/s)	Up to 6000	Up to 6 000	Up to 6000	Up to 6000*	Up to 6000*
Targeted flood extent (acres)	17,000	17,000	7000–10,000	7000–10,000	7000–10,000
Proposed duration	30–45 days or longer	30–45 days or longer	30 days	30 days	30 days
Targeted species for floodplain habitat (does not include passage)	Winter-run and spring-run Chinook salmon, Sacramento splittail	Fall-, winter-, and spring-run Chinook salmon, Sacramento splittail	Fall, spring, and Butte Creek spring-run Chinook salmon, and steelhead		Late fall-run Chinook salmon and steelhead

SOURCE: Based on Grimm and Lund (2016).
*Only in years with natural overflow.

compliance with federal and state endangered species acts, respectively (Yolo Habitat Conservancy 2016). The planning effort focuses on habitat for a range of terrestrial endangered species; although the planning is focused on areas outside the Bypass, some species, such as the giant garter snake and western pond turtle, are common in the Bypass. Plans and decisions for the Bypass will likely need to consider the potential for management actions for flooding and waterfowl to affect endangered species.

Managers are actively seeking solutions to balance these diverse uses of the Bypass. Potential conflicts over the timing, duration, and depth of flooding between fish, birds, and farmers illustrate these challenges. Grimm and Lund (2016) suggest that pursuing a reconciliation approach to floodplain ecosystems and management, rather than focusing on classic restoration, allows managers greater flexibility for finding balanced solutions within highly altered systems like the Yolo Bypass. To guide managers in finding such solutions, Grimm and Lund (2016) developed a multi-objective systems model with various options, quantified performance of those options across various objectives, and illustrated trade-offs. The model focuses on February–April, when most flooding occurs naturally and when most birds and fishes are likely to be using the Bypass. In addition to the current conditions, the model also considers potential future changes to management. For example, the addition of a gated notch in the Fremont Weir would give managers the ability to direct flows into the Bypass at times when the river stage was below the weir, increasing the frequency of flooding in the Bypass. Further, managers could control characteristics of this intentional flooding, such as timing and duration, to maximize benefits for native fish (table 13.2).

> **BOX 13.1** · Experimental Demonstration of Flooded Rice Fields to Rear Salmon
>
> Rice is an important summer crop on the Yolo Bypass. Because rice grows in standing water, areas farmed for rice are organized into paddies that can also be intentionally flooded after the growing season to attract migratory waterfowl. During natural inundation of the Bypass, rice fields provide rearing habitat for juvenile salmon, carried with flood waters onto the floodplain. These juvenile salmon utilize a range of natural and farmed habitats, and the fact that salmon have high growth rates in rice fields provides a clear illustration of reconciliation ecology.
>
> Katz (2015) used experimental treatments to explore how rice farming practices affect winter rearing habitat for juvenile Chinook salmon. Nine experimental rice fields, 0.08 ha each, were built in the Bypass and randomly assigned each field to one of three postharvest farming options: fallow, rice stubble, or tilled. Each field was stocked with 4500 juvenile salmon of hatchery origin and the fish were allowed to rear in the fields for 6 weeks, simulating the optimal flooding duration for multiple purposes as described in Suddeth (2014). After 6 weeks, the fields were drained and salmon were counted, measured, and released. Despite high densities of diverse zooplankters observed in all field treatments, salmon fed almost exclusively on the cladoceran *Daphnia magna*. Salmon growth (0.93 mm/day) and body condition were high across all treatments and, averaged across all treatments for the 41 day study, were the highest ever recorded in the Central Valley. This indicates that postharvest field treatment can be tailored to the needs of a farmer's operation, without affecting the ability of a field to provide productive rearing habitat for Chinook salmon and illustrates the potential to reconcile floodplain management for salmon with the needs of farmers in an intensely managed landscape.

(c) habitat requirements of waterfowl and fish; and (d) revenue and costs from different types of agricultural land use. The model illustrates the trade-offs between different flooding scenarios and/or management interventions. For example, the model demonstrates that, in dry years, just 3 weeks of flooding in January or February could significantly benefit fish and birds at no cost to farmers. Allowing longer flooding provides additional benefits to fish and birds, but also imposes costs on farmers.

The model also indicates that economic incentives for farmers to grow rice or wild rice would benefit fish and wildlife, because these crops function as de facto wetlands (box 13.1). Converting Bypass land currently in low-value agriculture (e.g., corn or pasture) to rice could have a minimum impact on farm income while providing large benefits to ducks and fish. Much of the low-value agricultural land cover is within the southern end of the Bypass, and this location makes it particularly valuable to be managed for annual flooding. This portion of the Bypass is adjacent to extensive tidal slough habitat, allowing easy access by spawning splittail, as well as promoting exit of juvenile salmon from the Bypass (Grimm and Lund 2016). Another potential management option would be to use managed flooding through a gated notch on Fremont Weir to create "miniature" seasonal floodplains along the Toe Drain (the east side of the Bypass), inundated at times that would most benefit native fish and birds. These possibilities underscore that the Bypass is not a uniform landscape; specific management interventions can be tailored to support the diverse land uses and habitats that occur in specific portions of the Bypass (figure 13.6). Indeed, active management that varies across the floodplain, calibrated to specific land uses and habitat types, will likely be a prominent feature of managing floodplains as reconciled landscapes.

Grimm and Lund's (2016) multi-objective model draws on a range of inputs including (a) hydrological and hydraulic data, such as timing, depth, and extent of flooding; (b) land use, including area devoted to different crops;

Conclusions

Although the Yolo Bypass occupies a portion of the historic floodplain and, in some years,

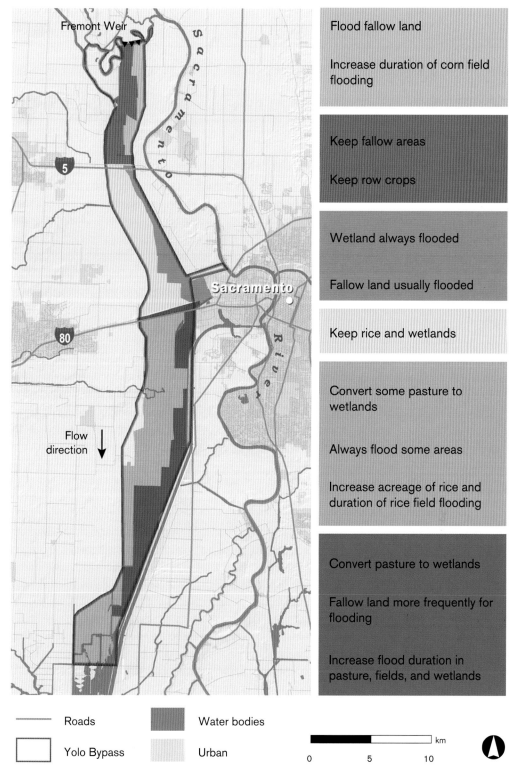

FIGURE 13.6 Suggested management interventions in subregions of the Yolo Bypass based on modeling results of Grimm and Lund (2016). Floods flow from Fremont Weir in the north to a drainage canal along the eastern edge and then to wetlands in the south, where ambient temperatures are cooler. Due to these differences in hydrology and temperature, the eastern and southern portions of the Bypass are less valuable for agriculture and the modeling results suggest that these areas have the most potential for providing the ecological benefits associated with floodplains (after Suddeth 2014; Gesch et al. 2002; CalAtlas 2012; Sacramento Area Council of Governments 2016).

experiences flooding that approximate historic conditions, there is very little that is natural about the Bypass. It is a novel floodplain ecosystem. Its surface is flattened and graded to drain rapidly, crisscrossed with ditches and levees to support farming and waterfowl operations. The timing and extent of flooding are limited by weirs (low walls) that prevent connectivity below a specific threshold of river flows, thus reducing the range and type of floods that can inundate the Bypass.

With these highly engineered or managed characteristics, it may seem quite surprising that the Bypass retains so much value as a floodplain ecosystem. Its value for waterfowl was long appreciated, as birds come to the Bypass in huge and highly visible flocks, prompting duck hunters to develop hunting clubs and, along with conservationists and agencies, to establish the YBWA. Because they were swimming literally beneath the surface and beyond easy observation, native fishes in the Bypass escaped equivalent recognition. Now, however, the Bypass receives a great deal of attention as the best remaining lowland floodplain habitat for native fishes. Its recognition as a valuable rearing area for juvenile salmon—a partial stand-in for the vast natural floodplains that no doubt helped maintain huge historical salmon runs—may spark a revolution in salmon management in the Central Valley. Floodplain rearing can increase the survivorship and size of smolts that enter the Pacific Ocean, ultimately improving returns to fisheries and ecosystems. Active management of flow, habitat, or both, in the Bypass and elsewhere in the Central Valley, can potentially maximize the value of floodplains to salmon populations. Although management to promote salmon can conflict with other uses of the Bypass—such as flood control, farming, and waterfowl—the same collaborative spirit that drove establishment of the YBWA can also catalyze the search for balanced solutions for these various values. A collaborative process is underway now to find these solutions. The Bypass can serve as a global model of a floodplain that is simultaneously managed for flood-risk reduction and a diverse range of ecosystem services, a reconciled floodplain system.

SACRAMENTO RIVER: PLANNING FOR RECONCILIATION

The Sacramento River, at 719 km long and draining 71,000 km^2, is California's longest and largest river. It is bounded by mountain ranges on the east (Californian Coast Ranges), west (Sierra Nevada), and north (Cascades), flowing from headwaters in the very northernmost part of California and draining through the Sacramento-San Joaquin Delta into San Francisco Bay. The Sacramento is a major source of water for agriculture, industry and municipal use and thus it has extremely high economic and social importance to California—a state that would have the eighth largest economy in the world if it were its own country. Providing this water as an engine of economic growth required significant alteration to the river's flow regime, channel, and floodplains. The river's historic conditions, changes, and current conditions are described in chapter 12. Here we discuss current plans for a reconciled floodplain-management system along the main channel.

Infrastructure

Water in California is intensely managed through a vast plumbing system, and storing and moving water from places with high water runoff to places with high water demands has necessitated a dramatic change to the state's rivers and their flow patterns. Early in the twentieth century, two major water engineering projects were initiated: The Central Valley Project (CVP) by the federal government and the California State Water Project (SWP) by the state government. Through these projects, and other smaller projects, the Sacramento River and its tributaries have been dammed, leveed, and otherwise altered.

Levees and other bank restraints were built on the Sacramento River beginning in the late

1800s, partly in response to the aggradation of the river as millions of tons of sediment were flushed to the system from hydraulic gold mining (chapter 12). The uppermost reach of the river, the first 300 km known locally as the "gorge reach," has few levees because the channel has a considerably steeper gradient than the rest of the river and is confined by steep sides; floodplains are much less expansive than in lower reaches. Below the gorge reach, much of the channel is lined with levees and other bank restraints (e.g., riprap) especially on sections of river that historically had extensive floodplains. The lower 390 km of the Sacramento runs downstream from the town of Red Bluff and is divided into two main reaches. The first runs from the town of Red Bluff to the town of Colusa (160 km) and the second starts at Colusa and ends at the Delta (230 km). The Red Bluff-Colusa reach has discontinuous levees and other bank restraints and the river is free to migrate laterally in some areas. The Colusa-Delta reach features continuous levees on both sides of the river, directly adjacent to the banks, preventing lateral movement.

Current Restoration

Numerous restoration projects have occurred or are being planned on the Sacramento River, including large-scale riparian restoration and levee setback projects. TNC and several conservation partners formed the Sacramento River Project in 1988 to pursue large-scale, process-based restoration of riparian and floodplain habitats of the Sacramento River (Golet et al. 2006, 2008; Larsen 2007). To date, the project has conserved approximately 5400 ha of riparian habitat along the Sacramento River, between the towns of Colusa and Red Bluff. Primary strategies include conservation of flood-prone land through acquisition or easement, active riparian restoration (i.e., planting), and restoration of natural river processes (Golet et al. 2008). Initial results suggested that, due to the altered hydrology of the Sacramento River, irrigation was necessary for successful riparian restoration (Alpert et al. 1999). Golet et al. (2008) reported that restored riparian sites supported a broad range of fauna, including birds, bats, and insects.

Planning for Floodplain Reconciliation

Drawing on lessons from existing projects, including levee setbacks and the multipurpose benefits of the Yolo Bypass, the CDWR has proposed a plan to expand the extent of connected floodplains along the Sacramento River (CDWR 2014), focused on balancing flood-risk management with ecological benefits. The plan seeks to identify areas with the greatest potential for ecological restoration of floodplains within an overall flood-management strategy.

The plan evaluates potential for reconciliation through two types of analyses: (1) floodplain inundation potential (FIP) and (2) the potential meander zone of the river. The FIP analysis identified floodplain areas that could be inundated by specific recurrence interval flood flows, considering both areas currently connected to the river and areas disconnected from the river by natural or built levees or other flow obstructions. This analysis drew on concepts of the "Frequently Activated Floodplain" (Williams et al. 2009, and discussed in detail in chapter 14). The FIP mapped areas of potential inundation, and the depth and duration of inundation, for three levels of flow (67%, 50%, and 10% exceedance probabilities). Each analysis included a specific water surface elevation and an associated season (e.g., spring) and duration (e.g., 7 days). For example, the "67% chance sustained spring FIP" refers to inundation of a floodplain area with 1 ft or more of water, during spring, from a flood level that has a 67% chance of occurring in any given year (i.e., two out of three chance) and is sustained for at least 7 days.

The meander zone analysis defined a "natural meander zone" that strives to capture potential future movements of the river, accounting for natural restraints such as geology. Together, these analyses showed where interventions could provide the most benefits in terms of

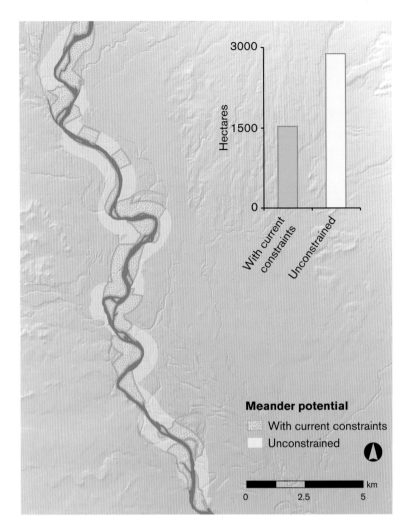

FIGURE 13.7 Currently constrained compared with unconstrained meander potential on the Sacramento River from river mile 201 to 222. "Current constraints" include levees, riprap, roads, and structures (Gesch et al. 2002; CDWR 2012; Sacramento River Forum 2016).

promoting or allowing natural geomorphic processes to occur. The plan assessed three types of potential floodplain interventions: (1) lowering or otherwise modifying the floodplain elevation to achieve longer periods of inundation; (2) eliminating flow restrictions, such as levees; and (3) connecting side channels during a greater range of flows. These actions can all increase flood capacity *and* restore floodplain habitat, including wetlands, oxbow lakes, and riparian forests. CDWR planners also considered the need to create additional resting and rearing habitat for salmon and steelhead within side channels and other floodplain features.

The estimated natural meander zone (figure 13.7) was used to establish potential levee setback areas within at least three active channel widths (Larsen, Fremier and Girvetz 2006) from the current channel edge. The evaluation of candidate sites for new setback levees focused on areas of higher ground, which would require lower levee heights. Further considerations for evaluating levee setback sites included (1) minimizing the length of the new setback levee to reduce maintenance costs, (2) prioritizing sites in upper reaches that could more effectively attenuate a flood wave before it reaches the more confined reaches downstream, (3) prioritizing sites adjacent to existing areas of wide levee separation to reduce bottlenecks in the levee alignments, and (4) considering sites' potential for habitat benefits and groundwater recharge.

The inundation potential and potential meander zone analyses were used to identify the most optimal areas for floodplain modification and levee setbacks. The Upper Sacramento was shown to have significantly more potential for floodplain modification (861 ha) than the Lower Sacramento (6 ha). Similarly, the upper river was determined to have 4092 ha of floodplain that could be reconnected with setback levees, while the lower river had 1688 ha.

The goals of this plan focus on attaining a flood-management system that also promotes key physical processes and ecological functions required to yield an improved ecosystem. In short, the plan seeks to reconcile functioning floodplains within the highly altered Sacramento River system. This plan illustrates how quantitative methods can be used to evaluate potential for reconciliation of floodplains that have been leveed and farmed for almost 150 years.

The planning process included several approaches to make the results of the plan more useful to decision makers and accessible to various stakeholders. First, the plan focused on identifying general areas for implementing actions rather than proposing specific actions. The plan used technically defensible methods and data but created simple end products. California is currently planning massive investments in its infrastructure and water-management systems to deal with drought, flood risk, climate change, and ecosystem restoration. This plan informs those decisions with a clear road map of where interventions could improve both flood risk and ecosystem services, such as habitat for fish and wildlife and groundwater recharge.

CHAPTER CONCLUSIONS

California has changed dramatically and continuously since the Gold Rush in the 1850s. As described in the next chapter, California's Central Valley will continue to face new challenges due to population growth, climate change, and other stressors on flood- and water-management systems. Floodplain reconciliation can help address many of these challenges. The Cosumnes River Preserve (CRP) floodplain, although still a novel ecosystem, provides the closest representation of a natural floodplain system that can inform management to benefit native species and ecosystems. The Yolo Bypass has evolved to become a highly managed floodplain that serves multiple purposes, including some—such as farming and fish—that were once regarded as being largely incompatible with each other. In many respects, the Bypass is a grand experiment in reconciliation ecology. Plans for the Sacramento River may advance this experiment further, exploring how an entire highly developed river system can include floodplain ecosystems as part of its flood-management system.

FOURTEEN

Reconciling Central Valley Floodplains

FLOODPLAINS OF CALIFORNIA'S central valley (the Central Valley) have changed dramatically over the past 150 years and will continue to change over the next century. In this chapter, we first discuss current flood risks that exist in the Central Valley, followed by a discussion of how risk is likely to increase with climate change. We then describe the potential benefits of large-scale floodplain reconciliation, much of it underway, followed by a conceptual model of floodplain ecology and management, to inform reconciliation efforts. The review of benefits and conceptual model also serve as a synthesis of the main theme of this book: how connections between river and floodplain and a range of biophysical processes produce multiple benefits for society.

FLOOD RISK

The human population of the Central Valley is growing rapidly and much of this population growth occurs in "deep floodplains," defined as areas where flooding following a levee breach would be 1 meter or higher in depth. As of 2007, approximately 1.8 million people in the Central Valley lived in areas protected by levees and this population is projected to increase to more than 2.3 million by 2020 (Galloway et al. 2007). The conversion of farmland to housing on deep floodplains greatly increases the number of people and the value of structures at risk from large floods (CDWR 2005).

Even without a growing human population, the Central Valley's geography, topography, climate, and land-use patterns all contribute to high flood risk. The Valley is surrounded by high, steep mountains that are relatively close to the ocean and thus intercept moisture-laden storms. Storm runoff, particularly from rain-driven or rain-on-snow events, drains quickly from the channel network to a broad, flat valley. The Valley's flood risk is rising due to a combination of population growth into flood-prone areas and climate change, and is exacerbated by levee design deficiencies and deterioration (CDWR 2005). Further, even absent trends that are increasing risk, current expectations of flood risk are based on calculations of the

100 year flood (1% exceedance probability) and these calculations are derived from relatively short records of hydrological data. As a result, the calculated magnitude of the 100 year event often rises following a new flood (Mount 1995). Despite the challenges in accurately calculating the 100 year flood, the designation of the "100 year floodplain" has great significance for planning, investment, and land-use patterns. Investments in flood protection are often designed to be just sufficient to offer 100 year flood protection. While investments to that level allow property owners to avoid carrying flood insurance, they do not provide protection from larger floods and, as described above, the 100 year floodplain is an uncertain and potentially shifting geographic reality (Davis 2007).

Levee maintenance in the Central Valley has been chronically underfunded (Leavenworth 2004), with 200 levee sections waiting for repair of erosion as of 2005 (Weiser 2005). A 2003 survey by the US Army Corps of Engineers found that 143 of the 1704 km of levees in the Sacramento River Control Project required repair, with an estimated cost of US$145 million (CDWR 2005). The California Department of Water Resources (CDWR) reported in a 2011 survey that half of 480 km of urban levees did not meet engineering design standards and 60% of rural levees had high potential for failure (CDWR 2011). Maintenance challenges in California reflect a broader trend in the United States where much of the flood-management infrastructure is aging and in need of repair or replacement, including an estimated US$12.5 billion for dams and US$50 billion for levees (American Society of Civil Engineers 2009).

The design and location of Central Valley levees add to the difficulty of their maintenance. Levees were initially built close to the channel in order to flush excess sediment from hydraulic mining out of the channel network. With much of the excess sediment now removed, levees in close proximity to the river are themselves vulnerable to erosion during high flows (CDWR 2005), which can lead to levee failure.

CLIMATE CHANGE

Climate change is predicted to alter the hydrology of the Central Valley in ways that could affect its river-floodplain ecosystems and increase flood risk for people. While models have a wide range of predictions about how climate change will influence total annual precipitation, the models agree that air temperatures will increase, resulting in higher water temperatures in rivers and floodplain wetlands, higher evapotranspiration rates, and an increase in the proportion of precipitation that falls as rain compared to snow in the Sierra Nevada and Cascades. Because atmospheric holding capacity for water vapor increases with air temperature, warmer air often produces more intense precipitation (Trenberth et al. 2003). Together, these factors indicate that climate change will result in an increase in flood magnitudes and frequencies in the Central Valley because of an increased frequency of storms, more intense precipitation, and increasing proportion of precipitation falling as rain. Models of Dettinger et al. (2011) predict an increase in flood magnitudes, with some models predicting that the largest floods will increase by up to 150% of historical magnitudes by the end of the twenty-first century. The models show that flood magnitudes will increase even under overall drier conditions, due to an intensification of the largest storms. Yoon et al. (2015) forecast that both intense droughts and severe floods will increase in frequency by 50% by the end of the century.

RESTORATION AND RECONCILIATION OF CENTRAL VALLEY FLOODPLAINS

Although the geographic extent of hydrologically connected floodplains in the Central Valley has declined dramatically in the past century, interest in, and funding for, floodplain restoration is increasing. While promoting floodplains' environmental benefits is clearly a primary objective driving these efforts, floodplain restoration can be integrated into broader water-management objectives and programs, under a reconciliation approach (chapter 10). As we emphasize in this book, floodplain reconciliation is a particularly useful approach for work in the Central Valley because of the need to integrate novel ecosystems, active environmental management, and natural ecosystem processes and services within a complex system of water and land management. As described in chapter 13, the ecosystem benefits from a reconciliation approach are diverse and large. In the next section, we briefly review how improving function of floodplain ecosystems can be linked to flood-risk management and other water-management objectives. Creating these linkages can to lead to a much greater extent of ecologically functioning floodplains than could be accomplished through environmental programs alone.

LINKING ECOSYSTEM BENEFITS TO OTHER WATER-MANAGEMENT OBJECTIVES

As described in chapter 10, a promising strategy for achieving floodplain reconciliation on a large scale is to integrate floodplain ecosystems into flood-management systems. This means integrating green infrastructure into a system of engineered infrastructure. California's flood-management system already relies on a mix of gray and green infrastructure. Thus, the Yolo and Sutter Bypasses are critical components of the system that combine gray (levees and weirs) with green (floodplain conveyance) infrastructure.

In addition to the system of bypasses, implemented nearly a century ago, new solutions that accomplish floodplain reconciliation through the integration of green and gray infrastructure are being proposed, such as planning for new management of the Yolo Bypass or levee setbacks, described in chapter 13. Recently, levee setbacks that have been implemented along the Bear, Feather, and Sacramento Rivers demonstrate these new solutions. The Bear River had levees set close to the channel, including at its confluence with the Feather River. When stage in both rivers was high, flood waters would back up into the Bear River channel, causing extensive flooding. Engineers determined that the best solution to reduce this backwater flooding was a levee setback project. The new setback levee is 3.2 km long and restores a large expanse of floodplain habitat (figure 10.3). The increased conveyance provided by the setback levee is projected to lower flood stages by 1 m during major floods, reducing flood risk along the lower Bear River. Because long-duration and frequent flooding can be particularly important for river-floodplain ecosystems and native fishes, a low wetland feature was added to ensure that portions of the reconnected floodplain experienced this type of flooding (Williams et al. 2009).

The Central Valley Flood Protection Plan (CDWR 2010) outlines a hybrid approach to flood-risk management, with improved gray infrastructure complemented by increased floodplain conveyance. The plan calls for increasing protection levels for urban areas up to the 200 year flood level (0.5% annual probability) and also describes potential expansion of existing bypasses (Sutter and Yolo) and construction of new bypasses on the Feather River and the lower San Joaquin River.

Floodplain restoration can be integrated into both levee setbacks and flood bypasses. Williams et al. (2009) describe how ecological restoration goals can be incorporated into flood projects, and a proposed setback levee on the Sacramento River at Hamilton City includes reconnection of 1200 ha of floodplain to the river. The combination of ecological and flood-risk objectives was critical for demonstrating the cost effectiveness of this project and gaining approval from the US Army Corps of Engineers (Opperman et al. 2010). Because improved conveyance is a key feature of flood management, restoration objectives and flood-management objectives need to be reconciled. For example, portions of the restored Bear River levee setback area will be maintained as grassland with low hydraulic roughness to maintain the desired conveyance levels (Williams et al. 2009).

> **BOX 14.1** · Linking Floodplain Management with Other Water-Management Objectives
>
> The Yolo Bypass illustrates the role that floodplains can play within an integrated water-management system (chapter 13). Although the Bypass was built decades before the present system of multipurpose reservoirs in the Sacramento Valley, it effectively illustrates how a geographically large, hydrologically connected floodplain can increase flexibility of multipurpose reservoirs. By conveying 80% of the volume of flood waters during major floods, the Bypass floodplain relieves pressure on levees and substitutes for billions of cubic meters of reservoir flood storage. Opperman et al. (2011) present an analysis of the role the Bypass played in 1986, when a 50–80 year recurrence interval flood occurred in the Sacramento River. During the 3 day peak of the flood (February 19–21) 3.3 billion cubic meter (bcm) flowed through the Bypass; this volume is approximately the same as the combined flood-control storage of the six largest reservoirs in the Sacramento River basin. During this peak, the multipurpose reservoirs' flood storage volume was nearly at capacity and the Sacramento River was essentially flowing at its channel capacity; higher flows would have led to levee overtopping. Thus, the 3.3 bcm that flowed through the Bypass during the peak 3 days could not have been stored or conveyed by any other element of the existing flood-control system. The "green infrastructure" of the Bypass floodplain provided a service that could not have been provided by the existing engineered infrastructure.
>
> Managing the volume of water associated with the 1986 flood without the Bypass floodplain, let alone a 100 or 200 year event, would require some combination of (a) additional flood-control infrastructure, including new or enlarged dams and extensive raising of levee height to increase channel capacity, and (b) considerable reallocation of existing reservoir storage from water supply to flood control. The first would require significant funding in a state where maintenance of the current levee system is chronically underfunded (Leavenworth 2004). The second would impinge upon California's already strained water supply resources. Thus, the hydrologically connected floodplain of the Yolo Bypass provides considerable economic value in the form of avoided costs for additional infrastructure and maintenance of the operational integrity and flexibility of the current system of water storage and allocation.
>
> In another example, an analysis of the nearby Mokelumne River shows the advantages of reducing flood storage in a major reservoir to 50% of current levels and adapting downstream agricultural land uses (e.g., crop selection) to be compatible with increased inundation frequencies. Together, these actions would increase reliability of water supplies in part by reducing the number of times the reservoir reached levels that triggered drought management. The changes to reservoir allocation would also result in a slight increase in hydropower revenue and a significant increase in the reservoir's ability to provide environmental flows for salmon. In total, the increased benefits outweigh the costs of adapting floodplain land use by several million dollars per year (Opperman, Hartmann, and Harirson 2015).

Proposals to reconnect floodplains, for either ecological restoration or flood-management objectives (or both), can lead to concerns from rural communities that loss of farmland will threaten the vitality of the agricultural economy. However, reconnected land doesn't necessarily mean a loss of private land or a cessation of revenue to landowners or local governments. Reconnected floodplain lands can continue to generate revenue through markets for ecosystem services or through flood-compatible agriculture (Opperman et al. 2009). Through these mechanisms, large-scale reconnection of floodplains can be consistent with a vibrant agricultural economy.

As a component of flood-risk-management projects, floodplain restoration can also have value for water-management objectives for water supply. Within multipurpose reservoirs, the need to leave reservation space to capture potential floods can compete with water storage for water supply and irrigation (Das et al. 2011). Within California, climate change will exacerbate this competition. California's large multipurpose reservoirs rely on refilling during April to July as the snowpack melts and, due to minimal risk of flooding, reservoir levels can be safely raised higher. Dettinger et al. (2011) predict that, under a warming climate, reduced snowpack could lead to a 14% decline in April-July flows for the Merced River, with a 7% reduction for the American River. Because of the interrelationship between flood management and water supply, the 2005 California Water Plan Update called for integrated flood management as one of the tools for improving water supply (see box 14.1).

Finally, by increasing populations of species with regulatory protection, floodplain reconciliation can also yield direct benefits to water-management systems in terms of lower regulatory conflicts and increased certainty and management flexibility. Because of this, water-management agencies have often funded ecosystem restoration projects in recent decades. Several species with regulatory protection benefit from floodplain reconciliation, including Chinook salmon, valley elderberry longhorn beetle, and giant garter snake (see table 13.1).

CONCEPTUAL MODEL OF CENTRAL VALLEY FLOODPLAINS

The first part of this chapter describes a range of benefits that can be achieved through floodplain reconciliation, such as promoting biodiversity and fisheries, improving flood-risk management, and increasing the overall flexibility of the water-management system. Through floodplain reconciliation, scientists and managers confront the challenge of designing projects and systems that can accomplish a specific management objective, such as flood-risk reduction, while also promoting desirable species and ecological processes. The engineering requirements of flood management are beyond the scope of this book, so here we provide a conceptual model of how physical processes interact with ecosystems to produce desired outputs (ecological benefits) at scales relevant for management (figure 14.1). Opperman (2012) presents a more extensive set of conceptual models for this purpose and a list of mechanistic tools and models that can be used to study or compare management or restoration alternatives.

Floodplains that can produce the various benefits described above require three primary characteristics: (1) hydrological connectivity; (2) a flow regime with a range of flow levels, including high flows that exceed bankfull; and (3) sufficiently large geographic area for benefits to accrue to a meaningful level.

Connectivity

The main prerequisite for an ecologically functional floodplain is hydrological connectivity between the river and floodplain (Amoros 1991; Tockner and Stanford 2002). Connectivity drives all hydrologic and geomorphic processes on the floodplain and strongly influences ecological processes. Connectivity can be achieved through multiple pathways including lateral overflow as river stage rises into a floodplain that fringes the main channel, through breaks in natural or constructed levees, and through sloughs or side channels into a flanking flood basin. Water on the floodplain can then perform geomorphic work (erosion and deposition), facilitate the exchange of organisms, nutrients, sediment, and organic material between the river and floodplain, and provide a medium in which biogeochemical processes and biotic activity (e.g., phytoplankton blooms, zooplankton and invertebrate growth and reproduction) can occur.

In addition to the natural connectivity found on some remnant floodplains in the Central Valley, connectivity has also been achieved through management actions. For example, connectivity in the Cosumnes River Preserve has been facilitated by numerous intentional levee breaches (Mount et al. 2003). Connectivity between the main-stem rivers and the Yolo and Sutter Bypasses is controlled by engineered weirs, although partial inundation in the Yolo Bypass can be achieved by inflow from smaller

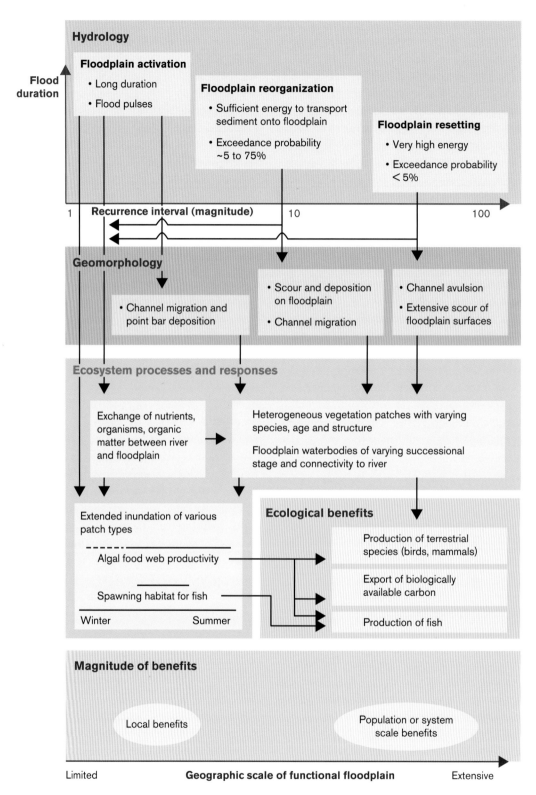

FIGURE 14.1 A conceptual model of floodplain processes, based on floodplains in California's Central Valley. Note the temporal scale bar (Winter → Summer) in the box "Extended inundation of various patch types," which indicates that the occurrence and magnitude of ecosystem processes vary with the season of inundation (after Opperman et al. 2010).

tributaries (Sommer, Harrell et al. 2001; Schemel et al. 2004).

It is worth noting that some ecological benefits can still be provided by Central Valley floodplains that are not hydrologically connected to rivers (other than through canals). For example, floodplain land converted to rice supports large numbers of migratory waterfowl and wading birds (see box 10.2; O'Malley 1999; Bird et al. 2000).

Flow Regime with a Range of Flow Levels

Various types of flows, mediated through connectivity, are ultimately responsible for the heterogeneous landforms and corresponding habitats and processes that occur on floodplains. Diverse flows ranging from baseflows to rare, large floods have ecological significance (Poff et al. 1997). Furthermore, numerous characteristics of any given flow, such as its seasonality, duration, and rate of change, influence its ecological effects. A long-duration flood with high residence time that occurs in January will result in lower levels of productivity than a flood with the same characteristics that occurs in April. To simplify, we will focus on four broad types of flows that each produce a characteristic suite of ecological benefits (figure 14.1): channel-forming flows, floodplain activation flows, floodplain reorganization flows, and floodplain resetting flows.

Channel-Forming Flows

Although bankfull discharge alone is often regarded as being most important for forming channels, river channel formation, and meander migration in particular, is correlated with the cumulative effect of a broad range of flows (chapter 3). The process of lateral migration removes older floodplain surfaces and mature vegetation at the cutbank and creates new depositional surfaces on point bars, facilitating colonization by riparian plants. Over time, channel-forming flows in meandering rivers rework a large portion of the floodplain provided the channel is free to migrate.

Floodplain Activation Flows

Floodplain activation flows provide ecological benefits, such as production of phytoplankton and rearing of juvenile fish, which can be exported to downstream portions of the watershed. The key elements of a floodplain activation flow are (a) *high frequency*, (b) *low magnitude*, and (c) *long hydrologic residence time*. Floodplain activation flows inundate floodplains for a sufficiently long time to allow ecologically significant processes to occur in the water column (productivity) and within substrates (decomposition, mineral recycling) and for fish, waterfowl, and other organisms to derive significant benefits from accessing the floodplain and/or to complete floodplain-dependent portions of their life history. Inundation must be relatively long for fish to effectively utilize floodplains, and must be sufficiently predictable for them to have adapted to using it, although many species can opportunistically use it. Because of their duration and frequency, floodplain activation flows promote several of the characteristics described in the Flood Pulse Concept (FPC): productivity, exchange of organisms and materials, and utilization of floodplain habitats by riverine organisms (Junk et al. 1989).

In the Central Valley, the primary ecological benefits provided by these types of floods are spawning (e.g., Sacramento splittail) and rearing (e.g., Chinook salmon and native cyprinids) for fish, feeding areas for waterfowl and wading birds, and production and export of biologically available carbon for rivers and downstream food webs, such as the Delta.

The characteristic benefits of floodplain activation flows can be produced in the Central Valley with a variety of types of flows interacting with riverine and floodplain topography. In the lower basin, floodplain activation flows occur through inundation of floodplains for extended periods of time. This includes flooding on the bypasses, which occurs frequently (60% of years) and results in large areas of shallow water habitat flooded for extended periods of time (up to a month or more; Schemel et al. 2004; Sommer, Harrell, Solger et al. 2004).

Floodplain activation flooding in higher gradient floodplains, such as along the upper Sacramento River, corresponds more closely to the "flow pulses" described by Tockner et al. (2000) than the overbank flood pulse of the FPC. Extended inundation (usually 2 weeks or more) of floodplain surfaces occurs much less frequently in this portion of the system (Williams et al. 2009). The ecological benefits provided by floodplain activation flows occur with flows that are largely contained within the bankfull channel. Although these flows are not out on an extended floodplain surface (in a geomorphic definition of floodplain), natural channels can have heterogeneous habitat forms within the bankfull channel. During a flow pulse, flow can be initiated in side channels and connections can form with other abandoned channel features (Tockner et al. 2000; Ward et al. 2001, 2002). These features will tend to have slower velocities and greater interaction with surrounding riparian vegetation and are thus likely to be more productive (Amoros 1991). Fish can use these connected habitats for feeding, rearing, and spawning. For example, Limm and Marchetti (2009) found that juvenile Chinook salmon grew faster with greater prey availability in aquatic habitats off the main channel of the Sacramento. Vegetated bars can become fully or partially inundated and flow can extend into annual vegetation or trees (e.g., willows) that grow on the bars and banks.

FLOODPLAIN REORGANIZATION FLOODS

Floodplain reorganization floods, the "erosive floods" of Tockner et al. (2000), contain sufficient energy to perform geomorphic work across the floodplain. While they occur less frequently than floodplain activation flows, they occur frequently enough (once to a few times during a decade) to continually create new landforms and habitat features on the floodplain. During floodplain reorganization floods, erosion occurs at channel cutbanks, reworking older floodplain surfaces and removing mature riparian vegetation, and erosion also occurs within side channels. Crevasses can form through levees, allowing new channels to form on the floodplain (Florsheim and Mount 2002). Sediment is deposited on bars within the channel, on crevasse splays, and in other locations on the floodplain. These deposition surfaces provide the necessary conditions for the regeneration of several species of riparian tree such as cottonwoods and willows (Trowbridge 2002). Floodplain reorganization floods can also scour side channels, abandoned channels, and other floodplain water bodies, "resetting" them to an earlier stage within hydrarch succession. These floods can also remove alluvial plugs between main channels and floodplain water bodies, increasing connectivity between these features (Amoros 1991). Due to the erosion of natural levees, banks, and alluvial plugs, floodplain reorganization flooding can move large amounts of wood and other organic material to be deposited in the channel or on the floodplain. During the draining or recession stage, floodplain reorganization floods will usually display characteristics of floodplain activation flows, with long residence times, high food-web productivity, and export of material to the river channel.

FLOODPLAIN RESETTING FLOODS

Very high magnitude, rare events can cause extensive changes to the morphology of river channels and floodplains (Trush et al. 2000). During these high-magnitude events, the floodplain conveys a large proportion of the overall flow, with high shear stress on the floodplain and in the channel. Thus, these rare flows can also cause extensive scouring of the floodplain. During high winter flood events (>40 year recurrence intervals), there is extensive geomorphic modification of the floodplain through channel avulsion, widespread scour and deposition, exchange of coarse particulate organic matter (CPOM) between channel and floodplain, and disturbance of riparian, wetland, and flood basin vegetation. These flows are associated exclusively with intense winter rainfall events, or rain-on-snow events ("atmospheric river storms"). Because these large events are able to scour and deposit large volumes of sediment, they exert the primary control on floodplain topography.

Flows that lead to channel migration and floodplain reorganization and floodplain resetting floods (collectively termed "erosive floods") create and maintain the template of landforms upon which floodplain ecosystems operate. The mix of vegetation types on the floodplain, including marshes, annual vegetation, and forests of varying successional stage, is largely controlled by the geomorphic work performed by erosive floods (figure 14.1; Stanford et al. 2005). Similarly, diversity of floodplain water bodies is also controlled by erosive floods. This heterogeneity of habitat types in turn influences the diversity of species that utilize these habitats (e.g., neotropical migrants, migratory fish). If erosive floods are eliminated or minimized due to flow regulation, floodplain habitats tend to simplify: riparian forests tend toward a dominance of mature structure and floodplain water bodies tend toward terrestrialization.

Floodplain activation flooding occurs upon this template of habitat types that is created and maintained by erosive flooding. The ecological benefits produced by floodplain activation flooding are strongly influenced by the types of habitats upon which the flooding occurs. For example, splittail and common carp spawn on flooded annual vegetation and thus require relatively open areas. Food-web productivity may be higher in open areas and initial evidence suggests that a mosaic of habitat types may be best for native fishes utilizing the floodplain (Crain et al. 2004); Hamilton et al. (1992) similarly suggest that a mosaic of forest, aquatic macrophytes, and open water is important for maintaining the diversity and productivity of the Orinoco River floodplain. Thus, both erosive floods (over longer time periods) and floodplain activation flows (during the flood event) influence the ecological processes that occur during periods of inundation.

SCALE

Spatial scale of floodplains is important for two reasons. First, the benefits provided by the floodplain habitat mosaic (e.g., habitat for riparian songbirds) are generally proportional to the size of the floodplain site and so, from a management perspective, the extent of functioning floodplain must be sufficiently large to produce benefits that are measurable and relevant for managing systems or populations. Relatively small areas of functioning floodplain will produce only localized benefits, whereas functioning floodplains throughout the Central Valley can produce benefits measured at the system or population scale (figure 14.1). The positive relationship between inundation of the Yolo Bypass and year class strength of Sacramento splittail provides a clear example of how current floodplains provide benefits measurable at the landscape or population scale (Sommer et al. 1997). As discussed in chapter 13, the extent of the bypasses is an order of magnitude smaller than the extent of the historic flood basins, and these massive areas of floodplain with prolonged flooding likely had significant influences on downstream ecosystems (e.g., export of productivity to the river and Delta) and population sizes of splittail, Chinook salmon, and other native fishes.

Second, many of the processes that maintain the habitat mosaic, such as meander migration and channel avulsion, can only occur on sites sufficiently large enough for these processes to occur. Small floodplains may not have sufficient geographic extent to encompass these dynamic processes (Tockner et al. 2000; Richards et al. 2002) and so restoration of small areas of floodplain may require active management to maintain heterogeneity.

MODEL SUMMARY

Floodplain habitats along the Sacramento River and throughout California's Central Valley have been greatly reduced from their historic extent, and key processes that create and maintain floodplains, such as flood flows and meander migration, have been greatly altered. These widespread alterations to habitats and processes have led to declines in many species' populations in California's Central Valley, creating challenges for both environmental and water management. To address these challenges, numerous entities and programs are now focused on restoring floodplains and other habitats. This chapter provides a conceptual model for floodplains that characterizes the key features and identifies the critical processes, drivers, and linkages that allow floodplains to produce a variety of functional ecological outputs. Ecologically functional floodplains—that is, those that can sustainably produce a full range of management outputs—require hydrological connectivity with their adjacent rivers across a wide range of flow levels. Meander migration and high-magnitude flood events create and maintain the habitat mosaic, including topographic features, such as side channels and oxbow lakes. Long-duration spring floods are essential for food-web productivity and the spawning and rearing of native fish. From a management perspective, meander migration, flow variability, and connectivity are necessary, but not sufficient to maintain production of desired floodplain outputs; functional floodplains must exist at sufficient geographic scales for these processes to operate and for outputs to be measurable and meaningful.

CONCLUSIONS

A broad range of benefits can be achieved through floodplain reconciliation—promoting ecosystem services of floodplains by integrating them more effectively within water and flood-management strategies. California is well-positioned to implement floodplain reconciliation projects that achieve multiple benefits because of the following:

- It has a flood season that usually does not overlap with the agricultural growing season.
- There is a strong need for floodplain ecosystem services such as groundwater recharge and habitat for fish and wildlife.
- Regulatory drivers exist for protecting and restoring floodplain-dependent species.
- There is unacceptably high flood risk for Valley urban areas.

A reconciliation approach is particularly fitting for a region that, through the Bypasses, provided such an early example of the benefits of incorporating hydrologically connected floodplains into its flood-management system.

FIFTEEN

Conclusions

MANAGING TEMPERATE FLOODPLAINS
FOR MULTIPLE BENEFITS

> The Sacramento Valley bioregion was formed by flooding, yet we often pretend flooding doesn't—or somehow shouldn't—happen.
>
> THAYER (2003, P.195)

IT IS PERHAPS IRONIC that the authors of this book all have a connection to the University of California, Davis, which is completely protected from flooding by traditional infrastructure projects. The campus is located on the former floodplain of Putah Creek, in the middle of some of some most productive and floodplain-flat farmland in the world. The creek formerly flowed from the highly erosive Coast Range and spread across the landscape, filling its valley deep with sediment. The indigenous Patwin people coexisted with the annual flood events by locating villages on the highest natural levees and leaving for high ground before major flood events. The first Anglo settlers decided to fight the floods rather than live with them, enhancing natural levees and directing the creek down a single channel. This strategy was remarkably successful because the creek quickly cut down into the gravelly sediment, creating a deeply incised channel that rapidly carried flood waters to the Sacramento River. In the twentieth century, incision was enhanced by state and federal agencies building higher levees and periodically scraping the channel free of vegetation. The final step that eliminated flooding, however, was construction of 93 m high Monticello Dam in 1957, which impounds Berryessa Reservoir on Putah Creek. The reservoir holds 1.976×10^9 m^3 of water (1.6 million acre-feet), over four times mean annual runoff from the watershed, so it rarely overflows.

The present-day Putah Creek below the dam is a classic novel ecosystem (Moyle 2013). It first flows as a blue-ribbon trout stream for 13 km before being mostly diverted to water prosperous farms and cities of Solano County. Below the Putah Creek Diversion Dam, flow remaining in the channel supports a rich assemblage of riparian plants and animals in a narrow trench or "canyon" that cuts through the agricultural landscape. Managed flow releases, or "environmental flows," from the

> **BOX 15.1** · The Putah Creek Sinks
>
> The Yolo Bypass was developed in response to repeated major flooding in Sacramento in the nineteenth century as well as failure of agriculture in the Bypass region due to flooding (Chapter 12). Part of the original land of the Bypass, near Davis, was called the Putah Creek Sinks on old maps because the creek's flow seemed to disappear into a vast morass of tules and cattails. In years of flood, however, the Sinks became a huge lake, filling with flood waters from Putah Creek, other small creeks, and the Sacramento River. The campaign to "reclaim" this land for farming ended in 1895, after the event described below.
>
>> On January 5, the raging waters of Putah Creek, with virtually no warning, roared through the break [in a levee] and poured into the canal on a collision course towards ... [Carey] ranch ... The "ocean of water" that came down from the mountains burst through the narrow channel and shot forward on to the [ranch] land, exploding over the tops of the levees which then "merely served to keep in the water" one of [the ranch hands] recalled. "You could not see the fences which were at least four feet high."
>>
>> VAUGHT (2006, p.173)
>
> The same channelization and downcutting of Putah Creek that reduced flooding upstream exacerbated the flooding in the Sinks, the lowermost part of the creek's floodplain. These alterations kept the water from spreading across the rest of the natural floodplain, which had been turned into farms. The Sinks were not reclaimed as farmland until after Monticello Dam on Putah Creek and other large dams on Central Valley rivers eliminated most extreme floods by capturing peak flows. The land that was the Sinks is now part of the managed floodplain of the Bypass, and some of it has been converted to wildlife habitat in the Yolo Bypass Wildlife Area.

is a vast novel ecosystem and a leading example of a large-scale hybrid flood-management project that integrates engineered and green infrastructure and provides multiple benefits. Although Putah Creek only occasionally experiences an unregulated flow event during periods of high precipitation in its watershed (Kiernan et al. 2012), it contributes to localized flooding in the Bypass.

But the complete success of Putah Creek infrastructure in eliminating floods from the lower watershed is an exception rather than the rule for California. The success is due to the combination of the highly erodible substrate on the valley floor and a huge reservoir that can contain potential flood events. The fact that Putah Creek flows into the Yolo Bypass, just east of Davis, gives us a front-door perspective on the challenges of flood management for the whole Central Valley, a region that typifies the river and flood challenges that confront regions around the world. These two very local, very novel ecosystems, along with the critical role the Bypass plays in the Sacramento Valley flood-management system, have strongly shaped our views of floodplain ecology and management.

As described in chapters 12 and 13, the Yolo Bypass is a central feature in the Sacramento Valley flood-management system. Its creation was a definitive response to a decades-long debate about whether the river could be completely contained between levees, or if the river required periodic access to large portions of its historic floodplain. Originally, the Bypass was devoted nearly entirely to agriculture, but in recent years there has been growing appreciation for its ability to support other ecosystem services, such as wildlife and fish habitat. In many ways, the evolution of the Bypass physically expresses the evolution of thinking about managing temperate floodplains and their ecosystems, starting with attempts to control floods with levees, to the decision to establish the Bypass, and thus reconnect a large area of floodplain, continuing through its transition into a feature managed for multiple benefits.

reservoir help to improve downstream fish habitat (Moyle et al. 1998).

The creek then flows into what was formerly the Putah Creek Sinks (box 15.1), now part of the Yolo Bypass (chapter 12). The Bypass itself

Our own thinking on floodplains and floodplain management has certainly been influenced by the evolution of the Bypass, so the rest of this chapter is a summary of our ideas on the role of green infrastructure and reconciliation ecology in floodplain management. We start with a brief summary of what we have learned about reconciling diverse uses of floodplains in California. We then summarize the key takeaway messages in this book as short maxims to guide sustainable floodplain management more broadly. We conclude by reiterating that a reconciliation approach, which looks at temperate floodplains as novel ecosystems, holds the most promise for floodplain management that not only protects people and structures but also enhances biodiversity and the delivery of ecosystem services.

Note that although we have relied heavily on the Yolo Bypass and the Central Valley to illustrate floodplains that produce multiple values, we are not suggesting the Valley has solved all of its floodplain challenges. Indeed both the river-floodplain ecosystems and flood-management systems of the Central Valley are under stress and agencies are currently trying to both increase the environmental benefits derived from functioning floodplains and improve management of growing flood risks, through integrated planning (see chapter 13).

LESSONS FROM CENTRAL VALLEY FLOODPLAINS

Agencies and decision makers are always confronted with a range of options for how to manage floodplains. Some options may focus on managing flood risk, others on maintaining or restoring floodplains ecosystems. The Central Valley provides a compelling case that, often, these objectives can best be accomplished by pursuing integrated strategies that can produce multiple benefits.

Consider managers seeking to increase ecosystem benefits from floodplains. Numerous options exist to increase connectivity and flow variability of floodplains, including levee setbacks and breaches, weirs and other control structures, and managed releases from reservoirs. Because ecosystem benefits derived from floodplains tend to be proportional to floodplain area, a fundamental management challenge is accomplishing these actions at a sufficient geographic scale to produce results measurable at the scales of species' populations or biological communities. Environmental goals alone can drive some floodplain restoration projects, such as the Cosumnes River Preserve. These projects can restore or protect rare resources, provide an arena for research, and raise awareness of the value of floodplain ecosystems. However, due to limited funds and other constraints, such projects tend to be relatively small. Large-scale restoration, better called reconciliation, will most likely be accomplished as part of integrated flood-management projects that incorporate elements of green infrastructure, including using floodplains for storage and conveyance. Today, the largest areas of remaining floodplain in the lowland Central Valley, the bypasses, were created as part of a flood-management system. Other than the Cosumnes River Preserve, most projects that have removed or set levees back have been driven by flood-management needs. For example, setback levee projects on the Bear and Feather Rivers were initiated as flood-risk-reduction projects, with ecosystem improvements seen as secondary benefits. The levee setback project approved for Hamilton City on the Sacramento River, on the other hand, integrated flood-risk reduction and floodplain ecosystem improvement from its inception and, in fact, both objectives were necessary to secure a project with a positive benefit-cost ratio (Golet et al. 2006).

For the objective of increasing benefits derived from functional floodplains, the scale of restoration possible through a comprehensive flood-management program is typically far greater than could be achieved strictly through environmental funding alone. However, environmental funding can support specific types of management or restoration that might not be part of green infrastructure projects.

Now consider agencies tasked with increasing safety from flooding. Producing a broader range of benefits is at best incidental to their primary goal to ensure public safety. But incorporating green infrastructure into management can potentially reduce flood risk and increase system resiliency, while providing a broader range of benefits (chapter 10). In the context of flood management, a system that supports these multiple benefits can potentially catalyze additional stakeholder support, avoid stakeholder or regulatory conflict, and draw on additional sources of funding. The Hamilton City project provides one specific example, in which the inclusion of floodplain restoration was key for generating support and funding to secure a flood-management project.

The potential for green infrastructure approaches to reduce risks and increase resiliency under future uncertainty can best be accounted for using the concept of residual risk (the risk that remains unaccounted for after project completion). Structural protection, such as from levees, usually results in high levels of residual risk because levees are always prone to failure at some level of flooding. Neighborhoods behind levees in cities like Sacramento may be considered outside of the "regulatory floodplain" (chapter 1) because they are protected from the so-called 100 year flood. But many of these areas behind levees are located within deep floodplains (e.g., where flood waters could rise to the level of a two-story building). Should those levees fail or be overtopped by a flood larger than the 100 year flood, the flooding can be rapid and catastrophic (Galloway et al. 2007). Raising levee heights or expanding the area defended by levees increases exposure to this residual risk. Flood management that relies on nonstructural measures, including providing flood conveyance on floodplains, is not vulnerable to the same kind of catastrophic failure, thus resulting in lower residual risk (Eisenstein and Mozingo 2013). Additionally, reliance on structural solutions can "lock in" decisions, while incorporating hydrologically connected floodplains leaves greater flexibility for future management decisions; this results in greater flexibility for water management in general, not just for flood management (Opperman et al. 2009; Cain and Unkel 2015). This so-called "option value" is thus much higher for green infrastructure solutions compared to structural solutions. Flood-management programs that incorporate hydrologically connected floodplains will have greater resiliency and flexibility for future conditions, and lower residual risks of catastrophic flooding (Eisenstein and Mozingo 2013).

A study of potential expansion of the Yolo Bypass illustrates how agencies and decision makers can compare alternative flood-management strategies. Flood-management agencies are seeking solutions to lower the risk of failure or overtopping for the levees of the Sacramento River near Sacramento. Two primary options are strengthening and raising the levees or increasing the area of the Yolo Bypass. Hydraulic modeling by the Sacramento Area Flood Control Agency (SAFCA 2008) predicted that the potential expansion of the Bypass would lower flood stages against levees near Sacramento by 1.3 m, significantly lowering risk. While raising levees can potentially decrease the risk of levee overtopping, this approach increases residual risk from catastrophic failure. Concern about risks of increasing levee heights is what prompted the Netherlands' "Room for the River" approach (chapter 11). In addition, levee improvement only provides flood-stage benefits at the local site and potentially contributes to higher flood stages elsewhere (Pinter et al. 2006). Conversely, the SAFCA study showed that bypass expansion would lower flood stages not just in the target area, but also up and downstream (Eisenstein and Mozingo 2013). Finally, while the levee improvement project would potentially have negative environmental impacts, bypass expansion would increase the area of hydrologically connected floodplain that provides a variety of benefits, such as habitat for native fish and birds, food-web productivity, and groundwater recharge.

As described in chapter 14, California is well positioned to explore reconciliation approaches to integrated flood management. Its flood season is largely discontinuous with the agricultural growing season, and seasonally inundated floodplains provide ecosystem services that provide considerable benefits to the state, such as groundwater recharge and habitat for the large annual influx of migratory fish and birds that use floodplains. Current flood risk is high for important urban areas and population growth will increase the number of people and value of real estate at risk. In addition, climate change models are showing that much of the state is going to be subject to larger and earlier floods, interspersed with long droughts. Thus, the need for improved, reconciliation-based approach flood management in California has never been greater.

MAXIMS FOR FLOODPLAIN MANAGERS

Drawing on the rest of the book and the sources cited within it, here we offer a set of basic guidelines or maxims for managing floodplains. Maxims should be able stand on their own, without further explanation, but the chapters that support each one are noted:

1. Most temperate floodplains are novel ecosystems (chapters 1, 5, and 8).
2. Floodplain ecosystems need flooding (chapter 5); the most biologically productive flooding occurs when inundation corresponds with increasing temperatures and photoperiods (chapter 7).
3. Improving connectivity is the key to improving altered floodplain ecosystems; floodplains and rivers function as one, integrated ecosystem (chapters 2, 3, and 5).
4. Just as flood waters change the floodplain, the floodplain changes flood waters (chapters 2, 3, and 4).
5. Rivers with more predictable timing of flood events will have more species adapted for use of floodplains (chapters 5 and 8).
6. Floodplains need to be dynamic and diverse physically to be diverse biologically (chapters 5, 7, and 8).
7. Let the river do the work for improving floodplains for geomorphic diversity (chapters 3 and 10).
8. Seasonal agriculture on floodplains can be compatible with flood management and biodiversity (chapters 9, 10, and 13).
9. Flood management is more likely to be successful than flood control (chapters 9 and 10).
10. A bigger flood is always possible than the biggest experienced so far (chapters 1, 2, and 10).
11. Although floods catalyze attention, intervals between floods are the best times to reassess floodplain-management strategies.
12. Work with, not against, natural processes: the best protection against a damaging flood is a large, well-managed floodplain (all).
13. Flood management is most effective when implemented at the scale of the entire river basin or watershed, relying on a diverse portfolio of management methods (chapter 10).
14. Flood-risk management that interweaves structural with nonstructural approaches can keep floods away from people and keep people away from floods (chapters 10)
15. Reconciliation is the approach that best incorporates floodplain ecosystems into human-dominated landscapes (chapters 1, 10, 11, and 13).

RECONCILIATION IN FLOODPLAIN MANAGEMENT

In a world with growing human populations, increased demand on resources, and rapidly changing climate, there has never been a greater need for floodplain-management strategies that reduce negative impacts of flooding on people while enhancing biodiversity, agriculture, and floodplain ecosystem services. We have stressed in this book that new management approaches can produce floodplains that provide habitat for some of the world's most spectacular fish and wildlife, while still protecting human communities from floods and providing clean water and open space for people. We reiterate that reconciliation is increasingly needed as an approach to floodplain management, because it recognizes that people are an integral part of the novel ecosystems that temperate floodplains support. Reconciliation is not an excuse to cease caring about, or managing for, native species or historic processes. Instead, it provides a flexible and adaptive approach for working within constraints and pursuing the possible.

Reconciliation approaches and management of novel ecosystems do suggest that we need to develop new baselines for floodplain ecosystems. Centuries of alteration, conversion, and management of floodplains have resulted in a general trend of their simplification (Peipoch et al. 2015). Historic conditions can inform management, but, realistically, if we want to reverse this simplification of biotic communities and prevent further loss of ecosystem services—and even restore them—we need baselines that are rooted in our understanding of present-day floodplain ecosystems (Kopf et al. 2015).

Establishing new, realistic baselines recognizes that we now live in the Anthropocene (Steffen et al. 2007) and human activities are continuing to rapidly change the global ecosystem, of which floodplain ecosystems are one small but important subset. Global climate change is predicted to increase variability and extremes in floods, droughts, and other climate patterns. While the changes in behaviors and policies required to avoid severe impacts from climate change are still possible, the future is uncertain and some change is already underway; the world needs to learn to adapt to the consequences, including large increases in flood risk, and improve the resiliency of both management systems and ecosystems. We believe that integrating floodplains as green infrastructure into flood management holds great promise for this needed process of adaptation, from both economic and ecological perspectives. Reconciled floodplains, managed to produce a range of benefits, can help society reconcile with an uncertain future.

Our time spent on rivers and floodplains has certainly shown us that much has changed and been lost over time. But we have seen more than just glimmers of hope in reconciled floodplains that are diverse and productive. We take heart from the huge flocks of migratory white geese and black ibis that congregate annually on California floodplains and from knowing that, beneath the flood waters, juvenile salmon are swimming, feeding, and growing among cottonwoods and rice stalks, before heading out to sea. We can envision greatly expanded floodplains that are centerpieces of many regions, protecting people but also featuring wildlands, wildlife, and floodplain-friendly agriculture. Connectivity among floodplains, people, and wild creatures is within reach, as is a future in which people work with natural processes rather than continually fighting them.

REFERENCES

Aalto, R., J. Lauer, and W. E. Dietrich. 2008. Spatial and temporal dynamics of sediment accumulation and exchange along Strickland River floodplains (Papua New Guinea) over decadal-to-centennial timescales. Journal of Geophysical Research: Earth Surface 113 (F1).

Aalto, R., L. Maurice-Bourgoin, T. Dunne, D. R. Montgomery, C. A. Nittrouer, and J. L. Guyot. 2003. Episodic sediment accumulation on Amazonian flood plains influenced by El Nino/Southern Oscillation. Nature 425 (6957): 493–497.

Abbe, T. B., and D. R. Montgomery. 1996. Large woody debris jams, channel hydraulics, and habitat formation in large rivers. Regulated Rivers: Research & Management 12: 201–221.

Acharya, G., and E. Barbier. 2002. Using domestic water analysis to value groundwater recharge in the Hadejia-Jama'are Floodplain, Northern Nigeria. American Journal of Agricultural Economics 84 (2): 415–426.

Aerts, J., W. Botzen, A. van der Veen, J. Krywkow, and S. Werners. 2008. Dealing with uncertainty in flood management through diversification. Ecology and Society 13 (1): 1–17.

Agostinho, A. A., L. C. Gomes, S. Verissimo, and E. K. Okada. 2004. Flood regime, dam regulation and fish in the Upper Parana River: effects on assemblage attributes, reproduction and recruitment. Reviews in Fish Biology and Fisheries 14: 11–19.

Ahearn, D. S., J. H. Viers, J. F. Mount, and R. A. Dahlgren. 2006. Priming the productivity pump: flood pulse driven trends in suspended algal biomass distribution across a restored floodplain. Freshwater Biology 51: 1417–1433.

Ahn, C., R. E. Sparks, and D. C. White. 2004. A dynamic model to predict the responses of millets (Echinochloa sp.) to different hydrologic conditions for the Illinois floodplain-river. River Research and Applications 20: 485–498.

Alexander, J., and S. B. Marriot. 1999. Introduction. Pp. 1–13. In Marriot, S. B. and Alexander, J., eds. Floodplains: Interdisciplinary Approaches. Special Publications. Geological Society of London, London, UK.

Alford, J. B., and M. R. Walker. 2013. Managing the flood pulse for optimal fisheries production in the Atchafalaya River Basin, Louisiana (USA). River Research and Applications 29(3):279–296.

Allan, J. D. 1995. Stream Ecology. Springer, Dordrecht, the Netherlands.

Allen, J. R. 1965. A review of the origin and characteristics of recent alluvial sediments. Sedimentology 5 (2): 89–191.

Alpers, C. N., C. Eagles-Smith, C. Foe, S. Klasing, M. C. Marvin-DiPasquale, D. G. Slotton, and L. Windham-Myers. 2008. Sacramento-San Joaquin Delta Regional Ecosystem Restoration Implementation Plan. US Geological Survey, Sacramento, CA.

Alpert, P., F. T. Griggs, and D. R. Peterson. 1999. Riparian forest restoration along large rivers: initial results from the Sacramento River project. Restoration Ecology 7 (4): 360–368.

Alsdorf, D. E., P. Bates, J. Melack, M. Wilson, and T. Dunne. 2007. Spatial and temporal complexity of the Amazon flood measured from space. Geophysical Research Letters 34 (8).

Alsdorf, D. E., J. M. Melack, T. Dunne, L. A. Mertes, L. L. Hess, and L. C. Smith. 2000. Interferometric radar measurements of water level changes on the Amazon flood plain. Nature 404 (6774): 174–177.

American Society of Civil Engineers. 2009. Report Card for America's Infrastructure. American Society of Civil Engineers, Washington, DC.

Amoros, C. 1991. Changes in side-arm connectivity and implications for river system management. Rivers 2 (2): 105–112.

Amoros, C., and G. Bornette. 2002. Connectivity and biocomplexity in waterbodies of riverine floodplains. Freshwater Biology 47 (4): 761–776.

Andersen, D. C., and P. B. Shafroth. 2010. Beaver dams, hydrological thresholds, and controlled floods as a management tool in a desert riverine ecosystem, Bill Williams River, Arizona. Ecohydrology 3 (3): 325–338.

Andersen, D. C., K. R. Wilson, M. S. Miller, and M. Falck. 2000. Movement patterns of riparian small mammals during predictable floodplain inundation. Journal of Mammalogy 81 (4): 1087–1099.

Anderson, M. G., D. E. Walling, and P. D. Bates, eds. 1996. Floodplain Processes. John Wiley & Sons, New York.

Araujo-Lima, C., B. Forsberg, R. Victoria, and L. Martinelli. 1986. Energy sources for detritivorous fishes in the Amazon. Science 234: 1256–1258.

Arthington, A. H. 2012. Environmental Flows: Saving Rivers in the Third Millennium. University of California Press, Berkeley.

Ashworth, P. J., and J. Lewin. 2012. How do big rivers come to be different? Earth-Science Reviews 114 (1–2): 84–107.

Association of State Floodplain Managers. 2007. Levees: The Double-Edged Sword. White Paper. Association of State Floodplain Managers, Madison, WI.

Baker, V. R., and J. E. Costa. 1987. Flood power. Pp. 1–21. In Mayer, L. and Nash, D., eds. Catastrophic Flooding. Allen and Unwin, Boston, MA.

Baki, A. B. M., and T. Y. Gan. 2012. Riverbank migration and island dynamics of the braided Jamuna River of the Ganges–Brahmaputra basin using multi-temporal Landsat images. Quaternary International 263 (0): 148–161.

Bakker, E. S. 1972. An Island Called California: An Ecological Introduction to Its Natural Communities. University of California Press, Berkeley.

Ballinger, A., and P. S. Lake. 2006. Energy and nutrient fluxes from rivers and streams into terrestrial food webs. Marine and Freshwater Research 57(1): 15–28.

Baptist, M. J., W. E. Penning, H. Duel, A. J. Smits, G. W. Geerling, G. E. Van der Lee, and J. S. Van Alphen. 2004. Assessment of the effects of cyclic floodplain rejuvenation on flood levels and biodiversity along the Rhine River. River Research and Applications 20 (3): 285–297.

Baranyi, C., T. Hein, C. Holarek, S. Keckeis, and F. Schiemer. 2002. Zooplankton biomass and community structure in a Danube River floodplain system: effects of hydrology. Freshwater Biology 47 (3): 473–482.

Barbier, E. B. 2003. Upstream dams and downstream water allocation: the case of the Hadejia-Jama'are floodplain, northern Nigeria. Water Resources Research 39 (11): 1311–1319.

Barbour, M., B. Pavlik, F. Drysdale, and S. Lindstrom. 1991. California vegetation: diversity and change. Fremontia 19 (1): 3–12.

Barnes, W. J. 1997. Vegetation dynamics on the floodplain of the lower Chippewa River in Wisconsin. Journal of the Torrey Botanical Society 124 (2): 189–197.

Barnett, A., J. Fargione, and M. P. Smith, M. P. 2016. Mapping trade-offs in ecosystem services from reforestation in the Mississippi alluvial valley. BioScience 66 (3): 223–237.

Barry, J. M., 1998. Rising Tide: The Great Mississippi Flood of 1927 and How It Changed America. Simon and Schuster, New York.

Batalla, R. J., C. M. Gomez, and G. M. Kondolf. 2004. Reservoir-induced hydrological changes in the Ebro River basin (NE Spain). Journal of Hydrology 290 (1): 117–136.

Batalla, R. J., and D. Vericat. 2013. River's architecture supporting life. Pp. 61–76. In Sabater, S., Elosegi, A., and Dudgeon, D., eds. River Conservation: Challenges and Opportunities. Fundacion BBVA, Bilbao, Spain.

Bathurst, J. C., A. Iroume, F. Cisneros, J. Fallas, R. Iturraspe, M. G. Novillo, A. Urciuolo et al. 2011. Forest impact on floods due to extreme rainfall and snowmelt in four Latin American environments 1: field data analysis. Journal of Hydrology 400 (3–4): 281–291.

The Bay Institute. 1998. From the Sierra to the Sea: The Ecological History of the San Francisco Bay-Delta Watershed. The Bay Institute, San Francisco, CA.

Bayley, P. B. 1989. Aquatic environments in the Amazon Basin, with an analysis of carbon sources, fish production, and yield. Pp. 110–127. In Dodge, D. P., ed. Proceedings of the International Large River Symposium. Canadian Special Publication of Fisheries and Aquatic Sciences 106, Department of Fisheries and Oceans, Ottawa, Canada.

Bayley, P. B. 1991. The flood pulse advantage and the restoration of river-floodplain systems. Regulated Rivers: Research and Management 6: 75–86.

Bayley, P. B. 1995. Understanding large river-floodplain ecosystems. BioScience 45 (3): 153–157.

Beechie, T., E. Beamer, and L. Wasserman. 1994. Estimating coho salmon rearing habitat and smolt production losses in a large river basin, and implications for habitat restoration. North American Journal of Fisheries Management 14: 797–811.

Beesley, L., A. J. King, B. Gawne, J. D. Koehn, A. Price, D. Nielsen, F. Amtstaetter, and S. N. Meredith. 2014. Optimising environmental watering of floodplain wetlands for fish. Freshwater Biology 59 (10): 2024–2037.

Beighley, R., K. Eggert, T. Dunne, Y. He, V. Gummadi, and K. Verdin. 2009. Simulating hydrologic and hydraulic processes throughout the Amazon River Basin. Hydrological Processes 23 (8): 1221–1235.

Beilfuss, R. 2010. Modeling trade-offs between hydropower generation and environmental flow scenarios: a case study of the Lower Zambezi River Basin, Mozambique. International Journal of River Basin Management 8 (3–4): 331–347.

Bellmore, J. R., C. V. Baxter, A. M. Ray, L. Denny, K. Tardy, and E. Galloway. 2012. Assessing the potential for salmon recovery via floodplain restoration: a multitrophic level comparison of dredge-mined to reference segments. Environmental management 49 (3): 734–750.

Bendix, J. 1992. Fluvial adjustments on varied timescales in Bear Creek-Arroyo, Utah, USA. Zeitschrift für Geomorphologie 36 (2): 141–163.

Bendix, J., and J. C. Stella. 2013. Riparian vegetation and the fluvial environment: a biogeographic perspective. Pp. 53–74. In Butler, D. and C. Hupp, eds. Treatise on Geomorphology: Ecogeomorphology. Vol. 12. Elsevier, San Diego, CA.

Benigno, G. M., and T. R. Sommer. 2008. Just add water: sources of chironomid drift in a large river floodplain. Hydrobiologia 600: 297–305.

Benke, A. C. 1990. A perspective on America's vanishing streams. Journal of the North American Benthological Society 9 (1): 77–88.

Benke, A. C. 2001. Importance of flood regime to invertebrate habitat in an unregulated river-floodplain ecosystem. Journal of the North American Benthological Society 20 (2): 225–240.

Benke, A. C., I. Chaubey, G. M. Ward, and E. L. Dunn. 2000. Flood pulse dynamics of an unregulated river floodplain in the southeastern US coastal plain. Ecology 81 (10): 2730–2741.

Benke, A. C., and J. B. Wallace. 2003. Influence of wood on invertebrate communities in streams and rivers. Pp. 149–177. In Gregory, S., Boyer, K., and Gurnell, A., eds. The Ecology and Management of Wood in World Rivers. American Fisheries Society, Bethesda, MD.

Bernal, B., and W. J. Mitsch. 2012. Comparing carbon sequestration in temperate freshwater wetland communities. Global Change Biology 18 (5): 1636–1647.

Beschta, R. L., and W. Ripple. 2006. River channel dynamics following extirpation of wolves in northwestern Yellowstone National Park, USA. Earth Surface Processes and Landforms 31 (12): 1525–1539.

Bird, J. A., G. S. Pettygrove, and J. M. Eadie. 2000. The impact of waterfowl foraging on the decomposition of rice straw: mutual benefits for rice growers and waterfowl. Journal of Applied Ecology 37 (5): 728–741.

Biron, P. M., T. Buffin-Bélanger, M. Larocque, G. Choné, C. A. Cloutier, M. A. Ouellet, S. Demers, T. Olsen, C. Desjarlais, and J. Eyquem. 2014. Freedom space for rivers: a sustainable management approach to enhance river resilience. Environmental Management 54 (5): 1056–1073.

Blackwell, M. S., E. Maltby, and A. L. Gerritsen, 2006. Ecoflood Guidelines: How to Use Floodplains for Flood Risk Reduction. European Union. Available from http://www.envia.bl.uk/handle/123456789/4274 and http://ec.europa.eu/ourcoast/download.cfm?fileID=951

Bond, N., J. Costelloe, A. King, D. Warfe, P. Reich, and S. Balcombe. 2014. Ecological risks and opportunities from engineered artificial flooding as a means of achieving environmental flow objectives. Frontiers in Ecology and the Environment 12(7): 386–394.

Booth, E. G., J. F. Mount, and J. H. Viers. 2006. Hydrologic variability of the Cosumnes River floodplain. San Francisco Estuary and Watershed Science 4 (2): 1–19.

Bormann, F. H., and G. E. Likens. 1979. Catastrophic disturbance and the steady state in northern hardwood forests. American Scientist 67 (6): 660–669.

Boulton, A. J., and L. N. Lloyd. 1992. Flooding frequency and invertebrate emergence from dry floodplain sediments of the River Murray, Australia. Regulated Rivers: Research and Management 7: 137–151.

Braatne, J., S. Rood, and P. Heilman. 1996. Life history, ecology and conservation of riparian cottonwoods in North America. Pp. 423–458. In

Stettler, R., Bradshaw, H., Heilman, P., and Hinckley, T., eds. Biology of Populus: Implications for Management and Conservation. National Research Council of Canada, Ottawa.

Bradley, C. E., and D. G. Smith. 1986. Plains cottonwood recruitment and survival on a prairie meandering river floodplain, Milk River, southern Alberta and northern Montana. Canadian Journal of Botany 64: 1433–1442.

Bradshaw, C. J. A., N. S. Sodhi, K. S. H. Peh, and B. W. Brook. 2007. Global evidence that deforestation amplifies flood risk and severity in the developing world. Global Change Biology 13 (11): 2379–2395.

Bravard, J., C. Amoros, and G. Pautou. 1986. Impact of civil engineering works on the succession of communities in a fluvial system. Oikos 47: 92–111.

Brawn, J. D., S. K. Robinson, and F. R. Thompson III. 2001. The role of disturbance in the ecology and conservation of birds. Annual review of Ecology and Systematics 32: 251–276.

Brettar, I., J. M. Sanchez-Perez, and M. Tremolieres. 2002. Nitrate elimination by denitrification in hardwood forest soils of the Upper Rhine floodplain :correlation with redox potential and organic matter. Hydrobiologia 469 (1–3): 11–21.

Bridge, J. S. 2003. Rivers and Floodplains: Forms, Processes, and Sedimentary Record. Blackwell Publishing, Malden, MA.

Bridge, J. S., and M. R. Leeder. 1979. A simulation model of alluvial stratigraphy. Sedimentology 26: 617–644.

Brierley, G., K. Fryirs, and V. Jain. 2006. Landscape connectivity: the geographic basis of geomorphic applications. Area 38 (2): 165–174.

Brinson, M. 1990. Human activities and ecological processes in bottomland hardwood ecosystems: the report of the ecosystem workgroup. Pp. 549–600. In Gosselink, J. G., Lee, L. C., and Muir, T. A., eds. Ecological Processes and Cumulative Impacts: Illustrated by Bottomland Hardwood Wetland Ecosystems. Lewis Publishers, Inc., Chelsea, MI.

Brizga, S. O., and B. L. Finlayson, 1990. Channel avulsion and river metamorphosis: the case of the Thomson River, Victoria, Australia. Earth Surface Processes and Landforms15 (5): 391–404.

Brooks, A. P., and G. J. Brierley. 2002. Mediated equilibrium: the influence of riparian vegetation and wood on the long term evolution and behaviour of a near pristine river. Earth Surface Processes and Landforms 27 (4): 343–367.

Brown, A. G. 1987. Holocene floodplain sedimentation and channel response of the lower River Severn, United Kingdom. Zeitschrift fur Geomorphologie 31: 293–310.

Brown, A. G. 1996. Floodplain paleoenvironments. Pp. 95–138. In Anderson, M. G., Walling D. E., and Bates P. D., eds. Floodplain Processes. John Wiley & Sons, New York.

Brown, A. G. 1997. Alluvial Geoarchaeology: Floodplain Archaeology and Environmental Change. Cambridge University Press.

Brown, D. W., S. M. Moin, and M. L. Nicolson. 1997. A comparison of flooding in Michigan and Ontario: "soft" data to support "soft" water management approaches. Canadian Water Resources Journal 22 (2): 125–139.

Brown, L. R. 1998. Assemblages of fishes and their associations with environmental variables, lower San Joaquin River drainage, California. Open-File Report 98-77, US Geological Survey, Sacramento, CA.

Brown, L. R., and M. L. Bauer. 2010. Effects of hydrologic infrastructure on flow regimes of California's Central Valley Rivers: implications for fish populations. River Research and Applications 26 (6): 751–765.

Brown, L. R., and P. B. Moyle. 2005. Native fishes of the Sacramento-San Joaquin drainage, California: a history of decline. American Fisheries Society Symposium 45:75–78.

Brown, T. G., and G. F. Hartman. 1988. Contribution of seasonally flooded lands and minor tributaries to the production of coho salmon in Carnation Creek, British Columbia. Transactions of the American Fisheries Society 117: 546–551.

Brunier, G., E. J. Anthony, M. Goichot, M. Provansal, and P. Dussouillez. 2014. Recent morphological changes in the Mekong and Bassac River channels, Mekong Delta: the marked impact of river-bed mining and implications for delta destabilisation. Geomorphology 224: 177–191.

Bryan, K. 1923. Geology and Ground-water Resources of Sacramento Valley, California. Paper 495, US Geological Survey Water-Supply, Washington, DC.

BryantMason, A., Y. Jun Xu, and M. A. Altabet. 2013. Limited capacity of river corridor wetlands to remove nitrate: a case study on the Atchafalaya River Basin during the 2011 Mississippi River Flooding. Water Resources Research 49(1): 283–290.

Buffin-Bélanger, T., P. M. Biron, M. Larocque, S. Demers, T. Olsen, G. Choné, M. A. Ouellet, C. A. Cloutier, C. Desjarlais, C., and J. Eyquem. 2015. Freedom space for rivers: an economically viable river management concept in a changing climate. Geomorphology 251:137–148.

Buijse, A. D., H. Coops, M. Staras, L. Jans, G. Van Geest, R. Grift, B. W. Ibelings, W. Oosterberg, and F. C. Roozen. 2002. Restoration strategies for river floodplains along large lowland rivers in Europe. Freshwater Biology 47 (4): 889–907.

Bunn, D. A., P. B. Moyle, and C. K. Johnson. 2014. Maximizing the ecological contribution of conservation banks. Wildlife Society Bulletin 38 (2): 377–385.

Bunn, S. E., P. M. Davies, and M. Winning. 2003. Sources of organic carbon supporting the food web of an arid zone floodplain river. Freshwater Biology 48 (4): 619–635.

Burt, T. P., and G. Pinay. 2005. Linking hydrology and biogeochemistry in complex landscapes. Progress in Physical Geography 29 (3): 297–316.

Burton, C., and S. L. Cutter. 2008. Levee failures and social vulnerability in the Sacramento-San Joaquin Delta area, California. Natural Areas Review 9 (3): 136–149.

Bustard, D. R., and W. Narver. 1975. Aspects of the winter ecology of juvenile coho salmon (Oncorhynchus kisutch) and steelhead trout (Salmo gairdneri). Journal of the Fisheries Research Board of Canada 31: 667–680.

Butler, D. R. 2006. Human-induced changes in animal populations and distribution and the subsequent effects on fluvial systems. Geomorphology 79 (3): 448–459.

C40 and CDC (C40 Cities Climate Leadership Group and Connecting Delta Cities). 2015. C40 Cities Homepage. Accessed April 10, 2017. http://www.deltacities.com/cities/tokyo/climate-change-adaptation.

Cabezas, A., F. A. Comín, S. Beguería, and M. Trabucchi. 2009. Hydrologic and landscape changes in the Middle Ebro River (NE Spain): implications for restoration and management. Hydrology and Earth Systems Sciences 13 (2): 273–284.

Cain, J. and C. Unkel. 2015. With new thinking, flood control projects can ease drought. In Sacramento Bee, April 3.

California Department of Fish and Wildlife. 2012a. Colusa Bypass Area. Accessed 2012. http://www.dfg.ca.gov/lands/wa/region2/colusabypass.html.

California Department of Fish and Wildlife. 2012b. Upper Butte Basin Wildlife Area. Accessed 2012. http://www.dfg.ca.gov/lands/wa/region2/upperbuttebasin.html.

California Land Stewardship Institute, Philip Williams and Associates, and Napa County Resource Conservation District. 2011. Napa River Sediment Reduction and Habitat Enhancement Plan: Oakville to Oak Knoll. State Water Resources Control Board and County of Napa, Napa, CA.

Carle, D. 2004. Introduction to Water in California. University of California Press, Berkeley.

Carling, P. A., and G. E. Petts, eds. 1992. Lowland Floodplain Rivers: Geomorphological Perspectives. John Wiley & Sons, New York.

CDWR (California Department of Water Resources). 2005. Flood Warnings: Responding to California's Flood Crisis. CDWR, Sacramento, CA.

CDWR. 2010. Central Valley Flood Protection Plan. Resources Agency, Sacramento, California.

CDWR. 2011. Flood Control System Status Report. CDWR, Central Valley Flood Management Planning Program, Sacramento, CA.

CDWR. 2012. Central Valley Flood Protection Plan, Attachment 9F: Floodplain Restoration Opportunity Analysis. CDWR, Sacramento, CA.

CDWR. 2014. Central Valley Flood Protection Plan Review Draft. California Department of Water Resources, Sacramento.

CDWR California Data Exchange Center (CDEC). 2016. Hydrologic Data. Accessed 2016. https://cdec.water.ca.gov/.

Chen, Y., K. Herzog, S. Shrestha, D. Grigas, J. Farrelly, C. Laskodi, and M. Skoog. 2015. Urban land use, water quality, and biological conditions in the Lower Mississippi River Basin bayous. Fisheries 40 (7): 334–335.

Chick, J. H., R. J. Cosgriff, and L. S. Gittinger. 2003. Fish as potential dispersal agents for floodplain plants: first evidence in North America. Canadian Journal of Fisheries and Aquatic Sciences 60: 1437–1439.

Chico State University. 2003. The Central Valley Historic Mapping Project. Department of Geography and Planning and Geographic Information Center, Chico, CA.

Clapp, W. B., and F. F. Henshaw. 1911. Surface Water Supply of the United States, 1909. Department of the Interior, United States Geological Survey, Washington, DC.

Clements, F. E. 1936. Nature and structure of the climax. Journal of Ecology 24 (1): 252–284.

Climatewire. 2012. How the Dutch make "Room for the River" by redesigning cities. Scientific American. Available from http://www.scientificamerican.com/article/how-the-dutch-make-room-for-the-river/.

Coates, D., O. Poeu, U. Suntornratana, N. T. Tung, and S. Viravong. 2003. Biodiversity and Fisheries in the Lower Mekong Basin. Mekong River Commission, Phnom Penh, Vietnam.

Colvin, R., G. R. Giannico, J. Li, K. L. Boyer, and W. J. Gerth. 2009. Fish use of intermittent water courses draining agricultural lands in the Upper Willamette River Valley, Oregon. Transactions of the American Fisheries Society 138 (6): 1302–1313.

Community Coalition for a Napa River Flood Management Plan. 1996. Goals and Objectives for a "Living" Napa River System Based on Geomorphic, Water Quality and Habitat Considerations. Community Coalition for a Napa River Flood Management Plan, Napa, CA.

Cook, R. A., B. Gawne, R. Petrie, D. S. Baldwin, G. N. Rees, D. L. Nielsen, N. S. P. Ning. 2015. River metabolism and carbon dynamics in response to flooding in a lowland river. Marine and Freshwater Research 66 (10): 919–927.

Cooley, H. 2006. Floods and Droughts. Island Press, Washington, DC.

Connell, J. H. 1978. Diversity in tropical rain forests and coral reefs. Science 199: 1302–1310.

Conner, W. H., J. G. Gosselink, and R. T. Parrondo. 1981. Comparison of the vegetation of three Louisiana swamp sites with different flood regimes. American Journal of Botany 63: 320–331.

Conrad, M. 2004. Flood peril we keep ignoring. Sacramento Bee, June 20, E2.

Constantine, J. A., T. Dunne, J. Ahmed, C. Legleiter, and E. D. Lazarus. 2014. Sediment supply as a driver of river meandering and floodplain evolution in the Amazon Basin. Nature Geosciences 7 (12): 899–903.

Cook, R. A., B. Gawne, R. Petrie, D. S. Baldwin, G. N. Rees, D. L. Nielsen, and N. S. Ning. 2015. River metabolism and carbon dynamics in response to flooding in a lowland river. Marine and Freshwater Research 66 (10): 919–927.

Cook, S. F. 1960. Colonial Expeditions to the Interior of California: Central Valley, 1800–1820. University of California Press, Berkeley.

Cooley, H. 2006. Floods and droughts. Pp. 91–116. In Gleick, P., ed. The World's Water 2006–2007: The Biennial Report on Freshwater Resources. Island Press, Washington, DC.

Cordes, L. D., F. M. R. Hughes, and M. Getty. 1997. Factors affecting the regeneration and distribution of riparian woodlands along a northern prairie river: the Red Deer River, Alberta, Canada. Journal of Biogeography 24 (5): 675–695.

Costa, J. E., and J. E. O'Connor. 1995. Geomorphically effective floods. Pp. 45–56. In Costa, J. E., Miller, A. J., Potter, K. W., and Willcock, P. R., eds. Natural and Anthropogenic Influences in Fluvial Geomorphology. Monograph 89. American Geophysical Union Geophysical, Washington, DC.

Costanza, R., R. d'Arge, R. de Groot, S. Farber, M. Grasso, B. Hannon, K. Limburg et al. 1997. The value of the world's ecosystem services and natural capital. Nature 387 (6630): 253–260.

County of Napa. 2008. Initial Study and Mitigated Negative Declaration: Napa River Rutherford Reach Restoration Project. Prepared by ICF Jones and Stokes, Napa, CA.

Crain, P. K., and P. B. Moyle. 2011. Biology, history, status and conservation of Sacramento perch, Archoplites interruptus. San Francisco Estuary and Watershed Science 9 (1): 1–37.

Crain, P. K., K. Whitener, and P. B. Moyle. 2004. Use of a restored Central California floodplain by larvae of native and alien fishes. Pp. 125–140. In Feyrer, F., Brown, L. R., Brown, R. L., and Orsi, J. J., eds. Early Life History of Fishes in the San Francisco Estuary and Watershed. American Fisheries Society, Bethesda, MD.

Cuffney, T. F. 1988. Input, movement and exchange of organic matter within a subtropical coastal blackwater river-floodplain system. Freshwater Biology 19: 305–320.

Cunningham, L. 2010. A State of Change: Forgotten Landscapes of California. Heyday Books, Berkeley, CA.

Cushing, C. E., and J. D. Allan. 2001. Streams: Their Ecology and Life. Academic Press, New York.

Das, T., M. D. Dettinger, D. R. Cayan, and H. G. Hidalgo. 2011. Potential increase in floods in California's Sierra Nevada under future climate projections. Climatic Change 109: 71–94.

Davidson, G., G. H. Mendell, W. C. Alberger, A. Miller, and C. E. Grunsky. 1896. Report of Board of Consulting Engineers to the Landowners of Reclamation District Number 108. Board of Consulting Engineers, San Francisco, CA.

Davis, D. W. 2007. In harm's way. Civil Engineering 77 (7): 60–65.

Dawson, R. J., T. Ball, J. Werritty, A. Werritty, J. W. Hall, and N. Roche. 2011. Assessing the effectiveness of non-structural flood management measures in the Thames estuary under conditions of socio-economic and environmental change. Global Environmental Change-Human and Policy Dimensions 21 (2): 628–646.

Decamps, H., M. Fortune, F. Gazelle, and G. Pautou. 1988. Historical influence of man on the riparian dynamics of a fluvial landscape. Landscape Ecology 1 (3): 163–173.

DeltaNet Project. 2015. DeltaNet Partners Ebro River web page. Accessed April 10, 2017. Available from http://www.deltanet-project.eu/ebro.

Dettinger, M. D., and D. R. Cayan. 1995. Large-scale atmospheric forcing of recent trends toward early snowmelt runoff in California. Journal of Climate 8 (3): 606–623.

Dettinger, M. D., F. M. Ralph, T. Das, P. J. Neiman, and D. R. Cayan. 2011. Atmospheric rivers, floods and the water resources of California. Water 3 (2): 445–478.

Dickson, D. 2015. Living river: The Napa Valley flood management plan. Pp. 186–199. In Bullock, J. A., Haddow G. D., Haddow K. S., and Coppola D. P., eds. Living with Climate Change: How Communities Are Surviving and Thriving in a Changing Climate. CRC Press, Boca Raton, FL.

Dietrich, W., and T. Dunne. 1978. Sediment budget for a small catchment in a mountainous terrain. Zeitschrift für Geomorphologie Supplement 29:191–206.

Dodds, W. K. 2009. Laws, Theories, and Patterns in Ecology. University of California Press, Berkeley.

Donath, T. W., N. Holzel, S. Bissels, and A. Otte. 2004. Perspectives for incorporating biomass from non-intensively managed temperate flood-meadows into farming systems. Agriculture Ecosystems and Environment 104 (3): 439–451.

Douglas, M. M., S. E. Bunn, and P. M. Davies. 2005. River and wetland food webs in Australia's wet-dry tropics: general principles and implications for management. Marine and Freshwater Research 56 (3): 329–342.

Douhovnikoff, V., J. R. McBride, and R. S. Dodd. 2005. Salix exigua clonal growth and population dynamics in relation to disturbance regime variation. Ecology 86 (2): 446–452.

Downs, P. W., and G. M. Kondolf. 2002. Post-project appraisals in adaptive management of river channel restoration. Environmental Management 29 (4): 477–496.

Dudgeon, D. 2000. Large-scale hydrological changes in tropical Asia: prospects for riverine biodiversity. BioScience 50 (9): 793–806.

Dunne, T., and L. B. Leopold. 1978. Water in Environmental Planning, W. H. Freeman, San Francisco.

Durand, J. R., R. A. Lusardi, D. M. Nover, R. J. Suddeth, G. Carmona-Catot, C. R. Connell-Buck, S. E. Gatzke et al. 2011. Environmental heterogeneity and community structure of the Kobuk River, Alaska, in response to climate change. Ecosphere 2 (4): 1–19.

Dutterer, A. C., C. Mesing, R. Cailteux, M. S. Allen, W. E. Pine, and P. A. Strickland. 2013. Fish recruitment is influenced by river flows and floodplain inundation at Apalachicola River, Florida. River Research and Applications 29(9):1110–1118.

Duvail, S. and O. Hamerlynck. 2003. Mitigation of negative ecological and socio-economic impacts of the Diama dam on the Senegal River Delta wetland (Mauritania), using a model based decision support system. Hydrology and Earth System Sciences Discussions, Copernicus Publications 7 (1): 133–146.

Dwyer, J. P., D. Wallace, and D. R. Larsen. 1997. Value of woody river corridors in levee protection along the Missouri River 1993. Journal of American Water Resources Association 33: 481–489.

Dybala, K. E., M. L, Truan, and A. Engilis, Jr. 2015. Summer vs. winter: examining the temporal distribution of avian biodiversity to inform conservation. The Condor 117(4): 560–576.

Dykaar, B. B., and P. J. Wigington, Jr. 2000. Floodplain formation and cottonwood colonization patterns on the Willamette River, Oregon, USA. Environmental Management 25 (1): 87–104.

Dynesius, M., and C. Nilsson. 1994. Fragmentation and flow regulation of river systems in the northern third of the world. Science 266: 753–762.

Easterling, D. R., J. L. Evans, P. Y. Groisman, T. R. Karl, K. E. Kunkel, and P. Ambenje. 2000. Observed variability and trends in extreme climate events: a brief review. Bulletin of the American Meteorological Society 81(3): 417–425.

Edwards, P. J., J. Kollman, A. Gurnell, G. E. Petts, K. Tockner, and J. V. Ward. 1999. A conceptual model of vegetation dynamics on gravel bars of a large Alpine river. Wetlands Ecology and Management 7: 141–153.

Egger, G., E. Politti, V. Garófano-Gómez, B. Blamauer, T. Ferreira, R. Rivaes, R. Benjankar, and H. Habersack. 2013. Embodying interactions between riparian vegetation and fluvial hydraulic processes within a dynamic floodplain model: concepts and applications. Pp. 407–427. In Maddock, I., Harby, A., Kemp, P., and Wood, P., eds. Ecohydraulics: An Integrated Approach. John Wiley & Sons, Oxford.

Eisenstein, W., and L. Mozingo. 2013. Valuing Central Valley Floodplains: a Framework for Floodplain Management Decisions. The Center for Resource Efficient Communities, University of California, Berkeley.

Esselman, P. C., and J. J. Opperman. 2010. Overcoming information limitations for the prescription of an environmental flow regime for a Central American river. Ecology and Society 15 (1): Article 6.

Faulkner, B. R., J. R. Brooks, K. J. Forshay, and S. P. Cline. 2012. Hyporheic flow patterns in relation to large river floodplain attributes. Journal of Hydrology 448: 161–173.

Faulkner, W. 1939. The Wild Palms. Random House, New York.

Felipe-Lucia, M., F. A. Comín, and E. M. Bennett. 2014. Interactions among ecosystem services across land uses in a floodplain agroecosystem. Ecology and Society 19 (1): 20.

Fenster, M. S., and C. B. Knisley. 2006. Impact of dams on point bar habitat: a case for the extirpation of the Sacramento Valley Tiger Beetle, C-hirticollis abrupta. River Research and Applications 22 (8): 881–904.

Fetherston, K. L., R. J. Naiman, and R. E. Bilbly. 1995. Large woody debris, physical process, and riparian forest development in montane river networks of the Pacific Northwest. Geomorphology 13: 133–144.

Feyrer, F., T. Sommer, and W. Harrell. 2006. Importance of flood dynamics versus intrinsic physical habitat in structuring fish communities: evidence from two adjacent engineered floodplains on the Sacramento River, California. North American Journal of Fisheries Management 26: 408–417.

Feyrer, F., T. R. Sommer, S. C. Zeug, G. O'Leary, and W. Harrell. 2004. Fish assemblages of perennial floodplain ponds of the Sacramento River, California (USA), with implications for the conservation of native fishes. Fisheries Management and Ecology 11 (5): 335–344.

Fiener, P., K. Auerswald, and S. Weigand. 2005. Managing erosion and water quality in agricultural watersheds by small detention ponds. Agriculture Ecosystems and Environment 110 (3–4): 132–142.

Fierke, M. K., and J. B. Kauffman, 2006. Riverscape-level patterns of riparian plant diversity along a successional gradient, Willamette river, Oregon. Plant Ecology 185(1):85–95.

Fisk, H. N. 1944. Geological Investigation of the Alluvial Valley of the Lower Mississippi River. Mississippi River Commission, Vicksburg, MS.

Fleskes, J. P., W. M. Perry, K. L. Petrik, R. Spell, and F. Reid. 2005. Change in area of winter-flooded and dry rice in the Northern Central Valley of California determined by satellite imagery. California Fish and Game 91 (3): 207–215.

Florsheim, J. L. 2004. Side-valley tributary fans in high-energy river floodplain environments: sediment sources and depositional processes, Navarro River basin, California. Geological Society of America Bulletin 116 (7–8): 923–937.

Florsheim, J. L., and M. D. Dettinger. 2007. Climate and floods still govern California levee breaks. Geophysical Research Letters 34 (22): L22403. doi:10.1029/2007GL031702.

Florsheim, J. L., and M. D. Dettinger. 2015. Promoting atmospheric-river and snowmelt-fueled biogeomorphic processes by restoring river-floodplain connectivity in California's Central Valley. Pp. 119–141. In Hudson, P. F. and Middelkoop, H., eds. Geomorphic Approaches to Integrated Floodplain Management of Lowland Fluvial Systems in North America and Europe. Springer, New York.

Florsheim, J. L., and J. F. Mount. 2002. Restoration of floodplain topography by sand-splay complex formation in response to intentional levee breaches, Lower Cosumnes River, California. Geomorphology 44 (1): 67–94.

Florsheim, J. L., and J. F. Mount. 2003. Changes in lowland floodplain sedimentation processes: pre-disturbance to post-rehabilitation, Cosumnes River, CA. Geomorphology 56 (3): 305–323.

Florsheim, J. L., J. F. Mount, and A. Chin. 2008. Bank erosion as a desirable attribute of rivers. BioScience 58 (6): 519–529.

Florsheim, J. L., J. F. Mount, and C. R. Constantine. 2006. A geomorphic monitoring and adaptive assessment framework to assess the effect of lowland floodplain river restoration on channel-floodplain sediment continuity. River Research and Applications 22 (3): 353–375.

Florsheim, J. L., B. Pellerin, N. H. Oh, N. Ohara, P. Bachand, S. M. Bachand, B. Bergamaschi, P. B. Hernes, and M. L. Kavvas. 2011. From deposition to erosion: spatial and temporal variability of sediment sources, storage, and transport in a small agricultural watershed. Geomorphology 132 (3): 272–286.

Ford, P. W., P. I. Boon, and K. Lee. 2002. Methane and oxygen dynamics in a shallow floodplain lake: the significance of periodic stratification. Hydrobiologia 485 (1–3): 97–110.

Forshay, K. J., and E. H. Stanley. 2005. Rapid nitrate loss and denitrification in a temperate river floodplain. Biogeochemistry 75: 43–64.

Förster, S., D. Kneis, M. Gocht, and A. Bronstert. 2005. Flood risk reduction by the use of retention areas at the Elbe River. International Journal of River Basin Management 3 (1): 21–29.

Francis, R. A., and A. M. Gurnell. 2006. Initial establishment of riparian trees within the active zone of a natural braided gravel-bed river (River Tagliamento, NE Italy). Wetlands 26 (3): 641–648.

Freitag, B., S. Bolton, F. Westerlund, and J. Clark. 2009. Floodplain Management: A New Approach for a New Era. Island Press, Washington, DC.

Fryirs, K. A., G. J. Brierley, N. J. Preston, and J. Spencer, 2007. Catchment-scale (dis) connectivity in sediment flux in the upper Hunter catchment, New South Wales, Australia. Geomorphology, 84 (3): 297–316.

Fuller, I. C., A. R. Large, D. J., and Milan. 2003. Quantifying channel development and sediment transfer following chute cutoff in a wandering gravel-bed river. Geomorphology 54 (3): 307–323.

Furst, D. J., K. T. Aldridge, R. J. Shiel, G. G. Ganf, S. Mills, and J. D. Brookes. 2014. Floodplain connectivity facilitates significant export of zooplankton to the main River Murray channel during a flood event. Inland Waters 4(4): 413–424.

Galat, D. L., L. H. Fredrickson, D. D. Humburg, K. J. Bataille, J. R. Bodie, J. Dohrenwend, G. T. Gelwicks et al. 1998. Flooding to restore connectivity of regulated, large-river wetlands: natural and controlled flooding as complementary processes along the lower Missouri River. Bioscience 48 (9): 721–733.

Galat, D. L., J. Kubisiak, J. Hooker, and L. Sowa. 1997. Geomorphology, distribution and connectivity of lower Missouri River floodplain waterbodies scoured by the flood of 1993. Verhandlungen des Internationalen Verein Limnologie 26: 869–878.

Gallardo, B., M. García, Á. Cabezas, E. González, M. González, C. Ciancarelli, and F. Comín. 2008. Macroinvertebrate patterns along environmental gradients and hydrological connectivity within a regulated river-floodplain. Aquatic Sciences 70 (3): 248–258.

Galloway, G. E. 1994. Sharing the Challenge: Floodplain Management into the 21st Century. Administration Floodplain Management Task Force, Washington, DC.

Galloway, G. E., J. J. Boland, R. J. Burby, C. B. Groves, S. Lien-Longville, L. E. Link, Jr., J. F. Mount et al. 2007. A California Challenge: Flooding in the Central Valley. A Report from an Independent Review Panel. Department of Water Resources, Sacramento, CA.

Garone, P. 2011. The Fall and Rise of the Wetlands of California's Great Central Valley. University of California Press, Berkeley.

Gasith, A., and V. H. Resh. 1999. Streams in Mediterranean climate regions: abiotic influences and biotic responses to predictable seasonal events. Annual Review of Ecology and Systematics 30: 51–81.

Geerling, G. W., H. Duel, A. D. Buijse, and A. J. Smits. 2013. Ecohydraulics at the landscape scale: applying the concept of temporal landscape continuity in river restoration using cyclic floodplain rejuvenation. Pp. 395–406. In Maddock, I., Harby, A., Kemp, P., and Wood, P., eds. Ecohydraulics: An Integrated Approach. John Wiley & Sons, Oxford.

Gergel, S. E., S. R. Carpenter, and E. H. Stanley. 2005. Do dams and levees impact nitrogen cycling? Simulating the effects of flood alterations on floodplain denitrification. Global Change Biology 11 (8): 1352–1367.

Gilbert, G. K. 1917. Hydraulic-mining Debris in the Sierra Nevada. Professional Paper 105, US Geological Survey, Washington, DC.

Girvetz, E. H. 2010. Removing erosion control projects increases bank swallow (Riparia riparia) population viability modeled along the Sacramento River, California, USA. Biological Conservation 143 (4): 828–838.

Gladden, J. E., and L. A. Smock. 1990. Macroinvertebrate distribution and production on the floodplains of two lowland headwater streams. Freshwater Biology 24: 533–545.

Golet, G. H., T. Gardali, C. A. Howell, J. Hunt, R. A. Luster, W. Rainey, M. D. Roberts, J. Silveira, H. Swagerty, and N. Williams. 2008. Wildlife response to riparian restoration on the Sacramento River. San Francisco Estuary and Watershed Science 6 (2): Article 1.

Golet, G. H., J. W. Hunt, and D. Koenig. 2013. Decline and recovery of small mammals after flooding: Implications for pest management and floodplain community dynamics. River Research and Applications 29 (2): 183–194.

Golet, G. H., M. Roberts, E. Larsen, R. Luster, R. Unger, G. Werner, and G. White. 2006. Assessing societal impacts when planning restoration of large alluvial rivers: a case study of the Sacramento River Project, California. Environmental Management 37 (6): 862–879.

Gomes, L. C., and A. A. Agostinho. 1997. Influence of the flooding regime on the nutritional state and juvenile recruitment of the curimba, *Prochilodus scrofa*, Steindachner, in upper Parana River, Brazil. Fisheries Management and Ecology 4: 263–274.

Gomez, B., D. N. Eden, D. M. Hicks, N. A. Trustrum, D. H. Peacock, and J. Wilmshurst. 1999. Contribution of floodplain sequestration to the sediment budget of the Waipaoa River, New Zealand. Geological Society, London, Special Publications 163 (1): 69–88.

Gómez, C. M., C. D. Pérez-Blanco, and R. J. Batalla. 2014. Tradeoffs in river restoration: flushing flows vs. hydropower generation in the Lower Ebro River, Spain. Journal of Hydrology 518, Part A: 130–139.

Gonzalez, E., M. Gonzalez-Sanchis, A. Cabezas, F. A. Comin, and E. Muller. 2010. Recent changes in the riparian forest of a large regulated Mediterranean river: implications for management. Environmental Management 45 (4): 669–681.

Górski, K., J. J. De Leeuw, H. V. Winter, D. A. Vekhov, A. E. Minin, A. D. Buijse, and L. A. Nagelkerke. 2011. Fish recruitment in a large, temperate floodplain: the importance of annual flooding, temperature and habitat complexity. Freshwater Biology 56 (11): 2210–2225.

Goulding, M. 1980. The Fishes and the Forest: Explorations in Amazonian Natural History. University of California Press, Berkeley.

Grantham, T. E., J. H. Viers, and P. B. Moyle. 2014. Systematic screening of dams for environmental flow assessment and implementation. BioScience 64: 1006–1018.

Greco, S. E., A. K. Fremier, E. W. Larsen, and R. E. Plant. 2007. A tool for tracking floodplain age land surface patterns on a large meandering river with applications for ecological planning and restoration design. Landscape Urban Planning 81: 354–373.

Greco, S. E., and E. W. Larsen. 2014. Ecological design of multifunctional open channels for flood control and conservation planning. Landscape and Urban Planning 131: 14–26.

Greco, S. E., and R. E. Plant. 2003. Temporal mapping of riparian landscape change on the Sacramento River, miles 196–218, California, USA. Landscape Research 28 (4): 405–426.

Gren, M., K.-H. Groth, and M. Sylvén. 1995. Economic values of Danube floodplains. Journal of Environmental Management 45 (4): 333–345.

Grift, R. E., A. D. Buijse, W. L. T. Van Densen, M. A. M. Machiels, J. Kranenbarg, J. P. K. Breteler, and J. Backx. 2003. Suitable habitats for 0-group fish in rehabilitated floodplains along the lower River Rhine. River Research and Applications 19 (4): 353–374.

Grime, J. P. 1979. Plant Strategies and Vegetation Processes. John Wiley and Sons, New York.

Grime, J. P. 1988. The C-S-R model of primary plant strategies: origins, implications and tests. Pp. 371–389. In Gottlieb, L. D. and Jain, S. K., eds. Plant Evolutionary Biology. Chapman and Hall, London.

Grimm, N. B., and S. G. Fisher. 1989. Stability of periphyton and macroinvertebrates to disturbance by flash floods in a desert stream. Journal of the North American Benthological Society 8: 293–307.

Grimm, R. S., and J. R. Lund. 2016. Multi-purpose optimization for reconciliation ecology on an engineered floodplain: Yolo Bypass, California. San Francisco Estuary and Watershed Science, 14 (1). Available from http://escholarship.org/uc/item/28j7r0hd.

Grosholz, E., and E. Gallo. 2006. The influence of flood cycle and fish predation on invertebrate production on a restored California floodplain. Hydrobiologia 568 (1): 91–109.

Grossinger, R. 2012. Napa Valley Historical Ecology Atlas: Exploring a Hidden Landscape of Transformation and Resilience. University of California Press, Berkeley.

Grunsky, C. E. 1929. Flood control with special reference to the Mississippi river: a symposium; the relief outlets and by-passes of the Sacramento Valley Flood-Control Project. Transactions of the American Society of Civil Engineers 93 (1): 791–811.

Guida, R. J., J. W. Remo, and S. Secchi. 2016. Tradeoffs of strategically reconnecting rivers to their floodplains: the case of the Lower Illinois River (USA). Science of the Total Environment 572: 43–55.

Gunderson, L. H. 2001. Panarchy: Understanding Transformations in Human and Natural Systems. Island Press, Washington, DC.

Guneralp, B., I. Guneralp, and Y. Liu. 2015. Changing global patterns of urban exposure to flood and drought hazards. Global Environmental Change-Human and Policy Dimensions 31: 217–225.

Günther-Diringer, D., and P. Weller. 1999. Danube Pollution Reduction Programme: Evaluation of Wetlands and Floodplain Areas in the Danube River basin. World Wildlife Fund (WWF) Danube-Carpathian Programme, Vienna, Austria, and WWF-Aueninstitut, Rastatt, Germany.

Gupta, N., P. M. Atkinson, and P. A. Carling. 2012. Decadal length changes in the fluvial planform of the River Ganga: bringing a mega-river to life with Landsat archives. Remote Sensing Letters 4 (1): 1–9.

Gurnell, A., K. Tockner, P. Edwards, and G. Petts. 2005. Effects of deposited wood on biocomplexity of river corridors. Frontiers in Ecology and the Environment 3 (7): 377–382.

Gurnell, A. M., and G. E. Petts. 2002. Island-dominated landscapes of large floodplain rivers, a European perspective. Freshwater Biology 47: 581–600.

Gurnell, A. M., G. E. Petts, D. M. Hannah, B. P. Smith, P. J. Edwards, J. Kollmann, J., J. V. Ward, and K. Tockner. 2001. Riparian vegetation and island formation along the gravel-bed: Fiume Tagliamento, Italy. Earth Surface Processes and Landforms 26 (1): 31–62.

Gutreuter, S., A. D. Bartels, K. Irons, and M. B. Sandheinrich. 1999. Evaluation of the flood-pulse concept based on statistical models of growth of selected fishes of the Upper Mississippi River system. Canadian Journal of Fisheries and Aquatic Sciences 56 (12): 2282–2291.

Haeuber, R. A., and W. K. Michener. 1998. Natural flood control. Issues in Science and Technology 15 (1): 74–80.

Haines, A., B. Finlayson, and T. McMahon. 1988. A global classification of river regimes. Applied Geography 8 (4): 255–272.

Hale, B. W. 2004. Conservation in Temperate River-floodplain Forests: A Comparative Analysis of the Lower Wisconsin State Riverway and the Middle Elbe Biosphere Reserve. PhD Dissertation, University of Wisconsin, Madison.

Hall, J. L., and R. C. Wissmar. 2004. Habitat factors affecting sockeye salmon redd site selection in off-channel ponds of a river floodplain. Transactions of the American Fisheries Society 133: 1480–1496.

Halyk, L. C., and E. K. Balon. 1983. Structure and ecological production of the fish taxocene of a small floodplain system. Canadian Journal of Zoology 61 (11): 2446–2464.

Hamilton, S. K., J. Kellndorfer, B. Lehner, and M. Tobler. 2007. Remote sensing of floodplain geomorphology as a surrogate for biodiversity in a tropical river system (Madre de Dios, Peru). Geomorphology 89 (1–2): 23–38.

Hamilton, S. K., and W. M. Lewis. Jr. 1987. Causes of seasonality in the chemistry of a lake on the Orinoco River floodplain, Venezuela. Limnology and Oceanography 32: 1277–1290.

Hamilton, S. K., W. M. Lewis, Jr., and S. J. Sippel. 1992. Energy sources for aquatic animals in the Orinoco River floodplain: evidence from stable isotopes. Oecologia 89: 324–330.

Harman, C., and M. Stewardson. 2005. Optimizing dam release rules to meet environmental flow targets. River Research and Applications 21 (2–3): 113–129.

Hauer, F. Richard, H. Locke, V. J. Dreitz, M. Hebblewhite, W. H. Lowe, C. C. Muhlfeld, C. R. Nelson, M. F. Proctor, and S. B. Rood. 2016. Gravel-bed river floodplains are the ecological nexus of glaciated mountain landscapes. Science Advances 2 (6): e1600026.

Hayhoe, K. 2004. Emission pathways, climate change, and impacts on California. Proceedings of the National Academy of Sciences of the United States of America 101 (34): 12422–12427.

Hayse, J. W., and T. E. Wissing. 1996. Effects of stem density of artificial vegetation on abundance and growth of age-0 bluegills and predation by largemouth bass. Transactions of the American Fisheries Society 125 (3): 422–433.

Hebblewhite, M., C. A. White, C. G. Nietvelt, J. A. McKenzie, T. E. Hurd, J. M. Fryxell, S. E. Bayley, and P. C. Paquet. 2005. Human activity mediates a trophic cascade caused by wolves. Ecology 86 (8): 2135–2144.

Heffernan, J. B. 2008. Wetlands as an alternative stable state in desert streams. Ecology 89 (5): 1261–1271.

Hein, T., C. Baranyi, W. Reckendorfer, and F. Schiemer. 2004. The impact of surface water exchange on the nutrient and particle dynamics in side-arms along the River Danube, Austria. Science of the Total Environment 328 (1–3): 207–218.

Helfield, J. M., and R. J. Naiman. 2001. Effects of salmon-derived nitrogen on riparian forest growth and implications for stream productivity. Ecology 82(9): 2403–2409.

Henery, R. E., T. R. Sommer, and C. R. Goldman. 2010. Growth and methylmercury accumulation in juvenile Chinook Salmon in the Sacramento River and its floodplain, the Yolo Bypass. Transactions of the American Fisheries Society 139 (2): 550–563.

Hermoso, V., M. Clavero, F. Blanco-Garrido, and J. Prenda. 2011. Invasive species and habitat degradation in Iberian streams: an analysis of their role in freshwater fish diversity loss. Ecological Applications 21: 175–188.

Hey, D. L., D. L. Montgomery, L. S. Urban, T. Prato, F. Andrew, M. Martel, J. Pollack, Y. Steele, and R. Zarwell. 2004. Flood Damage Reduction in the Upper Mississippi River Basin: An Ecological Alternative. The McKnight Foundation, Minneapolis, MN.

Hickey, J. T., R. F. Collins, J. M. High, K. A. Richardson, L. L. White, and P. E. Pugner. 2002. Synthetic rain flood hydrology for the Sacramento and San Joaquin river basins. Journal of Hydrologic Engineering 7 (3): 195–208.

Hickin, E. J., and G. C. Nanson. 1984. Lateral migration rates of river bends. Journal of Hydraulic Engineering 110:11: 1557–1567.

Hobbs, R. J., E. Higgs, C. M. Hall, P. Bridgewater, F. S. Chapin III, E. C. Ellis, J. J. Ewel, L. M. Hallett, J. Harris, and K. B. Hulvey. 2014. Managing the whole landscape: historical, hybrid, and novel ecosystems. Frontiers in Ecology and the Environment 12 (10): 557–564.

Hobbs, R. J., E. Higgs, and J. A. Harris. 2009. Novel ecosystems: implications for conservation and restoration. Trends in Ecology and Evolution 24 (11): 599–605.

Hobbs, R. J., E. S. Higgs, and C. Hall. 2013. Novel Ecosystems: Intervening in the New Ecological World Order. John Wiley & Sons, West Sussex, UK.

Hocking, M. D., and J. D. Reynolds. 2012. Nitrogen uptake by plants subsidized by Pacific salmon carcasses: a hierarchical experiment. Canadian Journal of Forest Research 42 (5): 908–917.

Hodges, J. D. 1997. Development and ecology of bottomland hardwood sites. Forest Ecology and Management 90 (2): 117–125.

Hodges, M. F., and D. G. Krementz. 1996. Neotropical migratory breeding bird communities in riparian forests of different widths along the Altamaha River, Georgia. The Wilson Bulletin: 496–506.

Hogan, Z. S., P. B. Moyle, B. May, M. J. V. Zanden, and I. G. Baird. 2004. The imperiled giants of the Mekong. American Scientist 92 (3): 228–237.

Holland, L. E. 1986. Distribution of early life history stages of fishes in selected pools of the Upper Mississippi River. Hydrobiologia 136 (1): 121–130.

Hooijer, A., F. Klijn, G. B. M. Pedroli, and A. G. Van Os. 2004. Towards sustainable flood risk management in the Rhine and Meuse river basins: synopsis of the findings of IRMA-SPONGE. River Research and Applications 20 (3): 343–357.

Hooke, J. M. 1980. Magnitude and distribution of rates of river bank erosion. Earth Surface Processes and Landforms 5:143–157.

Hudson, P. F., and Heitmuller, F. T. 2003, Local-and watershed-scale controls on the spatial variability of natural levee deposits in a large fine-grained floodplain: lower Pánuco Basin, Mexico: Geomorphology 56 (3): 255–269.

Hudson, P. F., and R. H. Kessel. 2000. Channel migration and meander-bend curvature in the lower Mississippi River prior to major human modification. Geology 28 (6): 531–534.

Hudson, P. F., and H. Middelkoop. 2015. Geomorphic Approaches to Integrated Floodplain Management of Lowland Fluvial Systems in North America and Europe. Springer, New York.

Huffman, J. 2014. Downtown investment tops $1 billion since 1996. Napa Valley Register. January 24, 2014.

Humphries, P., H. Keckeis, and B. Finlayson. 2014. The river wave concept: integrating river ecosystem models. BioScience 64 (10): 870–882.

Humphries, P., A. J. King, and J. D. Koehn. 1999. Fish, flows and flood plains: links between freshwater fishes and their environment in the Murray-Darling River system, Australia. Environmental Biology of Fishes 56 (1–2): 129–151.

Hundley, N. (1992) 2001. The Great Thirst: Californians and Water, 1770s–1990s. University of California Press, Berkeley.

Hupp, C. R., and W. R. Osterkamp. 1985. Bottomland vegetation distribution along Passage Creek, Virginia, in relation to fluvial landforms. Ecology 66 (3): 670–681.

Hupp, C. R., and W. R. Osterkamp. 1996. Riparian vegetation and fluvial geomorphic processes. Geomorphology 14 (4): 277–295.

Hupp, C. R., E. R. Schenk, D. E. Kroes, D. A. Willard, P. A. Townsend, and R. K. Peet. 2015. Patterns of floodplain sediment deposition along the regulated lower Roanoke River, North Carolina: annual, decadal, centennial scales. Geomorphology 228: 666–680.

Huston, M. A. 1979. A general hypothesis of species diversity. American Naturalist 113: 81–101.

Inbar, M. 1987. Effects of a high magnitude flood in a Mediterranean climate: a case study in the Jordan River basin. Pp. 333–353. In Mayer, L. and D. Nash, eds. Catastrophic Flooding. Allen and Unwin, London.

Irons, K., G. Sass, M. McClelland, and J. Stafford. 2007. Reduced condition factor of two native fish species coincident with invasion of non-native Asian carps in the Illinois River, USA. Is this evidence for competition and reduced fitness? Journal of Fish Biology 71: 258–273.

Jackson, C. R., and C. M. Pringle. 2010. Ecological benefits of reduced hydrologic connectivity in intensively developed landscapes. BioScience 60(1): 37–46.

Jacob, J. 2003. The response of small mammal populations to flooding. Mammalian Biology 68 (2): 102–111.

Jacobson, R. B., and D. J. Coleman. 1986. Stratigraphy and recent evolution of Maryland Piedmont floodplains. American Journal of Science 286: 617–637.

Jacobson, R. B., G. Lindner, and C. Bitner. 2015. The role of floodplain restoration in mitigating flood risk, Lower Missouri River, USA. Pp. 203–243.

In Hudson, P. F. and Middelkoop, H., eds. Geomorphic Approaches to Integrated Floodplain Management of Lowland Fluvial Systems in North America and Europe. Springer, New York.

James, L. A., and M. B. Singer. 2008. Development of the lower Sacramento Valley flood-control system: historical perspective. Natural Hazards Review 9 (3): 125–135.

Janauer, G. A., U. Schmidt-Mumm, and W. Reckendorfer. 2013. Ecohydraulics and aquatic macrophytes: assessing the relationship in river floodplains. Pp. 245–259. In Maddock, I., Harby, A., Kemp, P., and Wood, P., eds. Ecohydraulics: An Integrated Approach. John Wiley & Sons, Oxford.

Jassby, A. D., and J. E. Cloern. 2000. Organic matter sources and rehabilitation of the Sacramento: San Joaquin Delta (California, USA). Aquatic Conservation: Marine and Freshwater Ecosystems 10: 323–352.

Jassby, A. D., J. E. Cloern, and A. B. Mueller-Solger. 2003. Phytoplankton fuels Delta food web. California Agriculture 57 (4): 104–109.

Jeffres, C. A., J. J. Opperman, and P. B. Moyle. 2008. Ephemeral floodplain habitats provide best growth conditions for juvenile Chinook salmon in a California river. Environmental Biology of Fishes 83 (4): 449–458.

Jercich, S. A. 1997. California's 1995 water bank program: purchasing water supply options. Journal of Water Resources Planning and Management 123 (1): 59–65.

Johannesson, H., and G. Parker. 1989. Linear theory of river meanders. Pp. 181–213. In Ikeda, S. and Parker G., eds. River Meandering. American Geophysical Union, Washington, DC.

Johnson, B. L., W. B. Richardson, and T. J. Naimo. 1995. Past, present, and future concepts in large river ecology. BioScience 45 (3): 134–140.

Jolly, I. D. 1996. The effects of river management on the hydrology and hydroecology of arid and semi-arid floodplains. Pp. 577–609. In Anderson, M. G., Walling, D. E., and Bates, P. D., eds. Floodplain Processes. John Wiley & Sons, New York.

Jones, K. L., G. C. Poole, W. W. Woessner, M. V. Vitale, B. R. Boer, S. J. O'Daniel, S. A. Thomas, and B. A. Geffen. 2008. Geomorphology, hydrology, and aquatic vegetation drive seasonal hyporheic flow patterns across a gravel-dominated floodplain. Hydrological Processes 22 (13): 2105–2113.

Joyce, C. 2011. Can "carbon ranching" offset emissions in California? National Public Radio. Available from http://www.npr.org/2011/12/07/142947234/can-carbon-ranching-offset-emissions-in-calif.

Junk, W. J., and P. B. Bayley. 2008. The scope of the flood pulse concept regarding riverine fish and fisheries, given geographic and man-made differences among systems. Pp. 1907–1923. In Nielsen, J. L., ed. Reconciling Fisheries with Conservation: Proceedings of the Fourth World Fisheries Congress. American Fisheries Society, Bethesda, MD.

Junk, W. J., P. B. Bayley, and R. E. Sparks. 1989. The flood pulse concept in river-floodplain systems. Pp. 110–127. In Dodge, D. P., ed. Proceedings of the International Large River Symposium. Canadian Special Publication in Fisheries and Aquatic Sciences 106.

Kareiva, P., M. Marvier, and M. McClure. 2000. Recovery and management options for spring/summer chinook salmon in the Columbia River basin. Science 290: 977–979.

Katibah, E. F. 1984. A brief history of riparian forests in the Central Valley of California. Pp. 23–28. In Warner, R. E. and Hendrix, K. M., eds. California Riparian Systems: Ecology, Conservation, and Productive Management. University of California Press, Berkeley.

Katz, J. V. E. 2015. Restoring Ecological Function by Mimicking Natural Process. PhD Dissertation, University of California Davis.

Katz, J. V. E., P. B. Moyle, R. M. Quiñones, J. Israel, and S. Purdy. 2013. Impending extinction of salmon, steelhead, and trout (Salmonidae) in California. Environmental Biology of Fishes 96: 1169–1186.

Kay, C. E. 1995. An alternative interpretation of the historical evidence relating to the abundance of wolves in the Yellowstone ecosystem. Pp. 77–84. In Carbynm L., Fritts, S., and Seip, D., eds. Ecology and Conservation of Wolves in a Changing World. Canadian Circumpolar Institute, Edmonton, AB.

Kay, C. E. 1997. Viewpoint: ungulate herbivory, willows, and political ecology in Yellowstone. Journal of Range Management 50 (2): 139–145.

Kay, C. E., and S. Chadde. 1992. Reduction of willow seed production by ungulate browsing in Yellowstone National Park. Pp. 92–99. In Clary, W. B., McArthur, E. D., Bedunah, D., and Wambolt, C. L., eds. Proceedings-Symposium on Ecology and Management of Riparian Shrub Communities, Sun Valley, ID.

Keckeis, S., C. Baranyi, T. Hein, C. Holarek, P. Riedler, and F. Schiemer. 2003. The significance of zooplankton grazing in a floodplain system of the River Danube. Journal of Plankton Research 25 (3): 243–253.

Kelley, R. 1989. Battling the Inland Sea: Floods, Public Policy, and the Sacramento Valley. University of California Press, Berkeley.

Kenow, K. P., G. L. Benjamin, T. W. Schlagenhaft, R. A. Nissen, M. Stefanski, G. J. Wege, S. A. Jutila, and T. J. Newton. 2015. Process, policy, and implementation of pool wide drawdowns on the upper Mississippi River: a promising approach for ecological restoration of large impounded rivers. River Research and Applications 32(3):295–308.

Kiernan, J. D., P. B. Moyle, and P. K. Crain. 2012. Restoring native fish assemblages to a regulated California stream using the natural flow regime concept. Ecological Applications 22 (5): 1472–1482.

King, A. J., P. Humphries, and R. S. Lake. 2003. Fish recruitment on floodplains: the roles of patterns of flooding and life history characteristics. Canadian Journal of Fisheries and Aquatic Sciences 60: 773–786.

Kingsford, R. T. 2000. Ecological impacts of dams, water diversions and river management on floodplain wetlands in Australia. Austral Ecology 25 (2): 109–127.

Klasz, G., W. Reckendorfer, H. Gabriel, C. Baumgartner, R. Schmalfuss, and D. Gutknecht. 2014. Natural levee formation along a large and regulated river: the Danube in the National Park Donau-Auen, Austria. Geomorphology 215: 20–33.

Klijn, F., D. de Bruin, M. C. de Hoog, S. Jansen, and D. F. Sijmons. 2013. Design quality of room-for-the-river measures in the Netherlands: role and assessment of the quality team (Q-team). International Journal of River Basin Management 11 (3): 287–299.

Klijn, F., M. van Buuren, and S. A. M. van Rooij. 2004. Flood-risk management strategies for an uncertain future: living with Rhine river floods in the Netherlands. AMBIO: A Journal of the Human Environment 33 (3): 141–147.

Kneib, R. T. 1997. The role of tidal marshes in the ecology of estuarine nekton. Oceanography and Marine Biology 35: 163–220.

Knighton, D. 1998. Fluvial Forms and Processes: A New Perspective. John Wiley & Sons, New York.

Knox, J. C. 1987. Historical valley floor sedimentation in the Upper Mississippi Valley. Annals of the Association of American Geographers 77 (2): 224–244.

Knox, J. C. 1993. Large increases in flood magnitude in response to modest changes in climate. Nature 361 (6411): 430–432.

Kondolf, G. M. 1995. Five elements for effective evaluation of stream restoration. Restoration Ecology 3 (2): 133–136.

Kondolf, G. M. 1997. Hungry water: effects of dams and gravel mining on river channels. Environmental Management 21 (4): 533–551.

Kondolf, G. M., J. W. Webb, M. J. Sale, and T. Felando. 1987. Basic hydrologic studies for assessing impacts of flow diversions on riparian vegetation: examples from streams in the eastern Sierra Nevada, California. Environmental Management 11: 757–769.

Kopf, R. K., C. M. Finlayson, P. Humphries, N. C. Sims, and S. Hladyz. 2015. Anthropocene baselines: assessing change and managing biodiversity in human-dominated aquatic ecosystems. Bioscience 65: 798–811.

Kousky, C., S. Olmstead, M. Walls, A. Stern, and M. Macauley. 2011. The Role of Land Use in Adaptation to Increased Precipitation and Flooding: A Case Study in Wisconsin's Lower Fox River Basin. Resources for the Future, Washington, DC.

Kozak, J. P., M. G. Bennett, B. P. Piazza, and J. W. Remo. 2016. Towards dynamic flow regime management for floodplain restoration in the Atchafalaya River Basin, Louisiana. Environmental Science and Policy 64: 118–128.

Kozlowski, T. T. 2002. Physiological-ecological impacts of flooding on riparian forest ecosystems. Wetlands 22(3): 550–561.

Kranjcec, J., J. M. Mahoney, and S. B. Rood. 1998. The responses of three riparian cottonwood species to water table decline. Forest Ecology and Management 110 (1): 77–87.

Kreissman, B. and B. Lekisch. 1991. California: An Environmental Atlas. Bear Klaw Press, Davis, CA.

Kuchler, A. W. 1977. The map of the natural vegetation of California. Pp. 909–938. In Barbour, M. G. and Major J., eds. Terrestrial Vegetation of California. John Wiley & Sons, New York.

Kui, L., J. C. Stella, A. Lightbody, A., and A. C. Wilcox. 2014. Ecogeomorphic feedbacks and flood loss of riparian tree seedlings in meandering channel experiments. Water Resources Research 50 (12): 9366–9384.

Kundzewicz, Z. W., L. Mata, N. W. Arnell, P. Döll, B. Jimenez, K. Miller, T. Oki, Z. Şen, and I. Shiklomanov. 2008. The implications of projected climate change for freshwater resources and their management. Hydrological Sciences Journal 53 (1): 3–10.

Kundzewicz, Z. W., and L. Menzel. 2005. Natural flood reduction strategies: a challenge. International Journal of River Basin Management 3 (2): 125–131.

Kundzewicz, Z. W., and K. Takeuchi. 1999. Flood protection and management: quo vadimus? Hydrological Sciences Journal 44 (3): 417–432.

Kusler, J. A. 1996. Our National Wetland Heritage: A Protection Guide. 2nd ed. Environmental Law Institute, Washington, DC.

Kwadijk, J., and H. Middelkoop. 1994. Estimation of impact of climate change on the peak discharge probability of the river Rhine. Climatic Change 27 (2): 199–224.

Larsen, E. W. 2007. Sacramento River Ecological Flows Study: Meander Migration Modeling. Report for the Nature Conservancy and Stillwater Sciences Ecological Flows Study funded by California Bay-Delta Authority's Ecosystem Restoration Program (CALFED grant ERP-02D0P61, Chico.

Larsen, E. W., A. K. Fremier, and E. H. Girvetz. 2006. Modeling the effects of variable annual flow on river channel meander migration patterns, Sacramento River, California, USA. Journal of the American Water Resources Association 42 (4): 1063–1075.

Larsen, E. W., A. K. Fremier, and S. E. Greco. 2006. Cumulative effective stream power and bank erosion on the Sacramento River, California, USA. Journal of American Water Resources Association 42 (4): 1077–1097.

Larsen, E. W., E. H. Girvetz, and A. K. Fremier. 2006. Assessing the effects of alternative setback channel constraint scenarios employing a river meander migration model. Environmental Management 37 (6): 880–897.

Larsen, E. W., E. H. Girvetz, and A. K. Fremier. 2007. Landscape level planning in alluvial riparian floodplain ecosystems: using geomorphic modeling to avoid conflicts between human infrastructure and habitat conservation. Landscape and Urban Planning 81: 354–373.

Larsen, E. W., and S. E. Greco. 2002. Modeling channel management impacts on river migration: a case study of Woodson Bridge State Recreation Area, Sacramento River, California, USA. Environmental Management 30 (2): 209–224.

Larson, L., and D. Plasencia. 2001. No adverse impact: new direction in floodplain management policy. Natural Hazards Review 2(4): 167–181.

Larson, L. A., M. J. Klitzke, and D. A. Brown. 2003. No Adverse Impact: A Toolkit for Common Sense Floodplain Management. Association of State Floodplain Managers, Madison, WI.

Lasne, E., S. Lek, and P. Laffaille. 2007. Patterns in fish assemblages in the Loire floodplain: the role of hydrological connectivity and implications for conservation. Biological Conservation 139 (3–4): 258–268.

LASRL (Landscape Analysis and Systems Research Laboratory). 2003. Flood Frequency Analysis of the Sacramento River at Bend Bridge. Department of Environmental Design, University of California, Davis.

Lawrence and Houseworth (photographers). Ca. 1862. Sacramento during the Flood, K Street East from Fourth Street. UC Berkeley, Bancroft Library, Berkeley, CA.

Le, P. 2012. Battle of building in Puget Sound flood plains. The Seattle Times, March 27.

Leavenworth, S. 2004. Rising risk. p. A1. In Sacramento Bee, Sacramento, CA, March 4.

Lecce, S. A. 1997. Spatial patterns of historical overbank sedimentation and floodplain evolution, Blue River, Wisconsin. Geomorphology 18: 265–277.

Lehmann, E. 2011. Flood-prone land likely to increase by 45%: a major challenge to federal insurance program. Climate Wire, July 22.

Lemke, M., A. F. Casper, T. D. Van Middlesworth, H. M. Hagy, J. Walk, K. D. Blodgett, K. Dungey, and N. Schrader. 2014. Ecological response of floodplain restoration to flooding disturbance: a comparison of the effects of heavy and light flooding. Pp. 112–1127. In World Environmental and Water Resources Congress, Portland, OR, June 1–5.

Leopold, L. B., and M. G. Wolman. 1957. River channel patterns: braided, meandering, and straight. USGS PP 282-B: 39–85.

Leopold, L. B., M. G. Wolman, and J. P. Miller. 1964. Fluvial Processes in Geomorphology. 1st ed. W. H. Freeman and Company, San Francisco, CA.

Lewin, J. 1996. Floodplain construction and erosion. Pp. 203–220. In Petts, G. and Calow P., eds. River Flows and Channel Forms. Blackwell Science, London.

Lewis, W. M. 1988. Primary production in the Orinoco River. Ecology 69: 679–692.

Lewis, W. M., S. K. Hamilton, M. A. Rodríguez, J. F. Saunders III, and M. A. Lasi. 2001. Foodweb analysis of the Orinoco floodplain based on production estimates and stable isotope data. Journal of the North American Benthological Society 20 (2): 241–254.

Leyer, I., E. Mosner, and B. Lehmann. 2012. Managing floodplain-forest restoration in European river landscapes combining ecological and flood-protection issues. Ecological Applications 22 (1): 240–249.

Ligon, F. K., W. E. Dietrich, and W. J. Trush. 1995. Downstream ecological effects of dams. BioScience 45 (3): 183–192.

Limm, M. P., and M. P. Marchetti. 2003. Contrasting patterns of juvenile chinook salmon (*Oncorhynchus tshawytschaw*) growth, diet, and prey densities in off-channel and mainstem habitats on the Sacramento River. The Nature Conservancy, Chico, CA.

Limm, M. P., and M. P. Marchetti. 2009. Juvenile Chinook salmon (*Oncorhynchus tshawytscha*) growth in off-channel and main-channel habitats on the Sacramento River, CA using otolith increment widths. Environmental Biology of Fishes 85 (2): 141–151.

Llewellyn, D. W., G. P. Shaffer, N. J. Craig, L. Creasman, D. Pashley, M. Swan, and C. Brown. 1996. A decision-support system for prioritizing restoration sites on the Mississippi River Alluvial Plain. Conservation Biology 10: 1446–1455.

Lóczy, D., E. Pirkhoffer, and P. Gyenizse. 2012. Geomorphometric floodplain classification in a hill region of Hungary. Geomorphology 147–148: 61–72.

Lopes, C. A., G. L. Manetta, B. R. S. Figuieredo, L. A. Martinelli, and E. Benedito. 2015. Carbon from littoral producers is the major source of energy for bottom-feeding fish in a tropical floodplain. Environmental Biology of Fishes 98 (4): 1081–1088.

Loth, P. 2004. The Return of the Water: Restoring the Waza Logone Floodplain in Cameroon. IUCN, Gland, Switzerland.

Lott, D. F. 2002. American Bison: A Natural History. University of California Press, Berkeley.

Louisiana Department of Wildlife and Fisheries. 2015. Louisiana Black Bear Status and Range (map). Accessed April 10, 2017. http://www.wlf.louisiana.gov/louisiana-black-bear-status-range.

Lowe-McConnell, R. H. 1975. Fish Communities in Tropical Freshwaters. Longman, London.

Lowrance, R., R. Todd, J. J. Fail, J. O. Hendrickson, R. Leonard, and L. Asmussen. 1984. Riparian forests as nutrient filters in agricultural watersheds. BioScience 34 (6): 374–377.

Lytle, D. A., and D. M. Merritt. 2004. Hydrologic regimes and riparian forests: a structured population model for cottonwood. Ecology 85 (9): 2493–2503.

Maceda-Veiga, A., A. Monleon-Getino, N. Caiola, F. Casals, and A. De Sostoa. 2010. Changes in fish assemblages in catchments in north-eastern Spain: biodiversity, conservation status and introduced species. Freshwater Biology 55 (8): 1734–1746.

Mac Nally, R., S. C. Cunningham, P. J. Baker, G. J. Horner, and J. R. Thomson. 2011. Dynamics of Murray-Darling floodplain forests under multiple stressors: the past, present, and future of an Australian icon. Water Resources Research 47(12): 1–11.

Madsen, T., and R. Shine. 2000. Rain, fish and snakes: climatically driven population dynamics of Arafura filesnakes in tropical Australia. Oecologia 124: 208–215.

Magana, H. A. 2013. Flood pulse trophic dynamics of larval fishes in a restored arid-land, river-floodplain, Middle Rio Grande, Los Lunas, New Mexico. Reviews in Fish Biology and Fisheries 23 (4): 507–521.

Magner, J., and S. Alexander. 2008. Drainage and nutrient attenuation in a riparian interception-wetland: southern Minnesota, USA. Environmental Geology 54 (7): 1367–1376.

Mahoney, J. M., and S. B. Rood. 1998. Streamflow requirements for cottonwood seedling recruitment: an integrative model. Wetlands 18 (4): 634–645.

Malamud-Roam, F., M. Dettinger, B. L. Ingram, M. K. Hughes, and J. L. Florsheim. 2007. Holocene climates and connections between the San Francisco Bay estuary and its watershed: a review. San Francisco Estuary and Watershed Science 5 (1): 1–28.

Malamud-Roam, F. P., B. L. Ingram, M. Hughes, and J. L. Florsheim. 2006. Holocene paleoclimate records from a large California estuarine system and its watershed region: linking watershed climate and bay conditions. Quaternary Science Reviews 25 (13): 1570–1598.

Malmqvist, B. 2002. Aquatic invertebrates in riverine landscapes. Freshwater Biology 47 (4): 679–694.

Mandeville, J. W. 1857. Map of Public Surveys in California to Accompany Report of Surveyor General. Surveyor General's Office, San Francisco, CA.

Manfree, A. D. 2014. Historical Ecology of Suisun Marsh. Pp. 9–44. In Moyle, P. B., Manfree A. D., and Fiedler P. L., eds. Suisun Marsh: Ecological History and Possible Futures. University of California Press, Berkeley.

Maraseni, T. N., and C. Mitchell. 2016. An assessment of carbon sequestration potential of riparian zone of Condamine Catchment, Queensland, Australia. Land Use Policy 54: 139–146.

Marks, C. O., K. H. Nislow, and F. J. Magilligan. 2014. Quantifying flooding regime in floodplain forests to guide river restoration. Elementa: Science of the Anthropocene 2 (1): 31.

Márquez-Ferrando, R., J. Figuerola, J. C. Hooijmeijer, and T. Piersma. 2014. Recently created man-made habitats in Doñana provide alternative wintering space for the threatened Continental European black-tailed godwit population. Biological Conservation 171 (1): 127–135.

Marriott, S. B., and J. Alexander. 1999. Floodplains: Interdisciplinary Approaches. Geological Society of London Special Publication 163, London.

Martin, A. R., and V. M. E. da Silva. 2004. River dolphins and flooded forest: seasonal habitat use and sexual segregation of botos (*Inia geoffrensis*) in an extreme cetacean environment. Journal of Zoology 263: 295–305.

McClain, M. E., E. W. Boyer, C. L. Dent, S. E. Gergel, N. B. Grimm, P. M. Groffman, S. C. Hart et al. 2003. Biogeochemical hot spots and hot moments at the interface of terrestrial and aquatic ecosystems. Ecosystems 6 (4): 301–312.

McCluney, K. E., N. L. Poff, M. A. Palmer, J. H. Thorp, G. C. Poole, B. S. Williams, M. R. Williams, and J. S. Baron. 2014. Riverine macrosystems ecology: sensitivity, resistance, and resilience of whole river basins with human alterations. Frontiers in Ecology and the Environment 12 (1): 48–58.

McColl, C. J., K. Andrews, M. Reynolds, and G. H. Golet. 2016. Pop-up wetland habitats benefit migrating birds and farmers. ArcUser (ESRI) Summer 2016: 20–24.

McCoy, K. J., and M. Minnie. 1929. When the Levee Breaks. Columbia Records, New York City, NY.

McDonald, E. V., and A. J. Busacca. 1988. Record of pre-late Wisconsin giant floods in the Channeled Scabland interpreted from loess deposits. Geology 16 (8):728–731.

McIntyre, P. B., C. R. Liermann, and C. Revenga. 2016. Linking freshwater fishery management to global food security and biodiversity conservation. Proceedings of the National Academy of Sciences of the United States of America 113 (45): 12880–12885.

Mertes, L. A. K. 1997. Documentation and significance of the perirheic zone on inundated floodplains. Water Resources Research 33 (7): 1749–1762.

Mertes, L. A. K. 1994. Rates of floodplain sedimentation on the central Amazon River. Geology 22: 171–174.

Mertes, L. A. K. 2000. Inundation hydrology. Pp. 145–166. In Wohl, E. E., ed. Inland Flood Hazards: Human, Riparian, and Aquatic Communities. Cambridge University Press, Cambridge, UK.

Mertes, L. A. K. 2002. Remote sensing of riverine landscapes. Freshwater Biology 47 (4): 799–816.

Merz, J. F., and P. B. Moyle. 2006. Salmon, wildlife and wine: marine derived nutrients in human-dominated ecosystems of central California. Ecological Applications 16: 999–1009.

Michalková, M., H. Piégay, G. M. Kondolf, and S. E. Greco. 2011. Lateral erosion of the Sacramento River, California (1942–1999), and responses of channel and floodplain lake to human influences. Earth Surface Processes and Landforms 36 (2): 257–272.

Micheli, E. R., J. W. Kirchner, and E. W. Larsen. 2004. Quantifying the effect of riparian forest versus agricultural vegetation on river meander migration rates, Central Sacramento River, California, USA. River Research and Applications 20 (5): 537–548.

Micheli, E. R., and E. W. Larsen. 2011. River channel cutoff dynamics, Sacramento River, California, USA. River Research and Applications 27 (3): 328–344.

Milly, P. C. D., J. Betancourt, M. Falkenmark, R. M. Hirsch, Z. W. Kundzewicz, D. P. Lettenmaier, and R. J. Stouffer. 2008. Stationarity is dead: whither water management? Science 319 (5863): 573–574.

Min, S.-K., X. Zhang, F. W. Zwiers, and G. C. Hegerl. 2011. Human contribution to more-intense precipitation extremes. Nature 470 (7334): 378–381.

Mitsch, W. J., J. K. Cronk, X. Y. Wu, R. W. Nairn, and D. L. Hey. 1995. Phosphorus retention in constructed fresh-water riparian marshes. Ecological Applications 5 (3): 830–845.

Mitsch, W. J., and J. W. Day. 2006. Restoration of wetlands in the Mississippi-Ohio-Missouri (MOM) River Basin: experience and needed research. Ecological Engineering 26 (1): 55–69.

Mitsch, W. J., J. W. Day, J. W. Gilliam, P. M. Groffman, D. L. Hey, G. W. Randall, and N. Wang. 2001. Reducing nitrogen loading to the Gulf of Mexico from the Mississippi River Basin: strategies to counter a persistent ecological problem. BioScience 51 (5): 373–388.

Mitsch, W. J., C. C. Dorge, and J. R. Wiemhoff. 1979. Ecosystem dynamics and a phosphorous budget of an alluvial cypress swamp in southern Illinois. Ecology 60: 116–1124.

Mitsch, W. J., and J. G. Gosselink. 2000. Wetlands. 3rd ed. John Wiley and Sons, New York.

Modde, T., R. T. Muth, and G. B. Haines. 2001. Floodplain wetland suitability, access, and potential use by juvenile razorback suckers in the middle Green River, Utah. Transactions of the American Fisheries Society 130 (6): 1095–1105.

Moffatt, K. C., E. E. Crone, K. D. Holl, R. W. Schlorff, and B. A. Garrison. 2005. Importance of hydrologic and landscape heterogeneity for restoring bank swallow (*Riparia riparia*) colonies along the Sacramento River, California. Restoration Ecology 13 (2): 391–402.

Moghraby, A. I. 1977. A study on diapause of zooplankton in a tropical river: the Blue Nile. Freshwater Biology 7: 77–117.

Montgomery, D. R. 2003. King of Fish: the Thousand-Year Run of Salmon. Westview Press, Boulder, CO.

Montgomery, D. R. 2016. Emperor Yu's Great Flood. Science 353: 538–539.

Montgomery, D. R., B. D. Collins, J. M. Buffington, and T. B. Abbe. 2003. Geomorphic effects of wood in rivers. Pp. 21–47. In Gregory, S., Boyer K., and Gurnell A., eds. The Ecology and Management of Wood in World Rivers. American Fisheries Society, Bethesda, MD.

Morris, J., J. Beedell, and T. M. Hess. 2016. Mobilizing flood risk management services from rural land: principles and practice. Journal of Flood Risk Management 9(1): 50–68.

Morris, M. R., and J. A. Stanford. 2011. Floodplain succession and soil nitrogen accumulation on a salmon river in southwestern Kamchatka. Ecological Monographs 80 (1): 43–61.

Mosepele, K., P. B. Moyle, G. S. Merron, D. R. Purkey, and B. Mosepele. 2009. Fish, floods, and ecosystem engineers: aquatic conservation in the Okavango Delta, Botswana. BioScience 59 (1): 53–64.

Mount, J. F. 1995. California Rivers and Streams: The Conflict between Fluvial Process and Land Use. University of California Press, Berkeley.

Mount, J. F., W. Fleenor, B. Gray, and B. Herbold. 2014. The Draft Bay-Delta Conservation Plan: assessment of environmental performance and governance. West-Northwest Journal of Environmental Law and Policy 20: 245–376.

Mount, J. F., J. L. Florsheim, and W. B. Trowbridge. 2003. Restoration of dynamic flood plain topography and riparian vegetation establishment through engineered levee breaching. Pp. 142–148. In Faber, P. M., ed. California Riparian Systems: Processes and Floodplain Management, Ecology, and Restoration. Riparian Habitat Joint Venture, Sacramento, CA.

Moyle, P. B. 2002. Inland Fishes of California. University of California Press, Berkeley.

Moyle, P. B. 2013. Novel aquatic ecosystems: the new reality for streams in California and other Mediterranean climate regions. River Research and Applications 30 (10): 1335–1344.

Moyle, P. B., R. D. Baxter, T. R. Sommer, T. C. Foin, and S. A. Matern. 2004. Biology and population dynamics of Sacramento splittail (*Pogonichthys macrolepidotus*) in the San Francisco Estuary: a review. San Francisco Estuary and Watershed Science 2 (2): 1–47.

Moyle, P. B., P. K. Crain, and K. Whitener. 2007. Patterns in the use of a restored California floodplain by native and alien fishes. San Francisco Estuary and Watershed Science 5 (3): Article 1.

Moyle, P. B., P. K. Crain, K. Whitener, and J. F. Mount. 2003. Alien fishes in natural streams: fish distribution, assemblage structure, and conservation in the Cosumnes River, California, USA. Environmental Biology of Fishes 68: 143–162.

Moyle, P. B., M. P. Marchetti, J. Baldrige, and T. L. Taylor. 1998. Fish health and diversity: justifying flows for a California stream. Fisheries (Bethesda) 23 (7): 6–15.

Moyle, P. B., and J. F. Mount. 2007. Homogenous rivers, homogenous faunas. Proceedings of the National Academy of Sciences of the United States of America 104 (14): 5711–5712.

Moyle, P. B., and R. M. Yoshiyama. 1994. Protection of aquatic biodiversity in California: a five-tiered approach. Fisheries 19 (2): 6–18.

Muller, M., A. Biswas, R. Martin-Hurtado, and C. Tortajada. 2015. Built infrastructure is essential. Science 349 (6248): 585–586.

Muller-Solger, A. B., A. D. Jassby, and D. C. Muller-Navarra. 2002. Nutritional quality of food resources for zooplankton (Daphnia) in a tidal freshwater system (Sacramento-San Joaquin River Delta). Limnology and Oceanography 47 (5): 1468–1476.

Murray, B. C., W. A. Jenkins, R. A. Kramer, and S. P. Faulkner. 2009. Valuing Ecosystem Services from Wetlands Restoration in the Mississippi Alluvial Valley. Report NI R 09-02. Nicholas Institute for Environmental Policy Solutions, Duke University, Durham, NC.

Naiman, R. J., J. S. Bechtold, T. J. Beechie, J. J. Latterell, and R. Van Pelt. 2010. A process-based view of floodplain forest patterns in coastal river valleys of the Pacific Northwest. Ecosystems 13 (1): 1–31.

Naiman, R. J., and H. Decamps. 1997. The ecology of interfaces: riparian zones. Annual Review of Ecology and Systematics 28: 621–658.

Naiman, R. J., H. Decamps, and M. E. McClain. 2005. Riparia: Ecology, Conservation, and Management of Streamside Communities. Elsevier Academic Press, Amsterdam.

Naiman, R. J., C. A. Johnston, and J. C. Kelley. 1988. Alteration of North American streams by beaver. BioScience 38: 753–762.

Nakano, S., and M. Murakami. 2001. Reciprocal subsidies: dynamic interdependence between terrestrial and aquatic food webs. Proceedings of the National Academy of Sciences of the United States of America 98 (1): 166–170.

Namafe, C. M. 2004. Flooding in the context of the Barotse people of the upper Zambezi wetlands. Southern African Journal of Environmental Education 21: 50–60.

Nanson, G. C. 1986. Episodes of vertical accretion and catastrophic stripping: a model of disequilibrium flood-plain development. Geological Society of America Bulletin 97: 1467–1475.

Nanson, G. C., and J. Croke. 1992. A genetic classification of floodplains. Geomorphology 4 (6): 459–486.

Nanson, G. C., and E. J. Hickin. 1986. A statistical analysis of bank erosion and channel migration in western Canada. Bulletin of the Geological Society of America 97: 497–504.

Napa County Department of Public Works. 2008. Napa County post-construction runoff management requirements. Department of Public Works, Napa, CA.

Napa County Resource Conservation District. 2007. Napa Wetlands Monitoring Program Final Report. County of Napa, Napa, CA.

National Committee on Levee Safety. 2009. Draft Recommendations for a National Levee Safety Program: A Report to Congress. National Committee on Levee Safety. Accessed September 14, 2015. http://www.leveesafety.org/rec_reporttocongress.cfm.

National Research Council. 1996. Upstream: Salmon and Society in the Pacific Northwest. National Academy Press, Washington, DC.

Nehlsen, W. 1997. Pacific salmon status and trends: a coastwide perspective. Pp. 41–52. In Stouder, D. J., Bisson, P. A., Naiman, R. J., and Duke, M. G., eds. Pacific Salmon and Their Ecosystems: Status and Future Options. Chapman and Hall, New York.

Nehlsen, W., J. E. Williams, and J. A. Lichatowich. 1991. Pacific salmon at the crossroads: stocks at risk in California, Oregon, Idaho and Washington. Fisheries 16 (2): 4–21.

Nienhuis, P. H., and R. Leuven. 2001. River restoration and flood protection: controversy or synergism? Hydrobiologia 444 (1): 85–99.

Nienhuis, P. H., R. S. E. W. Leuven, and A. M. J. Ragas. 1998. General Introduction. Pp. 1–10. In Nienhuis, P. H., Leuven, R. S. E. W., and Ragas, A. M. J., eds. New Concepts for Sustainable Management of Rivers. Backhuys Publishers, Leiden, the Netherlands.

Nilsson, C., C. A. Reidy, M. Dynesius, and C. Revenga. 2005. Fragmentation and flow regulation of the world's large river systems. Science 308: 405–408.

Noe, G. B., and C. R. Hupp. 2005. Carbon, nitrogen, and phosphorus accumulation in floodplains of Atlantic Coastal Plain rivers, USA. Ecological Applications 15 (4): 1178–1190.

Notebaert, B., and H. Piégay. 2013. Multi-scale factors controlling the pattern of floodplain width at a network scale: the case of the Rhône basin, France. Geomorphology 200: 155–171.

Nyssen, J., J. Pontzeele, and P. Billi. 2011. Effect of beaver dams on the hydrology of small mountain streams: example from the Chevral in the Ourthe Orientale basin, Ardennes, Belgium. Journal of Hydrology 402 (1): 92–102.

Ogston, L., S. Gidora, M. Foy, M., and J. Rosenfeld. 2014. Watershed-scale effectiveness of floodplain habitat restoration for juvenile coho salmon in the Chilliwack River, British Columbia. Canadian Journal of Fisheries and Aquatic Sciences, 72(4): 479–490.

Olechnowski, B. F., and D. M. Debinski. 2008. Response of songbirds to riparian willow habitat structure in the Greater Yellowstone Ecosystem. The Wilson Journal of Ornithology 120 (4): 830–839.

Ollero, A. 2010. Channel changes and floodplain management in the meandering middle Ebro River, Spain. Geomorphology 117 (3): 247–260.

O'Malley, R. E. 1999. Agricultural wetland management for conservation goals: invertebrates in California ricelands. Pp. 857–885. In Batzer, D. P., Rader R. B., and Wissinger S. A., eds. Invertebrates in Freshwater Wetlands of North America. John Wiley & Sons, New York.

Omedas, M., R. Galvan, and C. M. Gomez. 2012. Water Planning Towards a Green Economy in the Ebro River Basin. UN-Water Decade Programme on Advocacy and Communication (UNW-DPAC) and Ebro River Basin Authority (Spain), Zargosa, Spain.

Omernik, J. M., A. R. Abernathy, and L. M. Male. 1981. Stream nutrient levels and proximity of agricultural and forest land to streams: some relationships. Journal of Soil and Water Conservation 36: 227–231.

Opperman, J. J. 2012. A conceptual model for floodplains in the Sacramento-San Joaquin Delta.

San Francisco Estuary and Watershed Science 10 (3): 1–28.

Opperman, J. J., G. E. Galloway, and S. Duvail. 2013. The multiple benefits of river-floodplain connectivity for people and biodiversity. Pp. 144–160. In Levin, S., ed. Encyclopedia of Biodiversity. 2nd ed. Academic Press, Waltham, MA.

Opperman, J. J., G. E. Galloway, J. Fargione, J. F. Mount, B. D. Richter, and S. Secchi. 2009. Sustainable floodplains through large-scale reconnection to rivers. Science 326 (5959): 1487–1488.

Opperman, J. J., G. Grill, and J. Hartmann, J., 2015. The Power of Rivers: Finding Balance between Energy and Conservation in Hydropower Development. The Nature Conservancy, Washington, DC.

Opperman, J. J., J. Hartmann, and D. Harrison. 2015. Hydropower within the climate, energy and water nexus. Pp. 79–107. In Pittock, J., Hussey, K., and Dovers, S., eds. Climate, Energy and Water: Managing Trade-Offs, Seizing Opportunities. Cambridge University Press, Cambridge.

Opperman, J. J., R. Luster, B. A. McKenney, M. Roberts, and A. W. Meadows. 2010. Ecologically functional floodplains: connectivity, flow regime, and scale. Journal of the American Water Resources Association 46 (2): 211–226.

Opperman, J. J., M. Meleason, R. A. Francis, and R. Davies-Colley. 2008. "Livewood": geomorphic and ecological functions of living trees in river channels. Bioscience 58 (11): 1069–1078.

Opperman, J. J., and A. M. Merenlender. 2000. Deer herbivory as an ecological constraint to restoration of degraded riparian corridors. Restoration Ecology 8 (1): 41–47.

Opperman, J. J., A. Warner, E. H. Girvetz, D. Harrison, and T. Fry. 2011. Integrated reservoir-floodplain management as an ecosystem-based adaptation strategy to climate change. Proceedings to American Water Resources Association 2011 Spring Specialty Conference on Climate Change and Water Resources, Baltimore, MD.

Osborne, L. L., and D. A. Kovacic. 1993. Riparian vegetated buffer strips in water-quality restoration and stream management. Freshwater Biology 29: 243–258.

Paillex, A., S. Dolédec, E. Castella, and S. Mérigoux. 2009. Large river floodplain restoration: predicting species richness and trait responses to the restoration of hydrological connectivity. Journal of Applied Ecology 46 (1): 250–258.

Pall, P., T. Aina, D. A. Stone, P. A. Stott, T. Nozawa, A. G. Hilberts, D. Lohmann, and M. R. Allen. 2011. Anthropogenic greenhouse gas contribution to flood risk in England and Wales in autumn 2000. Nature 470 (7334): 382–385.

Paller, M. H. 1987. Distribution of larval fish between macrophyte beds and open channels in a southeastern floodplain swamp. Journal of Freshwater Ecology 4 (2): 191–200.

Palmer, M., J. Liu, J. Matthews, M. Mumba, and P. D'Odorico. 2015. Manage water in a green way. Science 349 (6248): 584–585.

Pander, J., M. Mueller, and J. Geist. 2015. Succession of fish diversity after reconnecting a large floodplain to the upper Danube River. Ecological Engineering 75: 41–50.

Papworth, S., J. Rist, L. Coad, and E. Milner-Gulland. 2009. Evidence for shifting baseline syndrome in conservation. Conservation Letters 2 (2): 93–100.

Parkinson, A., R. Mac Nally, and G. Quinn. 2002. Differential macrohabitat use by birds on the unregulated Ovens River floodplain of southeastern Australia. River Research and Applications 18 (5): 495–506.

Pearsall, S. H., B. J. McCrodden, and P. A. Townsend. 2005. Adaptive management of flows in the lower Roanoke River, North Carolina, USA. Environmental Management 35 (4): 353–367.

Peipoch, M., M. Brauns, F. R. Hauer, M. Weitere, and H. M. Valett. 2015. Ecological simplification: human influences on riverscape complexity. BioScience 65: 1057–1065.

Peltzer, D. A., P. J. Bellingham, H. Kurokawa, L. R. Walker, D. A. Wardle, and G. W. Yeates. 2009. Punching above their weight: low-biomass non-native plant species alter soil properties during primary succession. Oikos 118 (7): 1001–1014.

Peterjohn, W. T., and D. L. Correll. 1984. Nutrient dynamics in an agricultural watershed: observations on the role of a riparian forest. Ecology 65 (5): 1466–1475.

Peters, D. L., D. Caissie, W. A. Monk, S. B. Rood, and A. St-Hilaire. 2016. An ecological perspective on floods in Canada. Canadian Water Resources Journal 41 (1–2): 288–306.

Petit, L. J. 1999. Prothonotary warbler (*Protonotaria citrea*) no. 408. P. 24. In Poole, A. and Gill, F., eds. The Birds of North America. The Academy of Natural Sciences, Philadelphia, PA.

Petrik, K., M. Petrie, A. Will, and J. McCreary 2012. Waterfowl Impacts of the Proposed Conservation Measure 2 for the Yolo Bypass: An Effects Analysis Tool. Ducks Unlimited, Rancho Cordova, CA.

Petts, G. E. 1998. Floodplain rivers and their restoration: a European perspective. Pp. 29–41.

In Bailey, R. G., José, P. V., and Sharwood, B. R., eds. United Kingdom Floodplains. Westbury Academic and Scientific Publishing, Ottley, UK.

Phelps, Q. E., S. J. Tripp, D. P. Herzog, and J. E. Garvey. 2015. Temporary connectivity: the relative benefits of large river floodplain inundation in the lower Mississippi River. Restoration Ecology 23 (1): 53–56.

Philip Williams and Associates, L. 2003. An Overview of the Potential for Achieving Large-Scale System-Wide Ecologic Restoration Benefits with a Modified Yolo Bypass. Prepared for Sacramento Flood Control Agency, San Francisco, CA.

Phillips, R. T. J., and J. R. Desloges. 2015. Alluvial floodplain classification by multivariate clustering and discriminant analysis for low-relief glacially conditioned river catchments. Earth Surface Processes and Landforms 40 (6): 756–770.

Piazza, B. P. 2014. The Atchafalaya River Basin: History and Ecology of an American Wetland. Texas A&M University Press, College Station.

Pielke, R. A., M. W. Downton, and J. B. Miller. 2002. Flood Damage in the United States, 1926–2000: A Reanalysis of National Weather Service Estimates. University Corporation for Atmospheric Research, Boulder, CO.

Pinay, G., V. J. Black, A. M. Planty-Tabacchi, B. Gumiero, and H. Decamps. 2000. Geomorphic control of denitrification in large river floodplain soils. Biogeochemistry 50 (2): 163–182.

Pingram, M. A., K. J. Collier, D. P. Hamilton, B. J. Hicks, and B. O. David. 2014. Spatial and temporal patterns of carbon flow in a temperate, large river food web. Hydrobiologia 729 (1): 107–131.

Pinter, N. 2005. One step forward, two steps back on US floodplains. Science 308 (5719): 207–208.

Pinter, N., B. S. Ickes, H. H. Wlosinski, and R. R. Van der Ploeg. 2006. Trends in flood stages: contrasting results from the Mississippi and Rhine River systems. Journal of Hydrology 331 (3): 554–566.

Pitlick, J. 1997. A regional perspective on the hydrology of the 1993 Mississippi River Basin floods. Annals of the Association of American Geographers 87 (1): 135–151.

Pittock, J. 2013. Lessons from adaptation to sustain freshwater environments in the Murray–Darling Basin, Australia. Wiley Interdisciplinary Reviews: Climate Change 4 (5): 429–438.

Pittock, J., J. Williams, and R. Grafton. 2015. The Murray-Darling Basin plan fails to deal adequately with climate change. Water: Journal of the Australian Water Association 42 (6): 26–30.

Poff, N. L. R., J. D. Allan, M. B. Bain, J. R. Karr, K. L. Prestegaard, B. D. Richter, R. E. Sparks, and J. C. Stromberg. 1997. The natural flow regime. BioScience 47 (11): 769–784.

Poff, N. L. R., J. D. Olden, D. M. Merritt, and D. M. Pepin. 2007. Homogenization of regional river dynamics by dams and global biodiversity implications. Proceedings of the National Academy of Sciences of the United States of America 104 (14): 5732–5737.

Pollock, M. M., R. J. Naiman, and T. A. Hanley. 1998. Plant species richness in riparian wetlands: a test of biodiversity theory. Ecology 79 (1): 94–105.

Pollock, M. M., G. R. Pess, T. J. Beechie, and D. R. Montgomery. 2004. The importance of beaver ponds to coho salmon production in the Stillaguamish River basin, Washington, USA. North American Journal of Fisheries Management 24: 749–760.

Poole, G. C., S. J. O'Daniel, K. L. Jones, W. W. Woessner, E. S. Bernhardt, A. M. Helton, J. A. Stanford, B. R. Boer, and T. J. Beechie. 2008. Hydrologic spiraling: the role of multiple interactive flow paths in stream ecosystems. River Research and Applications 24 (7): 1018–1031.

Poole, G. C., J. A. Stanford, S. W. Running, and C. A. Frissell. 2006. Multiscale geomorphic drivers of groundwater flow paths: subsurface hydrologic dynamics and hyporheic habitat diversity. Journal of the North American Benthological Society 25 (2): 288–303.

Postel, S. 2005. Liquid Assets: The Critical Need to Safeguard Freshwater Ecosystems. Worldwatch Institute, Washington, DC.

Postel, S., and S. Carpenter. 1997. Freshwater ecosystem services. Pp. 195–214. In Daily, G. C., ed. Nature's Services: Societal Dependence on Natural Ecosystems. Island Press, Washington, DC.

Postel, S., and B. D. Richter. 2003. Rivers for Life: Managing Water for People and Nature. Island Press, Washington, DC.

Power, M. E. 1995. Floods, food chains, and ecosystem processes in rivers. Pp. 52–60. In Jones, C. G. and Lawton, J. H., eds. Linking Species and Ecosystems. Chapman and Hall, New York.

Power, M. E., A. Sun, G. Parker, W. E. Dietrich, and J. T. Wootton. 1995. Hydraulic food-chain models: an approach to the study of food-web dynamics in large rivers. BioScience 45 (3): 159–166.

Puijalon, S., and G. Bornette. 2013. Multi-scale macrophyte responses to hydrodynamic stress and disturbances: adaptive strategies and biodiversity patterns. Pp. 261–273. In Maddock, I., Harby, A., Kemp, P., and Wood, P., eds. Ecohydraulics: An Integrated Approach. John Wiley & Sons, Oxford.

Purdy, S. E., P. B. Moyle, and K. W. Tate. 2012. Montane meadows in the Sierra Nevada: comparing terrestrial and aquatic assessment methods. Environmental Monitoring and Assessment 184 (11): 6967–6986.

Raibley, P. T., T. M. O'Hara, K. S. Irons, K. D. Blodgett, and R. E. Sparks. 1997. Largemouth bass size distributions under varying annual hydrological regimes in the Illinois River. Transactions of the American Fisheries Society 126 (5): 850–856.

Rantala, H., D. Glover, J. Garvey, Q. Phelps, S. Tripp, D. Herzog, R. Hrabik, J. Crites, and M. Whiles. 2016. Fish assemblage and ecosystem metabolism responses to reconnection of the Bird's Point-New Madrid Floodway during the 2011 Mississippi River Flood. River Research and Applications 32: 1018–1029.

Reckendorfer, W., H. Keckeis, G. Winkler, and F. Schiemer. 1999. Zooplankton abundance in the River Danube, Austria: the significance of inshore retention. Freshwater Biology 41 (3): 583–591.

Reese, E. G., and D. P. Batzer. 2007. Do invertebrate communities in floodplains change predictably along a river's length? Freshwater Biology 52 (2): 226–239.

Reffalt, W. C. 1985. A nationwide survey: wetlands in extremis. Wilderness 49 (171): 28–41.

Reiner, R., and T. Griggs. 1989. The Nature Conservancy undertakes riparian restoration projects in California. Ecological Restoration 7 (1): 3–8.

Reiner, R. J. 1996. The Cosumnes River Preserve, 1987–95: fertile ground for new conservation ideas. Fremontia 24 (1): 16–19.

Reinfelds, I., and G. Nanson. 1993. Formation of braided river floodplains, Waimakariri River, New Zealand. Sedimentology 40 (6):1113–1127.

Ribeiro, F., P. K. Crain, and P. B. Moyle. 2004. Variation in condition factor and growth in young-of-year fishes in floodplain and riverine habitats of the Cosumnes River, California. Hydrobiologia 527 (1): 77–84.

Ricciardi, A., and J. B. Rasmussen. 1999. Extinction rates of North American freshwater fauna. Conservation Biology 13 (5): 1220–1222.

Richards, K., J. Brasington, and F. Hughes. 2002. Geomorphic dynamics of floodplains: ecological implications and a potential modeling strategy. Freshwater Biology 47 (4): 559–579.

Richards, K., S. Chandra, and P. Friend. 1993. Avulsive channel systems: characteristics and examples. Geological Society, London, Special Publications 75 (1): 195–203.

Richter, B. 2014. Chasing Water: A Guide for Moving from Scarcity to Sustainability. Island Press, Washington, DC.

Richter, B. 2016. Water Share: Using Water Markets and Impact Investment to Drive Sustainability. The Nature Conservancy, Washington, DC. Available from www.nature.org/watershare.

Richter, B. D., D. P. Braun, M. A. Mendelson, and L. L. Master. 1997. Threats to imperiled freshwater fauna. Conservation Biology 11 (5): 1081–1093.

Richter, B. D., S. Postel, C. Revenga, T. Scudder, B. Lehner, A. Churchill, and M. Chow. 2010. Lost in development's shadow: the downstream human consequences of dams. Water Alternatives 3 (2): 14–42.

Rijke, J., S. van Herk, C. Zevenbergen, and R. Ashley. 2012. Room for the River: delivering integrated river basin management in the Netherlands. International Journal of River Basin Management 10 (4): 369–382.

Ripple, W. J., and R. L. Beschta. 2003. Wolf reintroduction, predation risk, and cottonwood recovery in Yellowstone National Park. Forest Ecology and Management 184 (1): 299–313.

Ripple, W. J., and R. L. Beschta. 2012. Trophic cascades in Yellowstone: the first 15 years after wolf reintroduction. Biological Conservation 145 (1): 205–213.

Risotto, S. P., and R. E. Turner. 1985. Annual fluctuations in abundance of the commercial fisheries in the Mississippi River and tributaries. North American Journal of Fisheries Management 5: 557–574.

Ritter, D. F., W. F. Kinsey, III, and M. E. Kauffman. 1973. Overbank sedimentation in the Delaware River Valley during the last 6000 years. Science 179: 374–375.

Ritter, D. F., R. C. Kochel, and J. R. Miller. 2011. Process Geomorphology. 5th ed. Waveland Press, Long Grove, IL.

River Partners. 2015. Model multiple benefit projects: Dos Rios Ranch and the San Joaquin River National Wildlife Refuge. Accessed April 10, 2017. http://www.riverpartners.org/where-we-work/projects-san-joaquin/.

Roach, K. A. 2013. Environmental factors affecting incorporation of terrestrial material into large river food webs. Freshwater Science 32 (1): 283–298.

Robbins, J. 2014. Paying farmers to welcome birds. The New York Times, April 14, 2014. Available from http://www.nytimes.com/2014/04/15/science/paying-farmers-to-welcome-birds.html?_.

Robertson, A. I., P. Bacon, and G. Heagney. 2001. The responses of floodplain primary production to flood frequency and timing. Journal of Applied Ecology 38 (1): 126–136.

Roegner, G. C., E. W. Dawley, M. Russell, A. Whiting, and D. J. Teel. 2010. Juvenile salmonid use of reconnected tidal freshwater wetlands in Grays River, Lower Columbia River Basin. Transactions of the American Fisheries Society 139 (4): 1211–1232.

Rood, S. B., J. H. Braatne, and F. M. R. Hughes. 2003. Ecophysiology of riparian cottonwoods: stream flow dependency, water relations and restoration. Tree Physiology 23 (16): 1113–1124.

Rood, S. B., and J. M. Mahoney. 1990. Collapse of riparian poplar forests downstream from dams in western Prairies: probable causes and prospects for mitigation. Environmental Management 14 (4): 451–464.

Rood, S. B., G. M. Samuelson, J. H. Braatne, C. R. Gourley, F. M. Hughes, and J. M. Mahoney. 2005. Managing river flows to restore floodplain forests. Frontiers in Ecology and the Environment 3 (4): 193–201.

Roos, M. 2006. Flood management practice in northern California. Irrigation and Drainage 55: S93–S99.

Rosenzweig, M. L. 2003. Reconciliation ecology and the future of species diversity. Oryx 37 (2): 194–205.

Ross, S. T., and J. A. Baker. 1983. The response of fishes to periodic spring floods in a southeastern stream. American Midland Naturalist 109 (1): 1–14.

Roth, B., S. Secchi, J. E. Garvey, G. G. Sass, J. Trushenski, D. Glover, P. M. Charlebois et al. 2012. Fishing Down the Bighead and Silver Carps: Reducing the Risk of Invasion to the Great Lakes. Project Completion Report, US Fish and Wildlife Service and Illinois Department of Natural Resources, Springfield, IL.

Rowland, J. C., W. E. Dietrich, G. Day, and G. Parker. 2009. Formation and maintenance of single-thread tie channels entering floodplain lakes: observations from three diverse river systems. Journal of Geophysical Research: Earth Surface 114 (F2).

Rowntree, K. M., and E. S. Dollar 1999. Vegetation controls on channel stability in the Bell River, Eastern Cape, South Africa. Earth Surface Processes and Landforms 24 (2): 127–134.

Rozo, M. G., A. C. R. Nogueira, and C. S. Castro. 2014. Remote sensing-based analysis of the planform changes in the Upper Amazon River over the period 1986–2006. Journal of South American Earth Sciences 51: 28–44.

Rude, N. P., J. T. Trushenski, and G. W. Whitledge. 2016. Fatty acid profiles are biomarkers of fish habitat use in a river-floodplain ecosystem. Hydrobiologia 773 (1): 63–75.

Rzoska, J. 1978. On the Nature of Rivers. Junk, The Hague.

SAFCA (Sacramento Area Flood Control Agency). 2008. Lower Sacramento River Regional Project Initial Report. Administrative Draft, Sacramento, SAFCA.

Salcido, R. E. 2012. Success and continued challenges of the Yolo Bypass Wildlife Area: a grassroots restoration. The Ecology Law Quarterly 39 (4): 1085–1134.

Salo, J., R. Kalliola, I. Häkkinen, Y. Mäkinen, P. Niemelä, M. Puhakka, and P. D. Coley. 1986. River dynamics and the diversity of Amazon lowland forest. Nature 322 (6076): 254–258.

San Joaquin River Restoration Program. 2015. Homepage. State and Federal Implementing Agencies. Accessed April 10, 2017. http://www.restoresjr.net.

Saunders, S., D. Findlay, T. Easley, and T. Spencer. 2012. Doubled trouble: more Midwestern extreme storms. The Rocky Mountain Climate Organization and the Natural Resources Defense Council, Louisville, CO and New York, 150: 409–419.

Sayers, P., G. Galloway, E. Penning-Rowsell, L. Yuanyuan, S. Fuxin, C. Yiwei, W. Kang, T. Le Quesne, L. Wang, and Y. Guan. 2015. Strategic flood management: ten "golden rules" to guide a sound approach. International Journal of River Basin Management 13 (2): 137–151.

Scheerer, P. D. 2002. Implications of floodplain isolation and connectivity on the conservation of an endangered minnow, Oregon chub, in the Willamette River, Oregon. Transactions of the American Fisheries Society 131 (6): 1070–1080.

Schemel, L. E., T. R. Sommer, A. B. Muller-Solger, and W. C. Harrell. 2004. Hydrological variability, water chemistry, and phytoplankton biomass in a large floodplain of the Sacramento River, CA, USA. Hydrobiologia 513: 129–139.

Schiemer, F., S. Keckeis, W. Reckendorfer, and G. Winkler. 2001. The "inshore retention concept" and its significance for large rivers. Algological Studies 135: 509–516.

Schipper, A. M., S. Wijnhoven, H. Baveco, and N. W. van den Brink. 2012. Contaminant exposure in relation to spatio-temporal variation in diet composition: a case study of the little owl (Athene noctua). Environmental Pollution 163: 109–116.

Schneider, K. S., G. M. Kondolf, and A. Falzone. 2003. Channel-floodplain disconnection on the Stanislaus River: a hydrologic and geomorphic perspective. Pp. 163–168. In Faber, P. M., ed. California Riparian Systems: Processes and Floodplain Management, Ecology and Restoration 2001 Riparian Habitat and Floodplain Conference. Riparian Habitat Joint Venture, Sacramento, CA.

Schramm, Jr., H. L., and M. A. Eggelton. 2006. Applicability of the Flood Pulse Concept in a temperate floodplain river ecosystem: thermal and temporal components. River Research and Applications 22: 543–553.

Schumm, S. A. 1977. The Fluvial System. John Wiley & Sons, New York.

Schuyt, K. D. 2005. Economic consequences of wetland degradation for local populations in Africa. Ecological Economics 53 (2): 177–190.

Scott, L. B., and S. K. Marquiss. 1984. An historical overview of the Sacramento River. Pp. 51–57. In Warner, R. E. and Hendrix, K. M., eds. California Riparian Systems: Ecology, Conservation, and Productive Management. University of California Press, Berkeley.

Seavy, N. E., T. Gardali, G. H. Golet, F. T. Griggs, C. A. Howell, R. Kelsey, S. L. Small, J. H. Viers, and J. F. Weigand. 2009. Why climate change makes riparian restoration more important than ever: recommendations for practice and research. Ecological Restoration 27 (3): 330–338.

Sedell, J. R., G. Reeves, R. Hauer, J. Stanford, and C. Hawkins. 1990. Role of refugia in recovery from disturbances: modern fragmented and disconnected river systems. Environmental Management 14: 711–724.

Sedell, J. R., J. E. Richey, and F. J. Swanson. 1989. The river continuum concept: a basis for the expected ecosystem behavior of very large rivers? Pp. 49–55. In Dodge, D. P., ed. Proceedings of the International Large River Symposium. Canadian Special Publication of Fisheries and Aquatic Sciences, Ottawa, ON.

Shafroth, P. B., G. T. Auble, J. C. Stromberg, and D. T. Patten. 1998. Establishment of woody riparian vegetation in relation to annual patterns of stream flow, Bill Williams River, Arizona. Wetlands 18 (4): 577–590.

Shafroth, P. B., J. C. Stromberg, and D. T. Patten. 2000. Woody riparian vegetation response to different alluvial water table regimes. Western North American Naturalist 60 (1): 66–76.

Shafroth, P. B., A. C. Wilcox, D. A. Lytle, J. T. Hickey, D. C. Andersen, V. B. Beauchamp, A. Hautzinger, L. E. McMullen, and A. Warner. 2010. Ecosystem effects of environmental flows: modelling and experimental floods in a dryland river. Freshwater Biology 55: 68–85.

Shaw, J. H. 1995. How many bison originally populated western rangelands? Rangelands 17(5): 148–150.

Sheaffer, J. R., J. D. Mullan, and N. B. Hinch. 2002. Encouraging wise use of floodplains with market-based incentives. Environment: Science and Policy for Sustainable Development 44 (1): 32–43.

Sheibley, R. W., D. S. Ahearn, and R. A. Dahlgren. 2006. Nitrate loss from a restored floodplain in the lower Cosumnes River, California. Hydrobiologia 571: 261–272.

Shelton, M. L. 1987. Irrigation induced change in vegetation and evapotranspiration in the Central Valley of California. Landscape Ecology 1 (2): 95–105.

Shields, Jr., F. D., A. Simon, and L. J. Steffen. 2000. Reservoir effects on downstream river channel migration. Environmental Conservation 27 (1): 54–66.

Singer, M. B. 2007. The influence of major dams on hydrology through the drainage network of the Sacramento River basin, California. River Research and Applications 23 (1): 55–72.

Singer, M. B., and R. Aalto. 2009. Floodplain development in an engineered setting. Earth Surface Processes and Landforms 34 (2): 291–304.

Singer, M. B., R. Aalto, and L. A. James. 2008. Status of the lower Sacramento Valley flood-control system within the context of its natural geomorphic setting. Natural Hazards Review 9 (3): 104–115.

Slingerland, R., and N. D. Smith. 1998. Necessary conditions for a meandering-river avulsion. Geology 26 (5): 435–438.

Slingerland, R., and N. D. Smith. 2004. River avulsions and their deposits. Annual Review of Earth and Planetary Sciences 32 (1): 257–285.

Smith, N. D., T. A. Cross, J. P. Dufficy, and S. R. Clough. 1989. Anatomy of an avulsion. Sedimentology 36 (1): 1–23.

Smith, N. D., and M. Perez-Arlucea. 1994. Fine-grained splay deposition in the avulsion belt of the lower Saskatchewan River, Canada.

Journal of Sedimentary Research 64 (2): 159–168.

Smock, L. A., J. E. Gladden, J. L. Riekenberg, L. C. Smith, and C. R. Black. 1992. Lotic macroinvertebrate production in three dimensions: channel surface, hyporheic, and floodplain environments. Ecology 73 (3): 876–886.

Snyder, W. D., and G. C. Miller. 1991. Changes in plains cottonwoods along the Arkansas and South Platte Rivers: eastern Colorado. Prairie Naturalist 23 (3): 165–176.

Sobczak, W. V., J. E. Cloern, A. D. Jassby, and A. B. Muller-Solger. 2002. Bioavailability of organic matter in a highly disturbed estuary: the role of detrital and algal resources. Proceedings of the National Academy of Sciences 99 (12): 8101–8105.

Sobczak, W. V., J. E. Cloern, A. D. Jassby, B. E. Cole, T. S. Schraga, and A. Arnsberg. 2005. Detritus fuels ecosystem metabolism but not metazoan food webs in San Francisco estuary's freshwater Delta. Estuaries 28 (1): 124–137.

Sommer, T., R. Baxter, and B. Herbold. 1997. Resilience of splittail in the Sacramento-San Joaquin estuary. Transactions of the American Fisheries Society 126: 961–976.

Sommer, T., B. Harrell, M. Nobriga, R. Brown, P. Moyle, W. Kimmerer, and L. Schemel. 2001. California's Yolo Bypass: evidence that flood control can be compatible with fisheries, wetlands, wildlife, and agriculture. Fisheries 26 (8): 6–16.

Sommer, T. R., W. C. Harrell, R. Kurth, F. Feyrer, S. C. Zeug, and G. O'Leary. 2004. Ecological patterns of early life stages of fishes in a large river-floodplain of the San Francisco estuary. Pp. 111–123. In Feyrer, F., Brown, L. R., Brown, R. L., and Orsi, J. J., eds. Early Life History of Fishes in the San Francisco Estuary and Watershed. American Fisheries Society, Bethesda, MD.

Sommer, T. R., W. C. Harrell, and M. L. Nobriga. 2005. Habitat use and stranding risk of juvenile Chinook salmon on a seasonal floodplain. North American Journal of Fisheries Management 25: 1493–1504.

Sommer, T. R., W. C. Harrell, A. M. Solger, B. Tom, and W. Kimmerer. 2004. Effects of flow variation on channel and floodplain biota and habitats of the Sacramento River, California, USA. Aquatic Conservation-Marine and Freshwater Ecosystems 14 (3): 247–261.

Sommer, T. R., M. L. Nobriga, W. C. Harrell, W. Batham, and W. J. Kimmerer. 2001. Floodplain rearing of juvenile Chinook salmon: evidence of enhanced growth and survival. Canadian Journal of Fisheries and Aquatic Sciences 58: 325–333.

Southwick, R., J. Bergstrom, and C. Wall. 2009. The economic contributions of human-powered outdoor recreation to the US economy. Tourism Economics 15 (4): 709–733.

Sparks, R. E., P. B. Bayley, S. L. Kohler, and L. L. Osborne. 1990. Disturbance and recovery of large floodplain rivers. Environmental Management 14 (5): 699–709.

Spink, A., R. E. Sparks, M. van Oorschot, and J. T. A. Verhoeven. 1998. Nutrient dynamics of large river floodplains. Regulated Rivers: Research and Management 14: 203–216.

Springborn, M., M. B. Singer, and T. Dunne. 2011. Sediment-adsorbed total mercury flux through Yolo Bypass, the primary floodway and wetland in the Sacramento Valley, California. Science of the Total Environment 412: 203–213.

Stammel, B., B. Cyffka, J. Geist, M. Müller, J. Pander, G. Blasch, P. Fischer, A. Gruppe, F. Haas, and M. Kilg. 2012. Floodplain restoration on the Upper Danube (Germany) by re-establishing water and sediment dynamics: a scientific monitoring as part of the implementation. River Systems 20 (1–2): 55–70.

Stanford, J. A., M. Lorang, and F. Hauer. 2005. The shifting habitat mosaic of river ecosystems. Proceedings of the International Association of Theoretical and Applied Limnology 29 (1): 123–136.

Stanford, J. A., and J. V. Ward. 1988. The hyporheic habitat of river ecosystems. Nature 335: 64–66.

Stanturf, J. A., E. S. Gardiner, P. B. Hamel, M. S. Devall, T. D. Leininger, and M. E. Warren. 2000. Restoring bottomland hardwood ecosystems in the Lower Mississippi Alluvial Valley. Journal of Forestry 98 (8): 10–16.

Stanturf, J. A., S. H. Schoenholtz, C. J. Schweitzer, and J. P. Shepard. 2001. Achieving restoration success: myths in bottomland hardwood forests. Restoration Ecology 9 (2): 189–200.

Steffen, W., P. J. Crutzen, and J. R. McNeill. 2007. The Anthropocene: are humans now overwhelming the great forces of Nature. AMBIO: A Journal of the Human Environment 36(8): 614–621.

Stella, J. C., J. J. Battles, J. R. McBride, and B. K. Orr. 2010. Riparian seedling mortality from simulated water table recession, and the design of sustainable flow regimes on regulated rivers. Restoration Ecology 18: 284–294.

Stella, J. C., J. J. Battles, B. K. Orr, and J. R. McBride. 2006. Synchrony of seed dispersal, hydrology and local climate in a semi-arid river reach in California. Ecosystems 9 (7): 1200–1214.

Stella, J., M. Hayden, J. Battles, H. Piegay, S. Dufour, and A. K. Fremier. 2011. The role of abandoned channels as refugia for sustaining pioneer riparian forest ecosystems. Ecosystems 14 (5): 776–790.

Stella, J. C., P. M. Rodríguez-González, S. Dufour, and J. Bendix, 2013. Riparian vegetation research in Mediterranean-climate regions: common patterns, ecological processes, and considerations for management. Hydrobiologia 719 (1): 291–315.

Stella, J., J. Vick, and B. Orr. 2003. Riparian vegetation dynamics on the Merced River. Pp. 302–314. In Faber, P. M. ed. California Riparian Systems: Processes and Floodplain Management, Ecology and Restoration. 2001 Riparian Habitat and Floodplain Conference. Riparian Habitat Joint Venture, Sacramento, CA.

Stillwater Sciences. 2013. 2012 Vegetation monitoring of the Napa River Flood Protection Project: Final Report. Stillwater Sciences, Berkeley, CA.

Stokes, D. 2014. The Fish in the Forest: Salmon and the Web of Life. University of California Press, Berkeley.

Strahan, J. 1981. Regeneration of riparian forests of the Central Valley. Pp. 58–67. In Warner, R. E. and Hendrix, K. M., eds. California Riparian Systems: Ecology, Conservation, and Productive Management. Proceedings of the California Riparian Systems Conference, Davis.

Stromberg, J. C., S. J. Lite, R. Marler, C. Paradzick, P. B. Shafroth, D. Shorrock, J. M. White, and M. S. White. 2007. Altered stream-flow regimes and invasive plant species: the Tamarix case. Global Ecology and Biogeography 16 (3): 381–393.

Stutler, L. L. 1973. An Historical Geography of the Yolo Bypass. MS Thesis, University of California, Berkeley.

Suddeth, R. J. 2014. Multi-Objective Analysis for Ecosystem Reconciliation on an Engineered Floodplain: The Yolo Bypass in California's Central Valley. PhD Dissertation, University of California Davis.

Svadlenak-Gomez, K. 2010. Biodiversity-Friendly Aquaculture on the Veta La Palma Estate, Spain. Ecoagriculture Partners, Washington, D.C.

Swales, S., R. B. Lauzier, and C. D. Levings. 1985. Winter habitat preferences of juvenile salmonids in two interior rivers in British Columbia. Canadian Journal of Zoology 64: 1506–1514.

Swanson, F. J. 2003. Wood in rivers: a landscape perspective. Pp. 299–313. In Gregory, S., Boyer, K., and Gurnell, A., eds. The Ecology and Management of Wood in World Rivers. American Fisheries Society, Bethesda, MD.

Swanson, K. M., E. Watson, R. Aalto, J. Lauer, M. T. Bera, A. Marshall, M. P. Taylor, S. C. Apte, and W. E. Dietrich. 2008. Sediment load and floodplain deposition rates: comparison of the Fly and Strickland rivers, Papua New Guinea. Journal of Geophysical Research: Earth Surface (2003–2012) 113 (F1).

Swenson, R. O., K. Whitener, and M. Eaton. 2003. Restoring floods on floodplains: riparian and floodplain restoration at the Cosumnes River Preserve. Pp. 224–229. In Faber, P. M., ed. California Riparian Systems: Processes and Floodplain Management, Ecology and Restoration. 2001 Riparian Habitat and Floodplain Conference. Riparian Habitat Joint Venture, Sacramento, CA.

Taft, O. W., M. A. Colwell, C. R. Isola, and R. J. Safran. 2002. Waterbird responses to experimental drawdown: implications for the multispecies management of wetland mosaics. Journal of Applied Ecology 39 (6): 987–1001.

Tappe, D. T. 1942. The Status of Beavers in California. Game Bulletin 3. State of California, Department of Natural Resources, Division of Fish and Game, Sacramento.

Taylor, C. 2013. A highway, a wetland and 250,000 bats. BayNature Magazine, July 25.

Teel, D. J., C. Baker, D. R. Kuligowski, T. A. Friesen, and B. Shields. 2009. Genetic stock composition of subyearling chinook salmon in seasonal floodplain wetlands of the Lower Willamette River, Oregon. Transactions of the American Fisheries Society 138 (1): 211–217.

Tharme, R. E. 2003. A global perspective on environmental flow assessment: emerging trends in the development and application of environmental flow methodologies for rivers. River Research and Applications 19 (5–6): 397–441.

Thayer, R. L. 2003. LifePlace: Bioregional Thought and Practice. University of California Press, Berkeley.

Thompson, K. 1961. Riparian forests of the Sacramento Valley, California. Annals of the Association of American Geographers 51: 294–315.

Thoms, M. C. 2003. Floodplain-river ecosystems: lateral connections and the implications of human interference. Geomorphology 56 (3–4): 335–349.

Thorp, J. H., and M. D. Delong. 1994. The riverine productivity model: an heuristic view of carbon sources and organic processing in large river ecosystems. Oikos 70 (2): 305–308.

Thorp, J. H., and M. D. Delong. 2002. Dominance of autochthonous autotrophic carbon in food webs of heterotrophic rivers. Oikos 96 (3): 543–550.

Thorp, J. H., M. D. Delong, K. S. Greenwood, and A. F. Casper. 1998. Isotopic analysis of three food web theories in constricted and floodplain regions of a large river. Oecologia 117: 551–563.

Tilman, D., J. Hill, and C. Lehman. 2006. Carbon-negative biofuels from low-input high-diversity grassland biomass. Science 314 (5805): 1598–1600.

Tobin, G. A. 1995. The levee love affair: a stormy relationship. Water Resources Bulletin 31 (3): 359–367.

Tockner, K., M. S. Lorang, and J. A. Stanford. 2010. River flood plains are model ecosystems to test general hydrogeomorphic and ecological concepts. River Research and Applications 26 (1): 76–86.

Tockner, K., F. Malard, and J. V. Ward. 2000. An extension of the flood pulse concept. Hydrological Processes 14 (16–17): 2861–2883.

Tockner, K., D. Pennetzdorfer, N. Reiner, F. Schiemer, and J. V. Ward. 1999. Hydrological connectivity, and the exchange of organic matter and nutrients in a dynamic river-floodplain system (Danube, Austria). Freshwater Biology 41 (3): 521–535.

Tockner, K., M. Pusch, D. Borchardt, and M. S. Lorang. 2010. Multiple stressors in coupled river-floodplain ecosystems. Freshwater Biology 55: 135–151.

Tockner, K., and F. Schiemer. 1997. Ecological aspects of the restoration strategy for a river-floodplain system on the Danube River in Austria. Global Ecology and Biogeography Letters 6: 321–329.

Tockner, K., and J. A. Stanford. 2002. Riverine flood plains: present state and future trends. Environmental Conservation 29 (3): 308–330.

Toral, G. M. and J. Figuerola. 2010. Unraveling the importance of rice fields for waterbird populations in Europe. Biodiversity and Conservation 19: 3459–3469.

Trenberth, K. E., A. Dai, R. M. Rasmussen, and D. B. Parons. 2003. The changing character of precipitation. Bulletin of the American Meteorological Society 84: 1205–1217.

Trigg, M. A., P. D. Bates, M. D. Wilson, G. Schumann, and C. Baugh. 2012. Floodplain channel morphology and networks of the middle Amazon River. Water Resources Research 48 (10): W10504. doi:10.1029/2012WR011888.

Trimble, S. W. 1999. Decreased rates of alluvial sediment storage in the Coon Creek basin, Wisconsin, 1975–93. Science 285: 1244–1246.

Trowbridge, W. B. 2002. The Influence of Restored Flooding on Floodplain Plant Distributions, PhD Dissertation, University of California, Davis.

Truan, M. L., A. English, Jr., and J. R. Trochet. 2010. Putah Creek Terrestrial Wildlife Monitoring Program Comprehensive Report, 1997–2009. Museum of Wildlife and Fish Biology. University of California, Davis.

Trush, W. J., S. M. McBain, and L. B. Leopold. 2000. Attributes of an alluvial river and their relation to water policy and management. Proceedings of the National Academy of Sciences of the United States of America 97 (22): 11858–11863.

Tu, M. 2000. Vegetation Patterns and Processes of Natural Regeneration in Periodically Flooded Riparian Forests in the Central Valley of California. PhD Dissertation, University of California, Davis.

UNEP (United Nations Environment Programme). 2010. Blue Harvest: Inland Fisheries as an Ecosystem Service. The WorldFish Center Working Papers, the WorldFish Center, Penang, Malaysia.

UNEP-DHI Partnership. 2014. Green Infrastructure Guide for Water Management: Ecosystem-based Management Approaches for Water-related Infrastructure Projects. United Nations Environment Programme, Nairobi, Kenya.

United States Congress. 1850. Arkansas Act. United States Statutes at Large, IX, chapter LXXXIV. Pp. 519–520.

Urban Design and Aesthetics Workgroup. 1997. Napa River Flood Management Plan Napa City Reach Overview and Recommendations. County of Napa, Napa, CA.

US Army Corps of Engineers. 1971. Floods of December 1946 and January 1965 in the Far Western States (photo pp. 100). US Army Corps of Engineers, Washington, DC. Accessed February 6, 2017. http://pubs.usgs.gov/wsp/1866a/report.pdf.

US Army Corps of Engineers. 1975. Napa River Flood Control Project Environmental Statement. US Army Corps of Engineers, San Francisco, CA.

US Army Corps of Engineers. 1993. Massachusetts Natural Valley Storage Investigation. US Army Corps of Engineers, New England Division, Waltham, MA.

US Army Corps of Engineers. 1995. Napa River, California, Draft Supplemental General Design Memorandum/Environmental Impact Statement and Environmental Impact Report. US Army Corps of Engineers, Washington, DC.

US Army Corps of Engineers. 2008. The Mississippi River and Tributaries Project: Designing the Project Flood. US Army Corps of Engineers, Washington, DC.

US Army Corps of Engineers. 2015. Mississippi River and Tributaries Project. US Army Corps of Engineers, Mississippi Valley Division, Vicksburg, MS. Accessed April 10, 2017. http://www.mvd.usace.army.mil/About/MississippiRiverCommission(MRC)/MississippiRiverTributariesProject(MRT).aspx.

US Army Corps of Engineers and Napa County Flood Control and Water Conservation District. 1998. Napa River and Napa Creek Flood Protection Project: Final Supplemental General Design Memorandum. Sacramento District, Napa, CA.

US Army Corps of Engineers and Napa County Flood Control and Water Conservation District. 1999. Napa River and Napa Creek Flood Reduction Project: Final Supplemental Environmental Impact Statement and Environmental Impact Report. Prepared by Design, Community, and Environment, Berkeley, CA.

US Fish and Wildlife Service. 2012. Butte Sink Wildlife Management Area. Accessed April 10, 2017. https://www.fws.gov/refuge/Butte_Sink/.

USDA (US Department of Agriculture). 2009. Vilsack announces nearly $42.3 million in recovery act funding for watershed projects: builds on USDA efforts to create jobs, help rural communities and conserve natural resources. Release No. 0190.09, June 2. Available from www.usda.gov.

Valett, H. M., M. A. Baker, J. A. Morrice, C. S. Crawford, M. C. Molles, C. N. Dahm, D. L. Moyer, J. R. Thibault, and L. M. Ellis. 2005. Biogeochemical and metabolic responses to the flood pulse in a semiarid floodplain. Ecology 86 (1): 220–234.

Van Cleve, G. W. 2012. Saving the Puget Sound wild salmon fishery. Seattle Journal of Environmental Law 2: 85.

van Groenigen, J. W., E. G. Burns, J. M. Eadie, W. R. Horwath, and C. van Kessel. 2003. Effects of foraging waterfowl in winter flooded rice fields on weed stress and residue decomposition. Agriculture Ecosystems and Environment 95 (1): 289–296.

van Stokkom, H. T., A. J. Smits, and R. S. Leuven. 2005. Flood defense in the Netherlands: a new era, a new approach. Water international 30 (1): 76–87.

Vannote, R. L., G. W. Minshall, K. W. Cummins, J. R. Sedell, and C. E. Cushing. 1980. The river continuum concept. Canadian Journal of Fisheries and Aquatic Sciences 37: 130–137.

Vaught, D. J. 2006. A swamplander's vengeance: R. S. Carey and the failure to reclaim Putah Sink 1855–1895. Sacramento History Journal 6: 161–176.

Volk, T. A., T. Verwijst, P. J. Tharakan, L. P. Abrahamson, and E. H. White. 2004. Growing fuel: a sustainability assessment of willow biomass crops. Frontiers in Ecology and the Environment 2 (8): 411–418.

Waananen, A. O., D. D. Harris, and R. C. Williams. 1971. Floods of December 1964 and January 1965 in the Far Western States. Part 1. Description. Geological Survey Water Supply Paper 1866-A. United States Government Printing Office, Washington, DC.

Walling, D. E., Q. He, and A. P. Nicholas. 1996. Floodplains as suspended sediment sinks. Pp. 399–440. In Anderson, M. G., Walling, D. E., and Bates, B. D., eds. Floodplain Processes. John Wiley & Sons, New York.

Walling, D. E., T. Quine, and Q. He. 1992. Investigating contemporary rates of floodplain sedimentation. Pp. 165–184. In Carling, P. A. and Petts, G. E., eds. Lowland Floodplain Rivers: Geomorphological Perspectives. John Wiley & Sons, New York.

Walsh, S. J., E. N. Buttermore, O. T. Burgess, and W. E. Pine, III. 2009. Composition of Age-0 Fish Assemblages in the Apalachicola River, River Styx, and Battle Bend, Florida. Open-File Report No. 2009-1145. US Geological Survey, Sacramento, CA.

Ward, B. R., and P. A. Slaney. 1988. Life history and smolt-to-adult survival of Keogh River steelhead trout (Salmo gairdneri) and the relationship to smolt size. Canadian Journal of Fisheries and Aquatic Sciences 45: 1110–1122.

Ward, J. V. 1989. The four-dimensional nature of lotic ecosystems. Journal of the North American Benthological Society 8 (1): 2–8.

Ward, J. V. 1997. An expansive perspective of riverine landscapes: pattern and process across scales. GAIA-Ecological Perspectives for Science and Society 6 (1): 52–60.

Ward, J. V., and J. A. Stanford. 1995. The serial discontinuity concept: extending the model to floodplain rivers. Regulated Rivers: Research and Management 10: 159–168.

Ward, J. V., K. Tockner, D. B. Arscott, and C. Claret. 2002. Riverine landscape diversity. Freshwater Biology 47 (4): 517–539.

Ward, J. V., K. Tockner, U. Uehlinger, and F. Malard. 2001. Understanding natural patterns and processes in river corridors as the basis for effective river restoration. Regulated Rivers: Research and Management 17 (4–5): 311–323.

Warner, A., J. J. Opperman, and R. Pietrowsky. 2011. A call to enhance the resiliency of the nation's water management. Journal of Water Resources Planning and Management 137 (4): 305–308.

Warner, J., and A. van Buuren. 2011. Implementing Room for the River: narratives of success and failure in Kampen, the Netherlands. International Review of Administrative Sciences 77 (4): 779–801.

Watkins (photographer). 1800s. Hydraulic Mining, Nevada County, California, No. 1417. UC Berkeley, Bancroft Library, Berkeley, CA.

Watson, K. B., T. Ricketts, G. Galford, S. Polasky, and J. O'Niel-Dunne. 2016. Quantifying flood mitigation services: the economic value of Otter Creek wetlands and floodplains to Middlebury, VT. Ecological Economics 130: 16–24.

Weiser, M. 2005. Years of neglect help put valley's levees at risk. Sacramento Bee, September 4.

Welcomme, R. L. 1979. Fisheries Ecology of Floodplain Rivers. Longman Group LTD, London.

Whelchel, A. W., and M. W. Beck. 2016. Decision tools and approaches to advance ecosystem-based disaster risk reduction and climate change adaptation in the twenty-first century. Pp. 133–160. In Renaud, F. G., Sudmeier-Rieux, K., Estrella, M., and Nehren, U., eds. Ecosystem-Based Disaster Risk Reduction and Adaptation in Practice. Springer International Publishing, Cham, Switzerland.

Wiens, J. A. 2002. Riverine landscapes: taking landscape ecology into the water. Freshwater Biology 47: 501–515.

Wijnhoven, S., A. J. M. Smits, G. van der Velde, and R. S. E. W. Leuven. 2006. Modeling recolonization of heterogeneous river floodplains by small mammals. Pp. 135–152. In Leuven, R. S. E. W., Ragas, A. M. J., Smits, A. J. M., and van der Velde, G., eds. Living Rivers: Trends and Challenges in Science and Management. Springer, Dordrecht, the Netherlands.

Wilcox, A. C., and P. B. Shafroth. 2013. Coupled hydrogeomorphic and woody seedling responses to controlled flood releases in a dryland river. Water Resources Research 49 (5): 2843–2860.

Williams, J. G. 2006. Central Valley salmon: a perspective on Chinook and Steelhead in the Central Valley of California. San Francisco Estuary and Watershed Science 4 (3), Article 2.

Williams, P. B. 1994. Flood control vs. flood management. Civil Engineering 64 (5): 51–54.

Williams, P. B., E. Andrews, J. J. Opperman, S. Bozkurt, and P. B. Moyle. 2009. Quantifying activated floodplains on a lowland regulated river: its application to floodplain restoration in the Sacramento Valley. San Francisco Estuary and Watershed Science 7 (1): 1–26.

Winemiller, K. O. 1989. Patterns of variation in life history among South American fishes in seasonal environments. Oecologia 81 (2): 225–241.

Winemiller, K. O. 1996. Factors driving temporal and spatial variation in aquatic floodplain food webs. Pp. 298–312. In Polis, G. and Winemiller, K., eds. Food Webs: Integration of Pattern and Dynamics. Chapman and Hall, New York.

Winemiller, K. O. 2004. Floodplain river food webs: generalizations and implications for management. Pp. 285–309. In Welcomme, R. L. and Petr, T., eds. Proceedings of the Second International Symposium on the Management of Large Rivers for Fisheries. Vol. 2. Food and Agricultural Organization (FAO) of the United Nations, Phnom Penh, Cambodia.

Winemiller, K. O., S. Tarim, D. Shormann, and J. B. Cotner. 2000. Fish assemblage structure in relation to environmental variation among Brazos River oxbow lakes. Transactions of the American Fisheries Society 129 (2): 451–468.

Winemiller, K. O., P. B. McIntyre, L. Castello, E. Fluet-Chouinard, T. Giarrizzo, S. Nam, I. G. Baird et al. 2016. Balancing hydropower and biodiversity in the Amazon, Congo, and Mekong. Science 351 (6269): 128–129.

Wohl, E. E. 2000. Geomorphic effects of floods. Pp. 167–193. In Wohl, E. E., ed. Inland Flood Hazards: Human, Riparian, and Aquatic Communities. Cambridge University Press, Cambridge, UK.

Wohl, E. E. 2010. Mountain Rivers Revisited. Water Resources Monograph 19. American Geophysical Union Geopress. p.573. doi:10.1029/2010WM001038.

Wohl, E. E. 2011. A World of Rivers: Environmental Change on Ten of the World's Great Rivers. University of Chicago Press, Chicago, IL.

Wolman, M. G., and L. B. Leopold. 1957. River Flood Plains: Some Observations on Their Formation. Professional Paper 282-C, US Geological Survey, Reston, VA.

Wolman, M. G., and J. P. Miller. 1959. Magnitude and frequency of forces in geomorphic processes. Journal of Geology 68: 54–74.

Wolsink, M. 2006. River basin approach and integrated water management: governance pitfalls for the Dutch space-water-adjustment management principle. Geoforum 37 (4): 473–487.

Woltemade, C. J. 2000. Ability of restored wetlands to reduce Nitrogen and Phosphorous concentrations in agricultural drainage water.

Journal of Soil and Water Conservation 55: 303–309.

Wood, J. K., N. Nur, C. A. Howell, and G. R. Geupel. 2006. Overview of Cosumnes Riparian Bird study and Recommendations for Monitoring and Management. A Report to the California Bay-Delta Authority Ecosystem Restoration Program. Point Reyes Bird Observatory (PRBO) Conservation Science, Petaluma, CA.

Wooton, J. T., M. S. Parker, and M. E. Power. 1996. Effects of disturbance on river food webs. Science 273: 1558–1561.

Wu, Q, Z. Zhao, L. Liu, D. E. Granger, H. Wang, D. J. Cohen, X. Wu et al. 2016. Outburst flood of 1920 BCE supports historicity of China's Great Flood and the Xia Dynasty. Science 353: 579–582.

WWF (WWF Danube-Carpathian-Programme and WWF-Auen-Institut). 1999. Evaluation of Wetlands and Floodplain Areas in the Danube River Basin: Final Report. WWF-Auen-Institut, Rastatt, Germany.

WWF. 2014. Living planet report 2014. Gland, Switzerland.

Yarie, J., L. Viereck, K. Van Cleve, and P. Adams. 1998. Flooding and ecosystem dynamics along the Tanana River. BioScience 48: 690–695.

Yarnell, S. M., G. E. Petts, J. C. Schmidt, A. A. Whipple, E. E. Beller, C. N. Dahm, P. Goodwin, and J. H. Viers. 2015. Functional flows in modified riverscapes: hydrographs, habitats and opportunities. BioScience 65 (10): 963–972.

Yarnell, S. M., J. H. Viers, and J. F. Mount. 2010. Ecology and management of the spring snowmelt recession. Bioscience 60 (2): 114–127.

Yolo Habitat Conservancy. 2016. Homepage. Accessed April 10, 2017. http://www.yolohabitatconservancy.org/.

Yoon, J., S. S. Wang, R. R. Gillies, B. Kravitz, L. Hipps, and P. J. Rasch. 2015. Increasing water cycle extremes in California and in relation to ENSO cycle under global warming. Nature Communications 6: 8657.

Young, M. P., G. W. Whitledge, and J. T. Trushenski. 2014. Fatty acid profiles distinguish channel catfish from three reaches of the Lower Kaskaskia River and its floodplain lakes. River Research and Applications 32 (3): 362–372.

Zedler, J. B. 2003. Wetlands at your service: reducing impacts of agriculture at the watershed scale. Frontiers in Ecology and the Environment 1 (2): 65–72.

Zehetner, F., G. J. Lair, and M. H. Gerzabek. 2009. Rapid carbon accretion and organic matter pool stabilization in riverine floodplain soils. Global Biogeochemical Cycles 23: 1–7.

Zeug, S. C., K. O. Winemiller, and S. Tarim. 2005. Response of Brazos River oxbow fish assemblages to patterns of hydrologic connectivity and environmental variability. Transactions of the American Fisheries Society 134 (5): 1389–1399.

Zeug, S. C., and K. O. Winemiller. 2008a. Evidence supporting the importance of terrestrial carbon in a large-river food web. Ecology 89 (6): 1733–1743.

Zeug, S. C., and K. O. Winemiller. 2008b. Relationships between hydrology, spatial heterogeneity, and fish recruitment dynamics in a temperate floodplain river. River Research and Applications 24 (1): 90–102.

Zevenbergen, C., J. van Tuijn, J. Rijke, M. Bos, S. van Herk, J. Douma, and L. van Riet Paap. 2013. Tailor-made Collaboration: A Clever Combination of Process and Content. Room for the River and UNESCO-IHE, Utrecht, the Netherlands.

Zhang M., K. Wang, Y. Wang, C. Guo, B. Li, and H. Huang 2007. Recovery of a rodent community in an agroecosystem after flooding. Journal of Zoology 272: 138–147.

Ziv, G., E. Baran, S. Nam, I. Rodríguez-Iturbe, and S. A. Levin. 2012. Trading-off fish biodiversity, food security, and hydropower in the Mekong River Basin. Proceedings of the National Academy of Sciences of the United States of America 109 (15): 5609–5614.

GEOSPATIAL DATA SOURCES

The Bay Institute. 1998. From the Sierra to the Sea: The Ecological History of the San Francisco Bay-Delta Watershed. The Bay Institute, San Francisco, CA. Available from http://www.thebayinstitute.org/page/detail/164.

CalAtlas. 2012. California Geospatial Clearinghouse. State of California. Accessed 2012.

CDWR (California Department of Water Resources). 2010. Levee Centerlines. California Department of Water Resources, Sacramento.

CDWR. 2012. Central Valley Flood Protection Plan. Attachment 9F: Floodplain Restoration Opportunity Analysis. California Department of Water Resources, Sacramento.

CDWR. 2016. California Levee Database. Accessed 2016. http://www.water.ca.gov/floodmgmt/lrafmo/fmb/fes/levee_database.cfm.

US Census. 2016. Cartographic Boundary Shapefiles – States. Accessed 2016. https://www.census.gov/geo/maps-data/data/cbf/cbf_state.html.

European Environment Agency. 2016. Lakes and Rivers. Accessed 2016. http://www.eea.europa.eu/data-and-maps.

Eurostat. 2016. European Union Digital Elevation Model. Hillshade, Netherlands. Accessed 2016. http://ec.europa.eu/eurostat/web/gisco/geodata/reference-data/elevation/eu-dem-laea.

Gesch, D., M. Oimoen, S. Greenlee, C. Nelson, M Steuck, and D. Tyler. 2002. The National Elevation Dataset. Photogrammetric Engineering and Remote Sensing 68 (1): 5–11.

GreenInfo Network. 2016. California Protected Areas Database. Accessed 2016. http://www.calands.org/.

Natural Earth. 2016. Land, Ocean, Rivers, Lake Centerlines, Lakes and Reservoirs, and Shaded Relief datasets. Accessed 2016. http://www.naturalearthdata.com/.

The Nature Conservancy. 2016. Freshwater Ecoregions of the World. Arlington, VA. Accessed 2016. http://maps.tnc.org/gis_data.html.

Sacramento Area Council of Governments. 2016. Flood Bypasses. Accessed 2016. http://www.sacog.org/gis-mapping-center.

Sacramento River Forum. 2016. Sacramento River Geonames. Accessed 2012. http://www.sacramentoriver.org/.

US Geological Survey. 2005, 2012, and 2014b. National Agricultural Imagery Program. Accessed 2016. https://lta.cr.usgs.gov/NAIP.

USGS (US Geological Survey). 2014a. Landsat imagery. Land Processes Distributed Active Archive Center. Sioux Falls, SD. Accessed 2016. http://lpdaac.usgs.gov.

US Geological Survey. 2016. Watershed Boundary Dataset. Accessed 2016. https://nhd.usgs.gov/wbd.html.

US Geological Survey. 2017. National Water Information System data available on the World Wide Web (USGS Water Data for the Nation). Accessed April 20. http://waterdata.usgs.gov/nwis/.

INDEX

Note: Page number followed by (f) and (b) indicates figure and box, respectively.

accretion, 22–27, 26(f), 30, 32–35
agriculture, 1, 2, 8, 29, 48, 66, 77, 94,99,101–6, 116, 130,132–33, 138, 144–45, 150, 153, 165–66, 169, 173–74, 179, 185, 194, 196, 206, 214, 217–18; flood-recession, 100, 113, 176,
agroecosystem, 111
algal mats, 39, 73, 77
alien species, 2, 47,82–87, 94, 99,105–112, 144, 149, 177, 179, 182–84, 190, 192–93
allochthonous, 52, 55, 70, 75
alluvial fan, 17(f), 18, 161, 185
alluvial, 4, 6, 16, 18, 22, 33, 35–36, 55, 61, 64, 99, 106, 137, 148, 210
Altamaha River, 92
Amazon River, 5, 6, 15, 27, 36, 52, 65, 70, 75, 80
American River, 169, 178, 185, 207
anoxic, 39–41
Apalachicola River, 80, 85
aquaculture, 103
Ara River, 117
arctic, 101. *See also* boreal
armored banks, 2(f), 7, 25, 50,117, 125, 144, 173–74. *See also* riprap
aseasonal, 5, 46, 83
assimilation, biological 37
Atchafalaya River, 4(f), 41, 65, 67, 146–48
Atchafalaya Floodway, 146–48
atmospheric river,159, 211
Australian Water Act, 154
authochthonous, 52, 70
avulsion, 3, 22–23, 29–32, 35, 46, 49, 54, 63, 65, 67, 94, 116, 162, 179, 208, 210–11

backswamp, 22, 34–35, 54,
Balonne River, 32
bank; armoring, 2(f), 7, 25, 50,117, 125, 144, 173–74; erosion, 26, 27, 32, 117, 149, 173, 210; stabilization, 152, 198. *See also* cutbank, riprap
bankfull flow, 12, 14, 16, 22–23, 43, 46; 126(f), 207, 209–10
bar, 23, 26, 27(f), 32–33, 50, 173, 210; braided channel, 25; point, 22(b), 24(f), 31, 61–63, 64(f), 118(f), 208(f), 209; scroll, 35
barbel, Ebro, 144
Barber, Dan, 103
Barmah Forest, 154
baseflow, 12, 14, 16, 53, 56, 61–62, 64, 144, 209
bass, largemouth, 81, 84, 86, 107, 111, 184; striped, 90; white, 81
bat, Mexican free-tailed, 187, 192
bayou, 67
Berryessa Reservoir, 213
Bear River, 87, 110, 127(f), 128, 131, 205
bear, black, 67
beaver, 35, 60, 65–66, 88, 89, 93–94, 109, 123, 164, 175, 192
beetle, tiger, 173; valley elderberry longhorn, 193, 207
Bidwell, John, 165
Big Lost River, 106
Bill Williams River, 109
biodiversity, 2, 3, 5, 6, 32, 45, 47–48, 50, 54, 57, 67, 100–102, 106, 111–12, 115, 137, 140,145,148, 175, 207, 215, 217–18
biofilm, 70, 75, 155

251

biogeochemical processes, 15, 37–44, 47, 57, 71, 94, 96–97, 100, 103, 207
Birch Creek, 18
birds, 19, 47, 56, 67, 78, 86, 90–98, 103, 111–12, 130, 142, 144, 147, 155, 179, 182, 184, 187, 191–92, 195–97, 199, 211, 216–17; wading, 92, 101, 103, 174, 184, 192, 208–9
Birds Point Floodway, 129
bison, 94–95
blackbird, tricolored, 19, 193
black fish, 83
blackfish, Sacramento, 84, 190
bluegill sunfish, 75, 81, 84, 86, 184
Bonnet-Carré Floodway, 130, 146
boreal rivers, 6
Boston, 124
Brahmaputra River, 25, 36
braided, 22–23, 24(f), 25–26, 27(f), 30, 32, 36, 49, 50(f), 95, 149, 151
braided channel bar. *See* bar
Brazos River, 4–5, 46, 82–83, 87
bulrush. *See* tules
Butte Basin, 161–63, 169, 174–75, 195
Butte Sinks. *See* Butte Basin
bream, common, 83
bypasses (*see also* Sutter Bypass, Yolo Bypass), 7, 8, 102, 122, 128–32, 167, 169, 172, 174–75, 190–91, 205, 211, 215

Cache Creek, 163, 185, 188, 191
Cache Slough, 185, 188
California (*see also* Central Valley), 177, 198, 201, 204–7, 212, 214–15; climate, 5, 121, 130, 134, 157–59, 160(f), 177; dams, 7, 88, 134, 198; history, 164–74; Department of Water Resources, 128, 187; State Water Project, 198
carbon, 11, 19, 32, 37, 44, 52–53, 55, 69, 70–78, 97, 113, 163, 174, 182, 189, 209; sequestration, 6, 100, 124, 133
carp, common, 84–85, 106–7, 111, 154, 182–84, 190, 211; bighead, 106; gibel, 83; grass, 106; silver, 106
catfish, blue, 80; bullhead, 84, 86, 107, 184; channel, 65, 75, 84; wels, 144
Central Valley; floodplains, 2(f), 3–4, 7, 9, 15, 55, 64(f), 72, 77, 102, 105(f), 106, 108, 130, 157–75, 177–201, 203, 215–17; conceptual models, 55, 207–12; ecosystem services, 101–6; endangered species 91, 193; extent, 174; fishes, 46, 79, 81–83, 87, 130, 101, 196; foodwebs, 96(f), 97; forests, 164–66; history, 160–69; hydrology, 169, 171–74; lessons from, 215; levees, 119, 127(f), 170(f), 171(f); mammals, 93; mercury in, 40; waterfowl, 92, 131. *See also* Bear River, Cosumnes River, Sacramento, San Joaquin, Yolo Bypass
Central Valley Project, 169, 198
channel, 11–19, 163, 209; cut-offs, 3, 63; fishes, 182; geomorphology, 21–36, 48; habitat, 6, 47, 50–57, 74, 83, 87–89, 106; incision 19; modification, 1, 2, 7, 49, 88, 105–6, 117, 125–29, 137–55, 161, 165; networks, 35; patterns, 22; types, 24(f)
Charles River, 124
Chilcotin River, 50(f)
Chilliwack River, 89
China, 104, 106; disasters, 116
Chironomidae, 74–78, 90, 97, 188–90
chlorophyll, 44, 72, 188
Chowchilla Floodplain, 53
chub, Catalan, 144; Oregon, 107; thicktail, 161
cladocerans, 76, 78, 188, 196
classification, fishes, 82–87
climate change, 5, 9, 13, 57, 62, 115, 119–20, 133, 134, 138, 148, 169, 201, 203–4, 206, 217–18
coarse particulate organic matter. *See* particulate organic matter
Columbia River, 88
conceptual models, 41, 46, 50, 54–57, 61, 74–76, 81, 113, 203, 207, 208(f), 212
Congo River, 6
conveyance, 110, 122, 127–28, 12, 130
connectivity, 11–14, 19, 30, 41, 43, 45–57, 60, 64–65, 74, 81, 95, 104–10, 126, 131, 144, 152–53, 175, 179–81, 184, 197, 207–9, 210, 212, 215, 217–18
cormorant, 97, 191
Cosumnes River, 7, 27–30, 42, 44, 163, 178, 188, 194, 215; algae, 38, 72, 73(f), 181; birds, 184; fish, 77, 83–87, 182–84, 190; hydrology, 105(f), 179, 181, 187; levee breach, 109, 181(f); novel ecosystem, 184; Preserve, 110, 178–85, 207, 215; reconciliation, 185; restoration, 110, 182; zooplankton, 77
cottonwood, 15, 46, 61–64, 66, 95, 106, 109–10, 179, 210, 218
crane, sandhill, 9, 92, 184, 193
crayfish, 86, 192
crevasse splay. *See* splay
cuckoo, yellow-billed, 101, 193
cutbank, 23, 32, 46, 63, 173, 209–10
cyanobacteria, 42, 71
cyclic floodplain rejuvenation, 64
cyprinid fishes, 75, 184, 209

dams, beaver, 65, 93, 109, 123; effects of, 1, 6, 7, 13, 16, 47, 57, 61–62, 82, 87, 104–12, 115, 117–21, 134, 138, 142, 144, 146, 149, 164, 169, 173, 178, 185, 206; ice, 116
dam, Alamo, 109; Friant, 104, Mequinença, 144–45, Monticello, 213, Oroville, 169, 172; Putah Creek, 213–14; Ribaroja, 144–45; Shasta, 169, 172–73, 185
Danube River, 4, 16, 18, 41, 43, 49, 72, 74, 76, 80, 82, 85, 101, 104, 106, 140–42
daphnia magna, 90, 189, 190, 196
Darling River. *See* Murray-Darling River
dead zone, 133, 145

denitrification, 39, 41–42, 100
deposition 3, 4, 11, 22–23, 25–35, 39, 43–50, 61–65, 67, 100, 105, 116, 138, 144, 151, 167, 173, 178–79, 191, 194, 207–10
de-poldering, 138, 139(f)
detritus, 38, 45, 55, 67, 69, 70, 74–78, 89, 96, 163, 182
diapause, 76, 189
diatoms, 71, 72
dissolved organic carbon (DOC), 37, 44, 189
disturbance, 27, 29, 47–49, 59–60, 62–64, 211
diversified portfolio approach, 121, 135, 217
diversity, *see* biodiversity
Dongting Lake, 94
Dos Rios Project 111, 113
drought, 60, 100–111, 120, 131, 133–34, 153–54, 201, 204, 206, 217–18
Drought, Millennium, 154
duck hunting clubs, 90, 10–2, 133, 174, 192, 194, 198
Dutch, 137–38, 140, 152

Ebro River, 4, 108, 111, 142–45
Economics, 3, 8, 81, 89, 97, 99, 102, 104, 111, 113, 116, 120, 122, 125, 129, 130, 132, 142, 145, 153, 157, 196, 198, 206, 218
ecosystems, floodplain, 45–57; novel, 1, 2, 6, 7, 47, 57, 103, 108, 112, 144, 178, 184, 187, 201, 205, 213–15, 217–18; resilience, 216; services, 37, 99–113, 214
endangered species, 48, 91, 97, 102, 107, 113, 194
Endangered Species Act (ESA), 101, 120, 192–93, 195
Elbe River, 129
elephants, 93–94
elk, 110, 164; tule, 93–95, 177
erosion, 3, 11, 19, 21, 23, 25–26, 29–32, 35, 45–49, 105, 117, 121, 123, 125–27, 149, 151, 169, 173, 194, 204, 207
Euphrates River, 103
eupotamon, 34
European Water Framework, 108, 111, 144
eurytopic fishes, 34, 83, 142
exceedance, percent, 4, 173, 179–80, 199, 204
Exe River, 26

fallout, 29
fatty acids (FA), 70, 75–76
Faulkner, William, 116
Feather River, 109, 127, 161–62, 167, 169, 174, 185, 205, 215
fine particulate organic matter. *See* particulate organic matter
fishes, classification, 79–91
fisheries, 6, 35, 45, 79–80, 90, 100–101, 103, 113, 115, 116, 118, 132, 144, 198, 207
Flathead River, 4, 48, 63
flood, 204; basin, 22, 35, 159–75, 190, 207, 211; Biblical, 116; control, 7, 8, 13, 60, 66, 82, 88, 103–5, 117–18, 120–21, 124, 128, 130, 140, 148, 152, 165–69, 175, 187, 198, 206, 217; duration, 5, 11–15, 19, 31–33, 38, 46, 53–54, 57, 74, 77, 80–82, 90, 97, 101, 109, 123–24, 131, 155, 157, 159, 163, 172, 174–75, 184, 188–89, 192, 194–95, 197, 199, 205, 209, 212; erosive, 54, 210–11; frequency, 2, 11–12, 14–15, 27, 31, 33, 46, 56–57, 64, 74, 77–78, 80–82, 93, 97, 105, 107, 116, 119, 130, 133, 144, 154, 165, 173–74, 182, 189, 195, 204, 209; great, 116, 129, 145; recurrence interval, 4, 12, 30, 78, 151, 169, 173, 199, 206, 210; insurance, 5, 118, 120, 122, 152, 204; management, 2, 8–9, 67, 100, 104, 106, 109, 111–13, 115, 117, 121–35, 137, 140, 145–52, 167, 170–72, 175, 185, 187, 194, 199, 201, 204–8, 212, 214–18; phases, 43–44, 72, 80, 181, 188; prevention, 48, 99, 100, 107; risk, 1, 2, 8–9, 48, 57, 67, 87, 99, 105, 108, 113, 115–35, 140–41, 145–46, 148, 151–52, 169, 172, 175, 198–99, 201–7, 212, 215–18; stage, 12–19, 22–23, 26, 30, 32, 34, 36, 43, 46, 72, 121, 128–29, 138, 146–47, 164, 195, 205, 207, 210, 216
Flood Control Act of 1917, 169
Flood Emergency Management Agency (FEMA), 5, 119–20, 134, 152
floodplain, activation, 172, 209–10; active, 4, 6, 110; classification, 23, 34–36; conceptual model, 208; definitions, 3–7; deltaic, 36; ecological, 142; fringing, 35; inundation potential (FIP), 199; 100-year, 4–5, 204; reorganization flood, 210; resetting flood, 210; temperate, 3
flood pulse concept (FPC), 14, 16, 43, 51(f), 52–54, 69, 71, 74, 209
flood-pulse advantage, 53–54, 80, 89, 101
floodwalls, 47, 104, 115, 117, 122, 148, 152
floodway, 122, 129–33, 141, 146–48
flow, 3, 11–36, 71, 209–10; activation, 209–11; characteristics, 12–16, 60, 209–11; functional, 47; overbank, 16–19, 22–23, 30–35, 46, 54, 60, 144, 161, 165, 167, 174, 178; regime, 7, 13, 37, 45–47, 49, 54, 56–57, 61–62, 91, 107–10, 144–145, 162, 198, 207, 209; regulation, 49, 50, 81–82, 104–6, 109–10, 112, 117, 120, 129, 131, 142, 144, 146–48, 153–55, 169, 172–75, 195, 214
flushing, 44, 72, 182, 189
flycatcher, willow, 101, 193
Fly River, 18
flyways, 92, 98, 101, 157
food webs, 9, 40–41, 47–48, 52, 55, 59, 67, 69–79, 89–91, 96(f), 97, 163, 182, 184, 188–89, 209–10, 211–12, 216
forest, 6, 11, 26, 29, 41, 43, 45, 47–48, 52, 59–67, 70, 89–90, 92–95, 98, 100–101, 106, 110–11, 121–23, 126, 128, 140–41, 144, 148, 163, 165, 173–74, 184, 199, 211, Barmah, 154; fishes, 65; hardwood bottomland, 5, 35, 41, 66–67, 91–92; red gum, 100, 153–55; Sacramento Valley, 163–64. *See also* tree, wood
Fremont Weir, 185, 187–88, 191, 195–96
frog, red-legged, 91, 193

INDEX 253

Gallatin River, 95
Ganges River, 16, 36, 103
gar, 82
geomorphology, 3, 5, 6, 9, 11, 13–14, 19–36, 46–47, 49–50, 54, 56–57, 60, 63–67, 93, 95, 109–10, 125, 146, 150–54, 158, 160–61, 163, 172–73, 199, 207–8, 210–11, 217
giant garter snake, 91, 193, 195, 207
godwit, black-tailed, 103
goldfish, 84–85,184
Gold Rush, 153, 157, 164–65, 201
goose, white-fronted, 192
gopher, pocket, 93
Gran Pantanal, 36
grapes, 65, 89
gravel, 19, 39, 50(f), 61, 87, 105, 142, 161, 173, 213
gray fish, 83
grazing, livestock, 40, 60, 164, 166, 179; zooplankton, 71, 72
groins, 138–39
groundwater, 11–12, 15–19, 23, 34, 37, 39–40, 43, 54–56, 60, 62, 106, 131, 133, 142, 178, 185; recharge, 60, 100, 113, 123, 149, 200–201, 212, 216–17
Guadalquivir River, 4, 103
Gulf of Mexico, 41, 100, 133, 145–46
guilds, fishes, 83–87, 182

habitat mosaic, 48–49, 54–55, 63, 91, 211–12
Hamilton City, 132, 205, 215–16
harrier, 191
hawk, Swainson's, 91, 111, 191, 193
hippopotamus, 93, 94
Huai River, 116
Huston dynamic equilibrium, 47–48
hydrarch succession, 49, 210
hydrologic connectivity, 29, 43, 45, 57, 65, 81, 95, 99, 104–6, 108–10, 124–25, 133, 146, 175, 188, 204, 206–7, 209, 212, 216; floodplain, 4, 33(f); phases, 42–44, 72
hydrology, 7, 11–19, 21, 31, 38, 49, 61, 71–72, 76, 83, 86, 154, 157, 163, 166, 169, 178, 181, 187, 196
hyporheic flow, 12–19, 39, 41, 45–46, 53–54. See also groundwater
hypoxic, 41, 74

ibis, white-faced, 191, 218
ide, 83
Illinois River, 4, 43, 52, 75, 106, 109, 110, 129
incision, channel, 7, 19, 31, 95, 105–6, 117, 144, 149, 173, 213
infrastructure, flood management (gray), 6, 7, 43, 57, 104–6, 110, 113, 115–20, 153, 198, 204–5, 213, 215; green, 9, 57, 120–35, 137, 140, 146, 148, 152, 199–201, 205–6, 214–18
insurance, 5, 118, 120, 122, 152, 204

inter-flood interval, 40
intermediate disturbance hypothesis, 47
invasive species. See alien species
invertebrates, aquatic, 19, 41, 45, 52, 65, 69, 74–78, 81, 86, 89–90, 96–97, 107, 144, 163, 174, 182, 184, 188–89, 192, 207
Inyo Mountains, 48
Isar River, 140
island formation, 23, 26, 30, 49–50, 65, 94

Jin Jiang Flood Area,129
Jordan River, 32

Kaskaskia River, 75
Kelley, Robert, 165–67, 177
Keyser Decision, 167
Knight's Landing, 185, 191
Kosi River, 30

larvae, fish, 56, 77, 81, 85–86, 89–90, 182, 189–90
levees, 2(f), 19, 22, 25, 30, 32–33, 47, 49, 66–67, 81, 92, 100, 104, 106, 111, 115–26, 128–31, 134, 138, 146–49, 151, 167, 169, 170(f), 171(f), 173, 185, 197–200, 203–4, 206, 216; breaches, 30, 85, 109–11, 119, 152, 179–80, 181(f), 203, 207; failure, 119, 129, 134, 138, 145–46, 167, 216; fuseplug,129, 146; natural, 18, 30, 32–33, 34(f), 35, 66, 161–63, 174,178, 185, 210, 213; setback, 87, 109–10, 122, 125, 127(f), 128, 131–32, 134–35, 141, 145, 151–52, 199–201, 205, 215
light, 53–54, 71, 78
limnophils, 83–84, 86, 142
littoral, 72, 76, 181
living river principle, 148, 151–52
Loire River, 19, 107
Lumber River, 66(f)

macrophytes, aquatic, 52, 70–75, 97, 155, 211
mallard, 192
mammals, 6, 41, 67, 90–91, 93–94, 96, 98, 142, 164, 192
management, 2, 137–55. See also flood management
McCoy, Kansas Joe, 116
meadows, 6, 56, 123
meander, 22–23, 24(f), 25–27, 28(f), 29–33, 35, 54, 63, 64(f), 66, 117–18, 125–26, 128–29, 140, 145, 152, 160–61,173, 178, 199, 200(f), 209, 211–12
Mekong River, 5, 6, 80, 105
Merced River, 161, 173, 207
mercury, 38, 40–41, 102, 191
mesic, 56
methane, 40
Meuse River, 4, 129, 137–38, 140
Mexico, Gulf of, 41, 100, 133, 145–46
mice, 93–94
mid-channel bar. See bar

Millennium Drought, 154
Mining, hydraulic, 29, 164–67, 168(f), 198, 204
minnow, Rio Grande silvery, 75
Mississippi River, 4, 18, 25, 27, 30, 41, 53, 65, 67, 80–81, 85–86, 92, 101–2, 104–6, 109, 116–17, 119, 121–22, 125, 128–30, 143(f), 145–48, 159, 183
Mississippi River and Tributaries Project (MR&T), 146–48
Missouri River, 4, 81, 85, 105–6, 120
mitigation, 102, 110, 112, 133
Modoc Plateau, 158
Mokelumne River, 65, 105, 163, 178, 180, 206
Mollicy Farms Project, 102, 109, 133
Monticello Dam, 213–14
Morganza Floodway, 146–47
mosquitofish, 84, 86, 107, 111, 183–84
multichannel, 7, 23
multiple benefits, ix, 2–3, 57, 110, 113, 115, 125, 131–32, 137, 175, 203, 212–18
Murray-Darling River, 4–5, 46, 53, 78, 82, 104, 153–55
Murrumbidgee River, 153
muskrat, 192

Napa River, 4, 148–53
nase, Ebro, 144
Nashua River, 124
Nature Conservancy, The (TNC), 108–10, 124, 132–33, 155, 178–79
Navarro River, 18
Netherlands, 93, 110, 119, 121, 128–29, 137–41, 216
New Madrid Floodway, 129–30, 133, 146–47
New Orleans, 146
Niger River, 6, 120
Nijmegen, 141
nitrate, 39, 41–42, 147
nitrogen cycle, 42–44; 42(f), 71
non-native species. *See* alien species
nutrients, 39–45, 47, 52–55, 57, 59, 63, 65, 71–72, 76, 89–90, 93, 97, 100, 103, 116, 123–24, 133, 145, 166, 174, 188, 207–8

oak, valley, 64, 110, 179
Oakville to Oak Knoll, Napa River, 150(f), 151
off-channel habitat, 34, 62, 88, 161
Ogeechee River, 53
Ohio River, 75, 106, 147(f)
Okavango Delta, 4, 6, 16, 36, 93–94
open water, 49, 76, 161, 164, 184, 211
organic material, 3, 21–22, 35, 38–44, 49, 52–57, 59–60, 63, 67, 69–71, 74–76, 96, 104, 123, 144, 188, 207, 210
Orinoco River, 36, 211
Oroville Dam, 169, 170(f), 172
Otter Creek, 124
otter, river, 97, 192

Ouachita River, 102, 109, 133
Ovens River, 93
overbank flows, 32, 46, 54, 56, 60, 144, 154, 161, 165, 174, 176, 178
overland runoff, 54
owl, little, 93
oxbow lake, 5, 6, 22, 31, 34, 36, 43, 49, 60, 71, 79, 80, 82, 87, 101, 141–42, 144–45, 149, 152, 173, 184, 199
oxygen, dissolved, 38–40, 54, 83
oxygenated zone, 39

paleopotamon, 34
Pantanal, 36
paradigm, 52, 138, 148, 185; new, 9, 87–89
Paraguay River, 36
parakeet, Carolina, 91
parapotamon, 34
particulate organic carbon (POC), 37, 43
particulate organic matter (POM), 43, 70, coarse, 210
perch; Sacramento, 107, 161; tule, 84
perennial marsh, 161
periphyton, 52, 55, 70–75, 78, 96(f), 97, 181
Peru, 36
phytoplankton, 37–39, 41, 43–44, 55, 70–75, 77–78, 96(f), 97, 103, 163, 181, 188–89, 207, 209. *See also* plankton
pikeminnow, 84, 182
picoplankton, 71
Piedra River, 4(f), 111
pintail, northern, 192
plankton. *See* phytoplankton, picoplankton, zooplankton
plant communities, 32–33, 46, 49, 57, 59, 63, 66, 128, 164, 181
plesiopotamon, 34
plover, snowy, 92, 192
point bar. *See* bar
polder, 129, 138–40
policy, 108, 121, 125, 134, 139, 151, 166
population, human, 1, 106, 116–17, 119–20, 134, 153, 157, 163, 164–65, 169, 201, 203, 217
precipitation, 9, 13, 15–16, 18–19, 34, 36, 40, 44, 46, 48, 56, 60, 67, 82, 92, 110, 119–20, 122–23, 134, 157–59, 160(f), 163, 185, 204, 214. *See also* rain, rainfall, runoff
production (primary, secondary), 43, 51(f), 52–54, 56–57, 69–78, 80, 89–91, 155, 163, 188–89, 208(f), 209, 212; microbial, 69
Puget Sound, 132
Putah Creek, 56, 163, 185, 188, 192, 213–14; sinks, 214

rain, 5, 19, 89–90, 134, 144; shadow, 158–59, 194, 203–4. *See also* precipitation, rainfall, runoff
rain-on-snow, 159, 203, 211
raindrops, 123

rainfall 5, 16, 17(f), 54, 105(f), 116, 119, 123, 159, 211. See also precipitation, rain, runoff
rainstorm, 82, 178
rat, 94, 184
rearing, fish, 41, 45, 80, 82–83, 85–86, 88–91, 112, 153, 183, 189, 191, 196, 198, 209, 212; habitat, 81, 93; ponds, 107, 161, 164, 172, 174, 190, 199, 210
reconciliation ecology, 1, 2, 6, 21, 92, 99, 108, 110–13, 115, 134, 137–55, 178, 185, 187, 195–99, 201, 203–12, 215, 217–18
reconnect, 9, 43, 47, 57, 87–88, 100–3, 108–10, 121, 128–29, 132–33, 135, 137, 140, 142, 145–46, 152, 177, 179, 185, 201, 205–6, 214
recreation, 1, 97, 100–102, 104, 113, 120, 123–24, 130–31, 133, 135, 137, 141, 151–52
recruitment; cottonwood, 61–62; plant, 45, 95
recruitment box model, 62(f)
Red River, 146
reed, giant, 107
reforestation, 100, 111, 123
refuge, 60, 67, 91, 130, 134–35, 174; aquatic species, 151; terrestrial species, 15, 93–94; fish, 45, 88; waterfowl, 92
refugee, 116
regulation, 97. See also flow regulation
reservoir, 53, 61, 80, 104–5, 117, 120, 124, 130, 134, 144–45, 153–54, 162, 169, 172–74, 178, 206–7, 213–14; releases, 47, 73, 87, 101, 109, 215
residence time, 11, 13–15, 29, 38, 40–44, 71–73, 76–78, 96(f), 97, 147, 162–63, 174, 181–84, 187–89, 194, 209–10; definition, 12
restoration, 7–9, 11, 13, 19, 30, 35, 40, 41, 47, 57, 64, 67, 87–89, 94, 97, 100, 102, 108–13, 115, 128, 132, 140–41, 148–51, 153–54, 177–78, 199, 204–7, 211, 215–16; of processes, 142
revetement. See armored banks, riprap
rheophilic, 83–85, 142
Rhine River, 4(f), 93, 104, 117, 119, 121, 128–29, 137–40
Rhone River, 4(f), 36, 107
rice, 92, 103, 144, 174, 179, 185, 187, 191–92, 194, 196, 197(f), 209, 228
ridge, 33, 66, 162
Rio Grande River, 29, 38, 41, 43, 74–75
riparian, 5, 8(f), 41, 43, 48, 55, 91, 93, 95, 110–11, 144–45, 149, 151, 153, 179, 199, 211; corridor, 152; forest, 6, 11, 29, 38, 45, 49, 52, 59–66, 92, 100, 125, 128, 163–65, 173–75, 178, 184, 187; restoration, 94; species, 15, 101, 106, 213; trees, 37, 42, 181; vegetation, 39, 109, 161, 209–10; wetland; zone, 56. See also forest
riprap, 57, 151, 173, 174, 198, 200(f). See also armored banks
risk, residual, 216. See also flood risk
river continuum concept (RCC), 50–52, 51(f), 55–56, 69, 74–75
riverine productivity model (RPM), 51(f), 55–56, 74–75

river wave concept, 51(f), 55–57
Roanoke River, 60
rodent, 93–95, 166
Room for the River, 121, 129, 131, 137–41, 145, 152, 216
roughness, hydraulic, 23, 43–44, 128, 130–31, 133, 173–74, 194, 205
rotifers, 76–78
runoff, 13, 57, 61, 66–67, 92, 116, 119–20, 122–23, 134, 144, 149, 159, 164, 198, 213; overland, 54; patterns, 5; storm, 123, 148, 203; timing, 82. See also precipitation, rain, rainfall
Rutherford Reach, Napa River, 149–51, 150(f)

Sacramento Area Flood Control Agency, 133, 216
Sacramento, City, 130–31, 158(f), 216; Valley, 8, 35, 113, 158, 177, 186–87, 206, 213; Weir, 197
Sacramento River, 2(f), 4(f), 18–19, 25, 28(f), 29, 41, 62, 75, 104, 110, 128–29, 131–32, 153, 158–74, 185, 187–91, 198–201, 206, 213–14; Control Project, 199, 204; floodplains, 94, 109, 161, 168–75, 177–79, 210–11; levees, 125, 205, 215–16
Sacramento-San Joaquin Delta, 178, 185, 188–89, 198, 209
Sacramento Valley flood management system, 214
St. Francis River, 146
salmon, 9, 19, 38, 56, 65, 79, 85, 87–89, 93, 101, 120, 132, 157, 163–64, 194–95, 198–99, 206, 218
salmon, Chinook, 38, 41, 79, 81–82, 84, 88–91, 97, 112, 161, 163, 173, 182, 183(f), 188–91, 193–98, 207, 209–11
salmon, coho, 88–89
San Francisco Estuary, 151–52, 158, 163, 167, 174, 187, 190, 198
San Joaquin, 4(f), 94, 158, 161, 169; Basin, 173; River, 104, 111–12, 159, 205; Valley, 172. See also Sacramento-San Joaquin Delta
sand, 27, 30, 33, 35, 61, 66, 105; splay, 161, 163, 181
sandpiper, 92
Saskatchewan River, 31
savannah, 89
Savannah River, 108, 124
Sawyer Decision, 167
scour, 19, 30, 48–49, 54, 61–62, 65, 74, 167, 208(f), 210–11
scroll bar. See bar
sculpin, prickly, 84, 190
seasonal, 5, 12, 38, 59, 80, 82, 93, 109, 157, 162, 178; agriculture 1, 185, 217; ecological processes, 15; flooding, 46, 60, 77, 103; floodplains, 6, 41, 47, 56, 79, 81, 83–86, 88–92, 182–84, 196; hydrology, 13, 16, 209; wetlands, 24, 35, 94, 154, 163. See also aseasonal
sediment, 3, 6–7, 12, 16, 18–19, 21–23, 28–32, 35, 37, 40, 43, 44, 49, 57, 70, 71, 76, 78, 100, 104–6, 123, 125, 144, 148–49, 204, 179, 181, 198, 211, 213;

capture, 173; deposition, 4, 26, 33, 36, 51, 60, 62–63, 65, 67, 116, 138, 151, 161, 164, 178, 180, 210; exchange, 207; hydraulic mining, 165, 167, 168(f), 169, 204; size, 59; transport, 13–14, 47, 152, 208(f)
sedimentation, 37, overbank, 29
Shasta Dam, 169, 170(f), 172–73, 185
shear stress, 19, 32, 210
shifting baseline syndrome, 106
shifting habitat mosaic, 49, 51(f), 54–55, 63
silverside, Mississippi, 84, 86, 183–84
silt, 27, 30, 35, 49, 66, 161
sinuosity, 22, 24(f)
Skagit River, 88–89
slough, 79, 83–90, 161–62, 179, 182–84, 191, 196, 207
Sly Park Reservoir, 178
snag removal, 65, 165
snake, 93; giant garter, 91, 193–95, 207
Snake River, 89
snow, 204, 144, 158–59, 204. *See also* precipitation, rain-on-snow
snowmelt, 5, 56, 61, 82, 90, 105, 109, 134, 158–59, 172–73, 178
snowpack, 159, 207
soil, 3, 16, 37–42, 56, 59–60, 63, 66, 77–78, 90, 96(f), 97, 99–100, 103, 107, 116, 122–23, 162, 165–66, 189–90; boreal, 6; development, 29, 32, 45; hydric, 39; levee, 117; moisture, 12, 61–62; salinity, 154; water, 54
Solano County, 213
songbirds, 92–93, 95, 98, 184, 191, 211
sparrow, song, 184, 191
spawning, 56, 80, 82–83, 85–86, 90, 153, 155, 161, 183, 190, 209–10, 212; habitat, 45, 81, 172, 174, 208(f); salmon, 89; splittail, 189, 193, 196
spiders, 53
spiraling, hydrologic, 14; nutrient, 52
splay, 179; crevasse, 22, 30, 33, 35, 181, 210; sand, 161, 163, 181
spruce, Sitka, 65
Stanislaus; County, 111; River, 173
steelhead, 161, 195, 199; Central Valley, 193
stickleback, threespine, 90
splittail, Sacramento, 78–79, 82, 84–85, 87, 161, 163, 182–83, 189–91, 193, 195–96, 209, 211
Stillaguamish River, 88, 164
storage, 123–24, 206; carbon, 100; dam, 47, 104, 105(f), 117, 144, 154, 169; flood, 15, 119, 129–31, 133–35, 146, 158, 162, 215; groundwater, 56; sediment, 29
straight channel, 22–23, 24(f), 30, 35
straightening, 1, 7, 117, 122, 126, 138
stranding, fish, 15, 79–80, 86–87, 191
strategy, fishes, 81–87; risk reduction, 8, 43, 57, 115, 137, 199, 217
stream power, 22, 25, 31–32, 35, 161
subside, subsidence, 27, 138

subtropical, 6, 21, 40, 52, 94
succession, fish, 85; hydrarch, 49, 210; vegetation, 26, 47, 49–50, 60, 62–64, 67, 74, 107, 144–45, 161, 163, 173, 184, 208(f), 211
Sumida River, 117
sunbleak, 83
Sutter Bypass/Basin, 8(f), 161–62, 169–70, 174–75, 185, 187, 205, 207
swale, 87, 110, 123; topography, 33
swallow, bank, 173; tree, 184
Swamp Land Commissioners, 166

Tagliamento River, 65
tamarisk, 48, 106–7, 109
Tana River, 120
teal, cinnamon, 191
temperate, 2, 3, 5, 46, 54, 57, 79, 87, 108; floodplains, 7, 9, 11, 21, 29, 45, 47, 60–61, 65, 69, 81, 90–91, 94, 97, 99, 106–7, 110, 113, 137, 140, 175, 177; forests, 66; management, 213–18; region, 1, 50, 52; rivers, 6, 40, 103, 119–20, 129, 153
temperature, 3, 53–54, 61, 71, 78, 90, 204, 217; air, 119; water, 18, 38–39, 44, 50, 55–56, 72, 80–83, 85–87, 107, 183–84, 194
Tensas River, 67
terrace, 33(f), 47, 64, 151–52, 174
territory, fluvial, 145
Thurra River, 31
tidal, 148, 152, 178, 188; influence, 36; marsh, 88, 90, 152, 185; slough, 83, 179, 196; transition zone, 151
Tigris River, 103
topsoil, 32. *See also* soil
tree, 50, 52, 59, 60–67, 70, 89, 93, 100–101, 110–11, 142, 164, 191, 210; alien, 112; corridor, 192; riparian, 15, 37, 42, 46, 95, 106, 179, 181. *See also* forest, wood
tropical, 5–6, 21, 52, 94; climate, 79; fishes, 83, 101; floodplains, 40, 65, 74–75; rivers, 33, 80–82, 102, 113; birds, 91–93, 98, 191
Truckee River, 4(f), 109
Tulare Basin, 94, 164
Tulare Lake, 94
tules, 93, 100, 163–66, 174, 178, 185, 214
Tuolumne River, 61, 111–12, 161
turtle, pond, 91, 193, 195

valley elderberry longhorn beetle, 193, 207
variable source area concept, 16
vegetation, 3, 15–16, 26–27, 31, 33, 38–39, 48–49, 52, 54, 59–61, 63–64, 66, 70–71, 73, 75, 89, 93–95, 96(f), 100, 106, 128, 142, 147, 161, 174, 179, 192; annual, 190; emergent, 72, 77; establishment, 23; flooded, 85; management, 108, 130, 133, 194, 213; native, 152; natural, 135; pattern, 12, 14, 46, 50, 55, 144, 173, 208(f), 209, 211; riparian, 41, 43, 56, 62, 151, 210; submerged, 163; terrestrial, 74, 97

Veta La Palma (VLP), 103
vireo, least Bell's, 91, 101, 193
vole, California, 93–94
Volga River, 4(f), 7, 83

Waal River, 93, 141
Waimakariri River, 26
Walla Walla River, 4, 118(f), 126
warbler, Bachman's, 91; prothonotary, 93
water, management, 2, 47, 57, 106, 112, 115, 119–20, 134–35, 147,153, 201, 205–7, 212, 216
waterfowl, 9, 45, 59, 78, 91–92, 94, 103, 109, 113, 131, 157, 163–64, 174–75, 179, 185, 187, 191–92, 194–98, 209
water market, 153
Water Sharing Investment Partnership (WSIP), 155
water table, 12, 14–16, 46, 54, 60–61, 66, 106, 110–11, 133, 163
weir, 129, 131, 167, 172, 174, 185, 187–88, 191, 195–97, 205, 207, 215
wetland, pop-up, 92
Wetland Reserve Program, 124, 133–34
white fish, 83
White River, 146
wildlife, 3, 67, 69, 91–92, 94, 101–3, 111–12, 130–31, 137, 140, 151–52, 157, 187, 194, 196, 201, 212, 214, 218
willow, 46, 61, 63–66, 93, 95, 109, 133, 163, 179, 210; sandbar, 106
Willow Slough, 185, 188

Willamette River, 2, 50, 88, 107, 184
windmill, 138
Wisconsin River, 49
wolves, 95
wood, large, 8, 13–14, 26, 30, 33, 35, 40, 48–49, 51, 64–65, 77, 123, 142, 165, 174, 210. *See also* forest, tree
woodpecker, ivory-billed, 91
woodrat, riparian, 91, 193
wren, winter, 78

Xia Dynasty, 116

Yangtze River, 4, 103–4, 116–17, 129–30
Yazoo River, 146
Yellow River, 104, 116–17, 129
Yellowstone National Park, 95; River, 7
Yolo Basin, 161, 163, 185, 194, 214
Yolo County Habitat Conservation Plan, 194
Yolo Bypass, 18, 185–98, 214, 216; birds, 191; fishes, 190; hydrology, 187; midge, 189; productivity,188–89
Yolo Bypass Wildlife Area (YBWA), 187, 194, 214
Yu, Emperor, 116
Yuba River, 161, 167, 169

Zambezi River, 113
zonation, 59, 63
zooplankton, 43, 45, 70–80, 90, 97, 103, 182, 184, 188, 190, 207. *See also* plankton